ABOUT ISLAND PRESS

Island Press is the only nonprofit organization in the United States whose principal purpose is the publication of books on environmental issues and natural resource management. We provide solutions-oriented information to professionals, public officials, business and community leaders, and concerned citizens who are shaping responses to environmental problems.

Since 1984, Island Press has been the leading provider of timely and practical books that take a multidisciplinary approach to critical environmental concerns. Our growing list of titles reflects our commitment to bringing the best of an expanding body of literature to the environmental community throughout North America and the world.

Support for Island Press is provided by the Agua Fund, The Geraldine R. Dodge Foundation, Doris Duke Charitable Foundation, The Ford Foundation, The William and Flora Hewlett Foundation, The Joyce Foundation, Kendeda Sustainability Fund of the Tides Foundation, The Forrest & Frances Lattner Foundation, The Henry Luce Foundation, The John D. and Catherine T. MacArthur Foundation, The Marisla Foundation, The Andrew W. Mellon Foundation, Gordon and Betty Moore Foundation, The Curtis and Edith Munson Foundation, Oak Foundation, The Overbrook Foundation, The David and Lucile Packard Foundation, Wallace Global Fund, The Winslow Foundation, and other generous donors.

The opinions expressed in this book are those of the author(s) and do not necessarily reflect the views of these foundations.

ABOUT THE SOCIETY FOR ECOLOGICAL
RESTORATION INTERNATIONAL

The Society for Ecological Restoration (SER) International is an international nonprofit organization comprising members who are actively engaged in ecologically sensitive repair and management of ecosystems through an unusually broad array of experience, knowledge sets, and cultural perspectives.

The mission of SER is to promote ecological restoration as a means of sustaining the diversity of life on Earth and reestablishing an ecologically healthy relationship between nature and culture.

The opinions expressed in this book are those of the author(s) and are not necessarily the same as those of SER International. SER, 285 W. 18th Street, #1, Tucson, AZ 85701. Tel. (520)622-5485, Fax (270)626-5485, e-mail, info@ser.org, www.ser.org.

RESTORING NATURAL CAPITAL

SOCIETY FOR ECOLOGICAL RESTORATION INTERNATIONAL

The Science and Practice of Ecological Restoration
James Aronson, EDITOR
Donald A. Falk, ASSOCIATE EDITOR

Restoring Natural Capital: Science, Business, and Practice

Edited by

James Aronson, Suzanne J. Milton, and James N. Blignaut

Foreword by

Peter Raven

SOCIETY FOR ECOLOGICAL RESTORATION INTERNATIONAL

ISLANDPRESS

Washington · Covelo · London

Island Press is a trademark of The Center for Resource Economics.

Library of Congress Cataloging-in-Publication Data

Aronson, James, 1953–
  Restoring natural capital : science, business, and practice / by James Aronson, Suzanne J. Milton, and James
N. Blignaut.
     p. cm.
  Includes bibliographical references and index.
  ISBN-13: 978-1-59726-076-3 (cloth : alk. paper)
  ISBN-10: 1-59726-076-2 (cloth : alk. paper)
  ISBN-13: 978-1-59726-077-0 (pbk. : alk. paper)
  ISBN-10: 1-59726-077-0 (pbk. : alk. paper)
  1. Restoration ecology. 2. Economic policy. 3. Natural resources—Management. 4. Sustainable
development. 5. Human capital. I. Milton, Suzanne J. (Suzanne Jane), 1952- II. Blignaut, J. N. III. Title.
  QH541.15.R45A76 2007
  333.71'53—dc22   2006100956

Printed on recycled, acid-free paper

Manufactured in the United States of America

10 9 8 7 6 5 4 3 2 1

CONTENTS

We could consider the earth as one big piece of real estate, as Bob Costanza has suggested through his idea of the Earth Inc. company. Then we the people of this earth are the shareholders and real estate managers, and we had better manage it well. But we are not only the real estate managers, we also form an integral and essential part of the real estate itself. We therefore have to come to grips with the fact that the earth's resources—*natural capital*—are finite and the limiting factor to development. Not only is natural capital finite, we are running out of it at an unprecedented scale and speed. This is because we're using up the earth's stocks of resources faster than we are replenishing them. We would need several more planets the size of this earth to support our insatiable consumption and conversion of natural capital in the medium term—and this is clearly untenable!

There is, however, a way out. Utilizing the earth's resources sustainably requires the restoration of natural capital at local, national, regional, and global scales, thus augmenting the natural capital stock. This book, investigating the business, science, and practice of restoring natural capital is thus a timely and valuable addition to the literature in this field, filling an important gap concerning environmental management, conservation, and resource use. Moreover, it is the first of its kind. A self-organizing group of seventy-one economists and ecologists from around the globe share their collective and individual experience, research, and skills around the core message: We can, and indeed must, supplement the earth's dwindling bioresources, even as we reexamine current patterns of consumption and the unjust distribution of those resources. Ecological restoration is proposed as part of the solution, but social issues, that is, the *restoration of social capital*, must be addressed as well. As the editors of this volume suggest, ecology and ecological restoration must be practiced *as if people matter*. The other half of the solution is that economics needs to refocus its analytic vision *as if ecology matters*. This requires a fundamental paradigm shift. We can no longer consume natural capital without restocking the earth's resources. By restocking the earth we recognize that we are part of the global ecosystem and that we have to take great care of this system, for our own well-being and for our children's sake.

We are at a precarious point in time. The decisions we make during the next few years will have an unprecedented effect on the future of life on Earth. This further emphasizes the importance of this book's message of a better tomorrow.

The challenge now lies in accomplishing what we know has to be done. We have to conserve biological diversity. This, however, cannot be achieved only by enlarging natural wilderness areas, valuable as that might be. We also have to adapt our outlook and learn to live within limits, within the matrix between humans and nature. We need to manage this delicate relationship to produce a better quality of life on Earth for all. One way of doing so, as this book so clearly illustrates, is to acknowledge and accommodate the fact that humans and other species are all fellow citizens on this planet, our one and only real estate.

The Missouri Botanical Garden is glad to be involved with this book and with the process of restoring natural capital. I believe we're entering a new era of nature conservation, an era that endorses the essence of restoring natural capital and an era that considers the prudent management of ecosystems as a key to economic development. While congratulating the contributors to this volume on a task well done, I would like to challenge them and everyone who reads (and uses) this book to take this message forward to a new level of implementation at all scales. The restoration of natural capital has to become mainstream science, policy, and business all around the globe.

Peter Raven, President
Missouri Botanical Garden, St. Louis, Missouri, USA
September 2006

Our precious planet is in peril. The economic overshoot of ecological thresholds is seemingly the order of the day. Yet, it is clearly not enough to lament the excessive economic use and human domination of ecosystems and sit as if in sackcloth and ashes while romanticizing the days gone by. How can the current process of environmental degradation be stopped? How can the process be reversed? And how can the damage already done, in part at least, be repaired? We believe that a two-pronged approach is required: first, a serious reduction in, and better management of, the demand for ecosystem goods and services, and second, an increase in the supply of these goods and services through what we call *restoring natural capital*. The restoration of natural capital, which is the exclusive focus of this book, is the shortest way to express an attempt to actively augment the stock of natural capital to yield more ecosystem goods and services, but in such a way that it contributes significantly to all aspects of human well-being.

It is the people who carry a vision, combined with a firm determination to accomplish it, that shape tomorrow's world, and change is imminent. Throughout this book the authors and editors wish to convey a message of a better tomorrow. There is another way forward. We are hopeful that, by acting collectively and bridging the ideological divide created by different disciplines, ecologists and economists, individuals and governments can achieve a different and positive outcome to the current economic-ecological crisis.

We believe that the market for the restoration of natural capital is now opening, albeit too slowly and on too small a scale. The good news is that a very wide range and a surprisingly great number of activities related to the restoration of natural capital are already happening in dozens of countries around the world. In the so-called developed world, this trend needs to be linked to halting, or indeed reversing, the substitution of natural capital with manufactured capital. In a developing world context, the restoration of natural capital has the potential to be incorporated within a larger-scale development strategy that includes food, water, and energy security programs. In light of this emerging market realization, this book focuses on the content and shape of strategies toward the restoration of natural capital to achieve the optimum and most desirable outcomes in the quickest possible time.

Part 1 of the book focuses on conceptual and theoretical issues, from both an economic and a restoration ecology perspective. Part 2 presents an array of case studies from around the globe. Part 3 deals with specific strategies to propel the restoration of natural capital forward.

With the help of the many people listed herein, we have worked hard to assure continuity and, above all, coherency throughout the book. We have not imposed our viewpoints or even our terminology on the seventy-one distinguished and experienced contributors to this volume. Indeed, the editors would like to very warmly thank and acknowledge the team of economists, restoration practitioners, land managers, and ecologists from around the world that participated in the two workshops (September 2004 in Prince Albert, South Africa, and October/November 2005 in St Louis, Missouri, USA) that led to this book, and who have so patiently cooperated in the long and detailed editing process. We also wish to thank all the workshop organizers, namely the staff at the University of Stellenbosch (Alta da Silva) and the Missouri Botanical Gardens (under the leadership of Peter Raven, and including Olga-Martha Montiel, Sandra Arango-Caro, and Jim Miller) for the work they have done to make the two workshops a success. The editors and contributors express their gratitude to all the chapter reviewers who generously gave of their time to help make this a better book. For each chapter, we sought reviews both internally, that is, from authors of other chapters in the book, as well as externally. Following is the list of reviewers: Martin Aguiar, Dean Apostol, Sean Archer, Philip Ashmole, Ricardo Bayon, Reinette Biggs, Joshua Bishop, Ivan Bond, Hugo Bottaro, Carina Bracer, Antje Burke, Peter Carrick, Pablo Cipriotti, Andre Clewell, Richard Coombe, Richard Cowling, John Craig, Dave Egan, Eugenio Figueroa B., James B. Friday, Brad Gentry, Becca Goldman, Eric Goldstein, Pamela Graff, Isla M. Grundy, Jim Harris, Matthew Hatchwell, Richard Hobbs, M. Timm Hoffman, Patricia M. Holmes, Tony Leiman, Malcom Hunter, Klaus Kellner, Ira Kodner, Roy Lubke, Porter P. Lowry II, Kathy MacKinnon, Jane Marks, Anthony Mills, Laszlo Nagy, Tim G. O'Connor, Jeff Opperman, Claire Palmer, Gunars Platais, P. S. Ramakrishnan, Rick Rohde, José Rotundo, Jan Sallick, Robert J. Scholes, Sjaak Swart, Colin Tingel, Simon Todd, David Tongway, Wessel Vermeulen, Nick Vink, Mathis Wackernagel, Christopher Ward, Cathy Waters, Adam Weltz, Martin de Wit, Paddy Woodworth, and Mike D. Young.

The editors extend a very warm and special thank you to Christelle Fontaine and Alex Chepstow-Lusty for their invaluable and unfailing help throughout the preparation of the book. Christelle coordinated the voluminous correspondence and helped with the technical preparation of all the chapters. Alex text-edited each chapter and had the unenviable task of reducing words without compromising content; he did a superb job. Andre Clewell also helped immensely by dialoging with the three editors. We also wish to thank Barbara Dean, Barbara Youngblood, Erin Johnson, Jessica Heise, and all the members of the team at Island Press. Finally, we gratefully acknowledge a two-year French (CNRS)–South African (NRF) research grant (No. 17370), the support of the National Science Foundation (USA), the Winslow Foundation, the Missouri Botanical Garden, and the Society for Ecological Restoration International (SER), without which the two workshops and the preparation of this book would not have been possible. In closing, we dedicate this book to all those bridge builders dedicating their lives to the interface between economics and environment, and to the children around the world who will inherit the natural, social, and human capital the present generation chooses to leave behind.

James Aronson, Sue Milton, and James Blignaut
April 2007

Restoring Natural Capital: The Conceptual Landscape

This book on the science, business, and practice of restoring natural capital aims to establish common ground between economists and ecologists, with respect to the ecological and socioeconomic restoration of degraded ecosystems and landscapes, and the still broader task of restoring natural capital.

Irrespective of theories and ideologies, all economists agree that if one wishes to be better off in the future, one's capital base needs to be expanded through investment. This principle applies across time and space and embraces everyone—from the smallest, poorest, and remotest household to the biggest multinational companies and nations. Concurrently, all ecologists would agree that it is the habitats and ecosystems on which individuals, populations, and species depend that maintain and nourish the diversity and, indeed, the vitality of life. If habitat destruction continues at the present pace, biological diversity, vitality, and resilience will decline; species will continue to disappear; and the flow of ecosystem goods and services will decline. By contrast, when damaged or degraded ecosystems are rehabilitated or restored, the marvelous diversity of organisms and the systems they form, with their enormous potential for adaptation, evolution, and self-organization, is much more likely to be conserved to the benefit of future human generations and, indeed, all life on the planet.

Both economists and ecologists therefore agree that any investment to broaden the (economic and ecological) base on which life depends will improve economic and ecological welfare and societal resilience. The challenge, however, is that though the principle of capital investment can be universally applied, the object of the investment is not the same. Economists focus predominantly on manufactured capital, and ecologists focus on what is broadly termed *nature* or what ecological economists call *natural capital*. Additionally, there is a problem of scale to overcome, in both time and space. At present, future and distant impacts on natural capital are discounted against present and proximal economic gains for people. This needs to change.

In this introductory section of six chapters, we provide a conceptual and contextual discussion of the new "leaping together" of pragmatically minded restoration ecologists and biologically aware economists. In the first chapter, we define natural capital and the restoration thereof. We make a special effort to explain the relationship between restoring natural capital and ecological restoration, as it is generally defined. Chapter 2 reflects on the restoration of natural capital from an ethical perspective, whereas chapters 3 and 4 consider the restoration

1

of natural capital from an ecological economics and a mainstream economic perspective, respectively. Chapter 5 assesses the restoration challenges ahead, on a gamut of scales, in the light of the work done by the Millennium Ecosystem Assessment (MA), and chapter 6 considers a practical tool to assess the level and extent of biodiversity loss and, hence, a way to estimate the restoration requirement. Together these chapters set the scene and provide the background to the various case studies presented in part 2, and to part 3, which is devoted to the development of strategies at local, regional, and global levels to promote restoration.

The fundamental notion we put forward is that the restoration of natural capital is a practical, realistic, and essential goal that requires the close collaboration of economists and ecologists. Through its application, lasting and mutually beneficial solutions can be obtained for all people and all of nature. In other words, for improved quality of life and greater hope for the future, it is vital to stop both the economic and ecological *rot* caused by the mismanagement and waste of biological resources and the failure to replenish our dwindling stocks of natural capital. The new vision we describe can be achieved only by natural scientists working in partnership with social scientists, forging a new path for ecologically sound, global and local economies. We call upon society's leaders to respond to the call for a radical paradigm shift and to help usher in a new era built upon twin conceptual pillars: *Economics as if nature matters, and ecology as if people matter*. This will allow us to move forward toward a sustainable and desirable future.

Chapter 1

Restoring Natural Capital: Definitions and Rationale

JAMES ARONSON, SUZANNE J. MILTON, AND JAMES N. BLIGNAUT

The restoration of natural capital is arguably one of the most radical ideas to emerge in recent years, because it links two imperatives—economics and ecology—whose proponents have been at loggerheads for decades. In economically developed and developing countries alike, however, we have to acknowledge that humans have transformed ecosystems to the extent that the supply of life-essential ecosystem goods and services is seriously threatened (Wackernagel and Rees 1997). This fact is summarized by two conclusions from the Millennium Ecosystem Assessment (MA 2005f):

> Over the past 50 years, humans have changed ecosystems more rapidly and extensively than in any comparable period of time in human history, largely to meet rapidly growing demands for food, fresh water, timber, fiber and fuel. This has resulted in a substantial and largely irreversible loss in the diversity of life on Earth.

> The changes that have been made to ecosystems have contributed to substantial net gains in human well-being and economic development, but these gains have been achieved at growing costs in the form of the degradation of many ecosystem services, increased risks of nonlinear changes, and the exacerbation of poverty for some groups of people. These problems, unless addressed, will substantially diminish the benefits that future generations obtain from ecosystems.

We argue that natural capital has become a limiting factor for human well-being and economic sustainability (Costanza and Daly 1992; Daly and Farley 2004; Aronson, Clewell, et al. 2006; Farley and Daly 2006; Dresp 2006) and advocate that the restoration of natural capital is the most direct and effective remedy for redressing the debilitating socioeconomic and political effects of its scarcity. Conservation, and reducing waste are indispensable, but likewise the investment in the restoration of natural capital that augments the pool of natural capital stock and hence stimulates the supply (or flow) of ecosystem goods and services (Repetto 1993; Cairns 1993; Jansson et al. 1994; Clewell 2000). The restoration of natural capital includes ecological restoration, but it goes further. The restoration of natural capital also considers the socioeconomic interface between humans and the natural environment. By functioning within this interface, the restoration of natural capital builds bridges between economists and ecologists and thereby offers new alternatives for ecologically viable economic development. It also offers new hope for bridging the worrisome gaps between scientists and nonscientists and between developed and underdeveloped countries.

Definitions of Terms and Concepts

Here we define a number of key terms pertinent to the concepts of restoration and natural capital, and explain how this focus complements related approaches to ecosystem repair and raises awareness of the need to make development ecologically, socially, and economically sustainable.

Natural Capital

Generally, development and the improvement of life quality are not possible without a growing asset, or capital, base. The concept *capital*, however, is not homogenous since one can distinguish between five principal forms of capital (Rees 1995; MA 2005f):

- *Financial capital* (money or its substitutes)
- *Manufactured capital* (buildings, roads, and other human-produced, fixed assets)
- *Human capital* (individual or collective efforts and intellectual skills)
- *Social capital* (institutions, relationships, social networks, and shared cultural beliefs and traditions that promote mutual trust)
- *Natural capital*, an economic metaphor for the stock of physical and biological natural resources that consist of renewable natural capital (living species and ecosystems); nonrenewable natural capital (subsoil assets, e.g., petroleum, coal, diamonds); replenishable natural capital (e.g., the atmosphere, potable water, fertile soils); and cultivated natural capital (e.g., crops and forest plantations)

Some clarification is required to distinguish between renewable, replenishable, and cultivated natural capital. *Renewable* natural capital is the composition and structure (stocks) of natural, self-organizing ecological systems that, through their functioning, yield a flow (or *natural income*) of goods and services. These flows are essential to life in general and are extremely useful to humans and all other species. *Replenishable* natural capital consists of stocks of nonliving resources that are continually recycled through their interaction with living resources over long periods (such as the interaction between surface mineral components and living organisms that produces fertile, stable soil). The condition of renewable natural capital stocks obviously influences the quality, quantity, and renewal rate of these essential, replenishable, natural capital stocks, and vice versa.

Cultivated natural capital arises at the dynamic interface of human, social, and natural capital. This interface produces agroecological systems and amenity plantings that may be more or less self-sustaining, depending on their design and management. Cultivated capital forms a continuum between renewable natural capital and manufactured capital and may be closer to one or the other, depending on the degree of transformation of the landscape, the genetic material, and the subsidies (e.g., energy, water, nutrients, seeding, weeding, pest control) required for maintaining the system. It is often forgotten that, in all cases, both cultivated resources and manufactured capital are derived from renewable, replenishable, and nonrenewable natural capital. This transformation of natural to human-made capital is "mining" the stock of renewable, replenishable, and nonrenewable natural capital, thereby reducing it for future use, unless it is restored where it has been used up or degraded.

Ecological Restoration and Restoration of Natural Capital

The *Society for Ecological Restoration International's Primer on Ecological Restoration* (SER 2002) defines ecological restoration as "the process of assisting the recovery of an ecosystem that has been degraded, damaged, or destroyed," but it is a much broader concept. The goal of ecological restoration, according to the SER *Primer*, is a resilient ecosystem that is self-sustaining with respect to structure, species composition, and function, while integrated into a larger landscape and congenial to "low impact" human activities. Ecological restoration "is intended to repair ecosystems with respect to their health, integrity, and self-sustainability" (SER 2002). An associated discipline is *ecological engineering*, which involves restoring and creating (thus, engineering) sustainable ecosystems "that have value to both humans and nature" (Mitsch and Jørgensen 2004). Lewis (2005) cogently adds that ecological engineers attempt to address both the restoration of damaged ecosystems and the creation of new sustainable systems "in a cost effective way."

The *restoration of natural capital* is any activity that integrates investment in and replenishment of natural capital stocks to improve the flows of ecosystem goods and services, while enhancing all aspects of human well-being. In common with ecological restoration, natural capital restoration is intended to improve the health, integrity, and self-sustainability of ecosystems for all living organisms. However, natural capital restoration focuses on defining and maximizing the value and effort of ecological restoration for the benefit of humans, thereby mainstreaming it into daily thought and action and promoting ecosystem health and integrity. Natural capital restoration activities may include but are not limited to (1) the restoration and rehabilitation of terrestrial and aquatic ecosystems; (2) ecologically sound improvements to arable lands and other lands that are managed for useful purposes; (3) improvements in the ecologically sustainable utilization of biological resources; and (4) the establishment or enhancement of socioeconomic activities and behavior that incorporate knowledge, awareness, conservation, and management of natural capital into daily activities.

Those motivated by a biotic rationale for restoration, as explained by Clewell and Aronson (2006), and whose concern lies with the perpetuation of biodiversity, may raise a concern here. They may argue that natural capital restoration's human-centered focus will obscure an essential insight of the restoration and conservation movements—that ecosystems and all the processes and species they contain are worth restoring and preserving "for their own sake," regardless of their economic (or other) value to humans. This is true (see chapter 2); however, in order to mainstream ecological restoration into the economy (chapter 34), it is also necessary to show how humans will benefit directly from it and how the interaction between economic and ecological systems could be improved through the restoration of natural capital.

Rehabilitation and Reallocation

In figure 1.1, *rehabilitation* is aligned with restoration in that both generally take an "original" (preanthropogenic era, sensu Crutzen and Stoermer 2000) or historic, culturally acceptable ecosystem or landscape as a reference for the orientation of interventions to halt

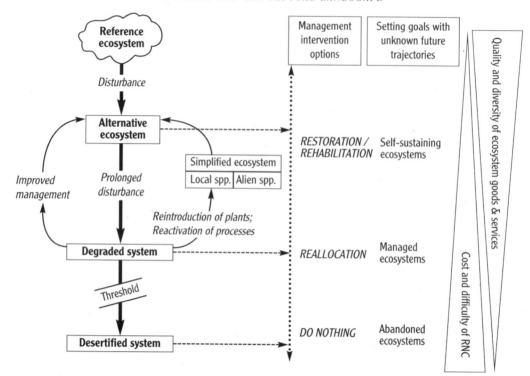

Figure 1.1. General model of ecosystem degradation and possible responses, modified from Aronson et al. (1993). In this scheme, disturbance refers to undesirable anthropogenically induced change. On the right-hand side of the figure, quantity and diversity of ecosystem goods and services refers to their availability, while cost and difficulty of restoration of natural capital are the relative financial and other expenditures and investments required for a continuum of management intervention options. The exact positions of transformed ecosystems, with a range of restoration inputs, depend on many variables, with the most plausible outcomes indicated.

degradation and initiate more sustainable ecosystem trajectories. By contrast, *reallocation* is a term that describes what happens when part of a landscape, in any condition is assigned a new use not necessarily bearing any relationship to the structure or functioning of the pre-existing ecosystems. Whereas, traditionally, restoration seeks a complete or near-complete return to a preexisting state (although this is being challenged as a result of the consequences of global climate change), by reassembling the species inventory, stresses, and disturbances, as far as possible, rehabilitation focuses on repairing ecosystem functions, in particular raising ecosystem productivity and services for the benefit of humans.

Where the spatial scale of damage is small and the surrounding environment is healthy in terms of species composition and function, amelioration of conditions in the damaged patch, together with ecological processes such as seed dispersal and natural recolonization by plants and animals can lead to full recovery of resilient, species-rich ecosystems that provide a range of services valued by humans (chapter 21)—including aesthetic, cultural, and what we may call "spiritual" services. However, in heavily modified ecosystems, which have crossed one or more thresholds of irreversibility (May 1977; Westoby et al. 1989; Aronson et al. 1993; Milton et al. 1994; Whisenant 1999; Hobbs and Harris 2001; Walker et al. 2002), restoration of the preexisting species inventory may no longer be feasible. In such cases, only rehabilitation and reallocation are likely to remain as viable, cost-effective alternatives, and any actions to reverse environmental damage should be determined by socioeconomic decision making that takes into account the spatial scale of the degradation, the present and future value of the resource to humans, and the condition and composition of the surrounding ecosystem.

Ecological restoration, rehabilitation, and reallocation can all contribute to the restoration of natural capital and be pursued simultaneously in different landscape units. Throughout this book, the term *restoration* (and hence, natural capital restoration) is often used so as to include rehabilitation, whereas *reclamation* is not employed because of prior connotations (Aronson et al. 1993; SER 2002).

Rationale for Restoring Natural Capital

We now present some basic principles, following Clewell and Aronson (2006), that collectively provide a rationale for the restoration, sustainable use, and enhanced protection of natural capital. They serve as a template that the editors and authors will use for the evaluation of the case studies, regional overviews, and other contributions in this volume.

Principle 1. In setting targets for the restoration of natural capital, our premise is that people of all cultures depend on the products and services derived from natural ecosystems to provide much of their sustenance and well-being (Daily 1997; Balmford et al. 2002). It follows that an improvement in the quantity or quality of natural ecosystems increases human well-being, while degradation causes the converse. We assert that self-sufficient, self-organizing natural ecosystems are appropriate restoration targets because, despite the deficiencies in our understanding of natural ecosystem functioning (Balmford et al. 2005), it would appear that they provide most ecosystem services (e.g., water purification, flood control) and some goods (e.g., natural pasture, marine fish) more cleanly, efficiently, and inexpensively than human-designed systems, such as "improved" pasture or aquaculture (Costanza et al. 1997; Balmford and Bond 2005). In the context of semicultural or cultural landscapes, and human-designed ecosystems (see, for example, chapter 16), the broader term of restoring natural capital is more readily applicable than ecological restoration, per se.

Principle 2. It has been remarked that anthropogenic global changes, including climate change, have profound implications for ecological restoration and biological conservation (Harris et al. 2006; Thomas et al. 2004), and the overlapping field of ecological engineering (Mitsch and Jørgensen 2004; Kangas 2004) that deals with the design and creation of ecosystems, as well as their restoration. However, we argue that the only durable way to increase ecosystem services is by restoring the functions (MacMahon 1987; Luken 1990; Falk 2006) and processes of self-sustaining ecosystems. Such systems will adapt to climate change and evolve as well or better than "designer ecosystems." Furthermore, restoring natural ecosystems on a large scale may actually help mitigate the effects of climate change (Clewell and Aronson 2006). Finally, climate change scenarios in no way alter the obvious benefits of restoring natural capital.

Principle 3. Costs of restoration of natural capital increase as a function of the spatial extent, duration, and intensity of environmental damage, and with the complexity of the target ecosystem or socioecological system (George et al. 1992; Aronson et al. 1993; Milton et al. 1994). This cost increase reflects the increasing number of interventions required to achieve restoration as damage initially depletes the plants and animals (for example, overfishing, deforestation), and then destroys the physical habitat (for example, through pollution, soil erosion, hydrological or climatic changes), not to mention the preexisting ties and links between people and the landscapes in which they lived and worked. Like ecological benefits, social and economic benefits from investments in restoring natural capital will generally take longer to be realized where not only ecological injuries but also adverse socioeconomic changes have been more profound and long lasting.

Principle 4. Natural capital and manufactured capital are complementary. Increasingly, the limiting factor for economic development is natural capital, and not manufactured capital, as it used to be.

Principle 5. Extinct species can never be recovered nor lost complexity fully understood or restored. Therefore, it is better to conserve or use resources sustainably than to restore, and better to invest in restoring natural capital during the earlier stages of resource degradation and loss of sustainability in managed systems than to postpone restoration activities.

Contribution

Here we have indicated that the restoration of natural capital includes ecological restoration, but it also considers the socioeconomic interface between humans and the natural environment, including managed systems such as food, fodder, tree fiber, and fish farms, and the awareness of the importance of natural capital in the daily lives of people. The recognition of the real possibility of restoring natural capital helps build bridges between economists and ecologists who can then develop a set of information and hypotheses to help develop new and sustainable economic pathways while also repairing some of the ecological and socioeconomic damage done in the past. As has been indicated, restoration and rehabilitation are not the only ways of developing these pathways. Conservation and revised management of resources and anthropogenic systems, as well as the reduction in consumer demand, among other things, are also vitally important. In the following chapters, various authors including, among others, economists and ecologists from various countries consider the theoretical, commercial, financial, and practical implications of restoring natural capital. The goal is a consilience of ecologists and economists offering practical strategies for redressing the debilitating socioeconomic and political effects of declining natural, social, and cultural capital worldwide. This poses an immense ethical challenge, as well as new conceptual approaches and revised strategy planning. In chapter 2, therefore, we reflect on the restoration of natural capital from an ethical vantage point before returning to economic, ecological, and political considerations.

Restoring Natural Capital: A Reflection on Ethics

JAMES N. BLIGNAUT, JAMES ARONSON, PADDY WOODWORTH,
SEAN ARCHER, NARAYAN DESAI, AND ANDRE F. CLEWELL

Over the past two centuries we have transformed natural capital to the extent that the supply of life-essential ecosystem goods and services—for us, and all other organisms on the planet—is quite seriously threatened. This calls for an urgent and active focus on and application of the science, business, and practice of the restoration of natural capital—the theme of this book. While the rest of this book deals with the restoration of natural capital from either a theoretical (conceptual), practical (experiential), or strategic (planning) perspective, this chapter reflects on the restoration of natural capital from an ethical vantage point. To do so, we will first demystify the prevailing economic ethic and then discuss sustainability and the contribution of restoring natural capital to the end of creating a new economic and socioecological ethic based on sustainability, fulfilled relationships, and social justice.

People and Nature: A Relationship Gone Astray

In conventional neoclassical economics the natural environment, though recognized as an essential production factor, is treated under the *ceteris paribus* (all other things being equal) assumption. Mainstream economics thus assumes no quantitative or qualitative change to stocks of natural resources due to substitutability and if there is no change to these stocks they are by definition infinite. Clearly this is an unrealistic proposition, but one with dangerous consequences.

On Economics, Values, and Ethics

In his penetrating book on ethics and economics, Wogaman (1986) states that it is important to distinguish between intrinsic and instrumental values or principles. An *intrinsic value* is something that is good in itself, and it requires no further justification. An *instrumental value*, however, is something that contributes to the fulfillment (or realization) of an intrinsic value. Instrumental values are, therefore, means to an end and not ends in and of themselves. The question with which we are concerned is this: what are the intrinsic and instrumental values in prevailing economic theory and thought?

The main theoretical construct of modern economics, generally called neoclassical economics, is based on Adam Smith's *The Wealth of Nations* (1776). Smith's central premise is that people seek the maximization of individual utility or satisfaction. This idea has been developed subsequently as the maximization of consumption, which has become the intrinsic value of neoclassical economics, the key value that requires no further explanation or justification. The instrumental value is self-interest. *Self-interest* is the basis upon which people compete with each other to achieve utility or consumption maximization. Self-interest per se is sometimes presented as a typical or "normal" element of Darwinian natural selection. In this context, however, the principle of self-interest is not applied to assure species survival, but to ensure domination—the domination of one individual over another in human society, and of *Homo sapiens* collectively over the rest of the natural world. While one can hardly argue against the application of the self-interest principle for the sake of species survival, the consequences of both individual and collective human domination are far reaching. Moreover, Kropotkin, a Russian aristocrat noted both for his libertarian politics and original contributions to evolutionary theory (Kropotkin 1902), qualified Darwinism with his insight that evolution involves mutual aid as well as competition. The zoologist Warder Clyde Allee developed this insight (Allee 1949; see also Bleibtrau 1970; Gould 1992), while Vermeij (2004) provided a significant counterpoint to the view that self-interested domination is the universal norm. In the same vein, we firmly endorse Vermeij's argument that humans, as top consumers, should provide corrective feedback to the economy. It is after all a self-organizing system, which needs to be allowed or even "pushed" to adapt in such a way as to sustain system stability and survival.

For a variety of reasons, these feedback mechanisms are failing to function: human society, and economies, are not getting a vital, life-sustaining feedback message or are not registering it strongly enough (figure 2.1). The limited feedback, or information blockage, is a result of "the market" not recognizing that humans are part of a larger ecosystem. As long as our species sees constantly increasing consumption for one and all as its supreme goal, the market will provide all the right signals and information to this end, ignoring the ecological and spiritual consequences. That maximization of, or growth in, consumption is a principle focus of modern economics, a notion based on the faith in neoclassical or neoliberal economic theory. An outcome of this "growth machine" model is increased polarization in global and national economics and politics (as per divergent outcomes in figure 2.1). In some developed countries there is far too much consumption, based as it is on clearly unsustainable levels of material and energy output, while in most developing countries, there is too little per capita consumption, leading to an increase in human vulnerability and chronic loss of dignity and well-being. Whatever the situation, the flow of information from environmental indicators back to the economy is "filtered" out.

Changing the prevailing ideology will require a new paradigm in which the outcome of "the market process" is redefined toward a new end (we reflect on this in more detail in the next section). This is not impossible since, as we noted earlier, the economy is a self-organizing system. The market acts and reacts to information and is based on the premise that people have the ability to reflect, analyze, and reinterpret the data that the market provides. It is the general failure of humans to absorb and act on the environmental information that inhibits the much-needed change in values, behavior, and lifestyle. This failure to absorb information is due to a variety of reasons, including (1) an overload of information, (2) a distrust of information sources, (3) an inability to comprehend the information, and/or (4) because essential bits of information are being withheld by governments and others with

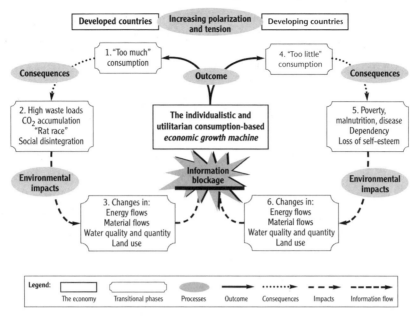

FIGURE 2.1. Simplified state and transition model of the global, consumption-based economic growth machine with indications of its various outcomes and environmental impacts. Overexploitation (often begun during colonial periods) combined with inequity and maldistribution confounds the problem of underconsumption.

power. However, as ecologists and economists, we understand that making information available is not enough. It is essential that we also contribute to communicating it to the wider society (which constitutes the *market*) through education, the media, culture, and all the means available within democratic politics.

That message is that our world sets natural limits to how much we can consume and that we are pushing way past those limits, blindly and recklessly. Ironically and tragically, the human drive to domination, as opposed to survival, threatens not only many other species but also, conceivably in the near future, the survival of the human species itself. Since consumption acts as the intrinsic economic value, consumption gains control over persons who desire to continually consume more. The supporters of "value-free," neoclassical economics deny this. "Normative economics is speculative and personal," as Friedman famously wrote; it is "a matter of values and preferences that are beyond science. Economics as science, as a tool for understanding and prediction, must be based solely on positive economics, which is in principle independent of any particular ethical position or normative judgments" (1953, 4).

Within *positive economics*, the efficiency criterion (calculation) gains supremacy over all other values, such as fairness, obligation, prudence, honesty, loyalty, sustainability, and practicability (Bromley 2000). Some would argue (see Von Hayek 1993) that it is unethical to restrict the market by introducing ethical guidelines, since the outcome of the market process is ethically desirable by definition. According to this view, the moral solution to all economically related problems would be to extend the boundaries of private property rights to be all-encompassing (Coase 1960), in other words, the systematic commodification of all public goods, including ecosystem goods and services of all kinds.

The procommodification view is in stark contrast with the belief that ethical guidelines should modify social actions and behavior. These guidelines, which Kant and Habermas call "a context-relevant common moral denominator," find their expression in the concepts of equality and human rights (Kant 1956; Habermas 1993). Commodification and consumerism are biased toward inequality and contribute to the marginalization of the weak in favor of "progress" and the self-interest of the strong. Commodification also disregards the fact that the natural environment, in all its diversity and complexity, is valuable. This is because of the difficulty of its valuation in monetary terms, and because some elements within this diversity have no direct value to humans. Many ecologists argue that ecosystems are indeed valuable in themselves, quite apart from their human-use value (Jordan 2003). That is a strong argument ethically, but it has not made a large enough impact on our societies to save the earth's natural capital from degradation. The restoration of natural capital argument, by making strong and interlinked economic and ecological cases, should have a much broader appeal and therefore a much deeper impact on public opinion and policymakers, globally and locally.

People, however, have to have a sense of purpose to build and maintain dignity, self-esteem, and meaning in their lives, which are of course much wider and deeper concepts than the maximization of consumption (Monod 1971). This implies that humans, as relational beings, have a fundamental need for "fulfilled relationships" at many levels that include subsistence, protection, affection, understanding, participation, recreation or leisure, capabilities, creativity, identity, and freedom (Max-Neef 1989). Therefore, people are not to be seen apart from, but rather as part of, the natural environment. This relationship requires restoration as well.

A Divided World: People Versus Nature

The maximization of consumption, as a prevailing value for society, has lead to the establishment of an ideology of economic growth. Heilbronner (1985, 62) eloquently describes this process as leading us "to the larger picture that [Adam] Smith had in mind. We would call it a growth model, although Smith used no such modern term himself. What we mean by this is that Smith shows us both a propulsive force that will put society on an upward growth path [consumption] and a self-correcting mechanism [self-interest] that will keep it there."

To grow economically requires an accumulation of manufactured capital. Manufactured capital is by and large converted or transformed natural capital (see chapter 1). This implies an asymmetrical application of the self-interest principle. As a result, not only do people fight each other for resources but collectively perceive their self-interest as being in conflict with the natural world, or as Schumacher (1973,13) wrote,

> Modern man does not experience himself as a part of nature but as an outside force destined to dominate and to conquer it. He even talks of a battle with nature, forgetting that if he won the battle, he would find himself on the losing side. Until quite recently, the battle seemed to go well enough to give him the illusion of unlimited powers, but not so well as to bring the possibility of total victory into view. This has now come into view, and many people, albeit only a minority, are beginning to realize what this means for the continued existence of humanity.

People are fighting nature because they consider themselves to be outside of, or set aside from, nature. A new ethic is required that insists that economics and political economic policies must take account—monetarily and otherwise—of the cost of consumption of natural capital and ecosystem services. Humans, in an increasingly crowded world, can no longer permit the ideology of consumption maximization to take precedence over the need for ecosystem resilience and human justice (Blignaut 2004a).

In sum, economic growth is important, but so are social relationships and relationships with nature, education, law, justice, and so on. In a holistic approach (Smuts 1926, 86), where ecology and economics are integrated, a new kind of scientifically and ethically based consensus is necessary to address current world problems. It is in this context that sustainability emerges as the signpost of the way forward and the means to restore healthy relationships both within and among human societies, and between people and nature.

Sustainability

We argue that fulfilled relationships, with oneself, with others, and with the natural world, are the most desirable of all ethical values frameworks. We consider sustainable development as the most effective instrument for building this framework. Within sustainable development, however, there are strong differences of emphasis. Some proponents stress human well-being, whereas others stress the maintenance of natural processes, sometimes known as ecosystem well-being. These differences are reflected in two prevailing and often cited definitions of *sustainability*: (1) "Providing for the needs of the current generation without compromising the ability of future generations to provide for their own needs" (Brundtland Report 1987), and (2) "The capacity to create, test and maintain the adaptive capability [of natural ecosystems]" (Holling et al. 2002). Though these definitions may seem incompatible, in reality they are not. The Holling definition can be interpreted as an ecological prerequisite for the Brundtland definition. In other words, if we do not maintain the adaptive capability of natural ecosystems, we compromise the ability of future generations to provide for their own needs. The converse is also true.

Not only is there a plethora of definitions for sustainable development, the matter is further complicated by the fact that many economists argue that natural capital (including untransformed natural capital) is not directly required for economies to function and that natural capital can largely be substituted by the growth of manufactured capital (figure 2.2a). This notion of substitutability is known as "weak sustainability," the first scenario discussed here (Van Kooten and Bulte 2000). Weak sustainability presupposes that all forms of capital are completely interchangeable in the process of production, in the estimation of total wealth, in tracking changes in asset values, and in calculating sustainable income (Pearce and Turner 1991; Solow 1991; Dorfman 1997; Pezzey and Toman 2002). The value system employed is unabashedly anthropocentric and utilitarian.

Second, the *strong sustainability* notion of ecological economists recognizes that natural and human-made capital are complementary but not substitutable. This is because natural capital is broader than just natural resources of direct use to humans or natural commodities that can be manufactured (Daly 1990; Ekins 2003; Ekins, Simon, et al. 2003) (see also chapters 1 and 3 in this regard). Strong sustainability is a concept currently more favored by ecologists, and their allies among economists, than by most politicians and mainstream

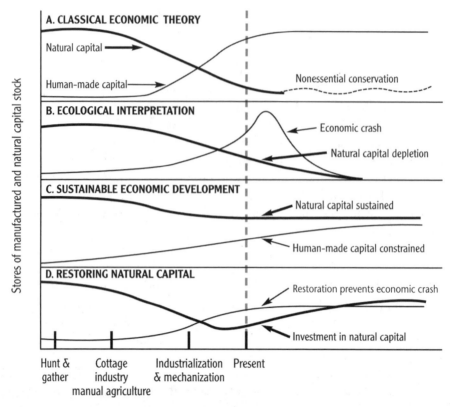

FIGURE 2.2. Four contrasting sustainability paradigms:
(A) Neoclassical perspective: economic growth can continue after natural capital has been depleted.
(B) Environmentalist pessimistic perspective: predicting an economic crash following natural capital depletion.
(C) Ecological economic vision wherein tradeoffs to growth are accepted in favor of maintaining natural capital and a more sustainable economy through qualitative improvement of ecosystem and resource management.
(D) The possible effects of restoration of natural capital on quantity and stability of human-made capital.
Panels A and B are redrawn from Folke et al. (1994), with permission from the authors and the publisher. Panels C and D are original and previously unpublished.

economists. Its central contention is that the internal substitution of different components of natural capital and, above all, its replacement by the other forms of capital, is only possible to a limited degree. This is based on the premise of the *precautionary principle* and the application of *safe-minimum standards* for prudent management. The proponents of strong sustainability affirm that economic growth based on the destruction of natural capital will be unsustainable, because economies require natural, social, and manufactured capital to survive (figure 2.2b). Most of those who promote the search for strong sustainability do nevertheless acknowledge the inevitability of tradeoffs, namely that economic objectives often have environmental and social costs (figure 2.2c), and vice versa (Duchin and Lange 1994). In this third scenario quality of goods and services is given equal or greater importance than quantity. This corresponds to the difference between development and growth of socioeconomic systems. The restoration of natural capital has economic costs, but these are greatly outweighed by its benefits, because it increases the prospects for sustainable development and reduces the threat of economic, social, and ecological disaster (figure 2.2d).

A compromise between weak and strong concepts of sustainability is offered by the notion of *critical natural capital* (chapter 3). This term is the product of a conceptual partitioning of natural capital between components that are irreplaceable, and therefore critical, and the remaining kinds that can indeed be replaced. This compromise allows bridge building between ecologists and economists to go forward in a consensual fashion.

More than twenty years ago, Norgaard (1985) highlighted two incontrovertible facts of relevance. First, environmental systems are not divisible, a fact that invalidates the neoclassical assumption that all resources are divisible and can be owned (which makes the procommodification view implausible). Second, "environmental systems almost never reach equilibrium positions, and . . . changes [in them] are frequently irreversible" (Norgaard 1985, 382–83). Furthermore, Norgaard wrote,

> Critical natural capital cannot be defined. It is not that there are no thresholds but that there are many, many thresholds that are interdependent, spatially and historically, and sensitive to a history of intertwined perturbations, etc., and we can predict few, if any, of the thresholds, or the consequences when they are crossed. To base the argument on the existence of something we cannot define puts one in the position of then having to define it if challenged. (Norgaard, 2 January 2006, personal communication)

Here Norgaard implicitly emphasizes the difference between the limiting aspects of positivist science, which focuses only on the predictive capacity of our epistemological knowledge, and the need for an ethical framework. We need a context-relevant, generally accepted moral determinant to guide us where positivist science cannot take us. The fact that we cannot determine the various threshold limits exactly does not imply that they do not exist. Indivisibility, disequilibria, irreversibility, uncertainty, and the existence of (ambiguous) "critical" components of natural capital lead proponents of strong sustainability to place an ever greater importance on natural capital. They see natural capital not only as a complement to manufactured capital but in fact vital for all life worthy of the name "life" (Ekins, Simon, et al. 2003). In more sober words, substitutability of different forms of capital is limited and inelastic. Therefore, the need to preserve critical natural capital imposes severe constraints on economic growth that depends on the transformation, pollution, or destruction of natural capital. Failure to respect these fundamental principles has in the past, and could again in the future, lead to ecosystem collapses and, what's more, "economic collapse might be brought about by ecosystem collapse" (Van Kooten and Bulte 2000, 240). The collapse of the Easter Island community, brought about by ecosystem collapse, is a good example (Diamond 2004). Neoclassical economists consider this view unnecessarily pessimistic and sometimes dub it "neo-Malthusian." Yet to ecologists and ecological economists it seems obvious, or self-evident, especially in a crowded world like the one we live in today.

Contribution

It is evident from the discussion presented that there are definite social, ecological, and ethical parameters within which economic development and growth has to take place. The prevailing ideology of growth is based on a false and seriously dangerous misconception that it is possible for humans to alienate themselves from the natural world and to dominate it, "as if from the outside," and yet continue to live comfortably "on the inside." When people decide to restore natural capital, they not only set out to physically and biologically repair degraded, damaged, or destroyed natural capital, they also—more or less consciously—seek to repair and restore a healthy psychological, social, and spiritual relationship with nature. Investing time, energy, and financial capital in the restoration of natural capital is an acknowledgement that people are part of an intricate web or matrix of relationships that encompasses nature, including other people near and far, other forms of life, other material objects, the economy, and science. The act of restoring natural capital might be costly in financial terms, but in addition to the augmentation of ecosystem goods and services, it adds both value and meaning to all of these relationships. It contributes to the development or reinforcement of dignity, self-esteem, and restored relationships to past, present, and future generations. These values are not all quantifiable, yet it is possible to determine the monetary value, implicitly or explicitly, of the value of the ecosystem goods and services delivered by restoration. Such monetary values will, however, always be partial and subject to the underlying ethical framework. To illustrate this point, the next two chapters provide a theoretic framework for both the valuation and the restoration of natural capital from two completely different ideological perspectives.

Restoration provides an essential alternative to the prevailing paradigm of maximizing consumption that can be achieved through an ethical framework based on fulfilled relationships. We recognize the daunting scale of the task of putting these principles into practice in a global societal context where the ideology of growth dominates economics, politics, culture, the media, and education. These practical issues are the focus of the rest of this book. It should be kept in mind that politics, which is the practical social expression of our ethics and our collective values system, plays a key role in shaping our societies and conduct. In all of these contexts, the strategy of restoring natural capital can, and should, make a difference. We believe that that difference is crucial to the outcome of current debates about the future of our societies and the biosphere within which we live, and without whose goods and services we could no longer exist.

Chapter 3

Restoring Natural Capital: An Ecological Economics Assessment

Joshua Farley and Erica J. Brown Gaddis

In this chapter we show how the application of the basic principles of ecological economics can provide concrete and practical guidelines for deciding if, when, and where to restore natural capital. A mainstream economic approach to the same question is presented in the next chapter. We begin by defining distinct categories of natural capital: source, service, sink, and site. We then propose corresponding strategies for measuring and valuing natural capital as they relate to the restoration thereof and distinguish between critical and noncritical natural capital. Measurement strategies are evaluated according to their usefulness in attaining the goals of ecological sustainability, social justice, and economic efficiency through the restoration of natural capital.

Important Concepts in Ecological Economics

Whereas neoclassical economics focuses on the microallocation of scarce aspects of natural capital among different market products, ecological economics focuses also on macroallocation, the apportionment of finite ecosystem structure between *economic production* (economic goods and services) and *ecological production* (ecosystem goods and services). Such an analysis is useful in answering the following types of questions related to the restoration of natural capital:

1. How much restoration is required to support life-sustaining ecosystem functions?
2. When is restoration imperative, and when should restoration be considered based on marginal costs and benefits?
3. When should restoration focus on restoration of ecosystem function or on ecosystem structure?
4. How should the costs and benefits of restoration be distributed within society and between generations?

Addressing these questions in depth requires that we first explore some ecological economic concepts describing the framework with which we approach the restoration of natural capital.

Ecological Economics: The Imperatives

As mentioned in chapter 2, neoclassical economics elevates the efficiency criterion above all other criteria for resource allocation. Ecological economics also considers efficient allocation important, but it is secondary to the issues of scale and distribution (Daly 1992).

Scale concerns the macroeconomic question of how large the economic system can be relative to the ecological system that sustains and contains it (sustainable scale), as well as how large it should be (desirable scale). If we exceed sustainable scale, we must restore natural capital until our ecosystems regain their ability to reliably generate critical life support functions and supply our economy with the raw materials and waste absorption capacity it requires. Efficient allocation entails sacrificing the least valuable ecological services in exchange for the most valuable economic ones. Economic growth beyond the point where ecological costs outweigh economic benefits is inefficient and exceeds desirable scale.

Distribution addresses equity or the apportionment of resources among different individuals. In a market economy, different distributions result in different allocations, so the desirability of a given allocation depends on the desirability of the distribution that generated it. With respect to restoration, a just distribution requires that those responsible for exceeding desirable scale and/or those who benefit from restoration pay for restore it (see chapter 26).

Natural Capital

In chapter 1, the various categories of natural capital were identified, and capital was defined as a stock that yields a flow of benefits. Defining natural capital requires that we evaluate the role of ecosystem structure and function in delivering a variety of benefits to society. First, natural capital in the form of ecosystem structure provides *sources* of raw materials for economic production, such as timber, fish, and fossil fuels. Equally important are the functions provided by natural capital as ecosystem *services* for climate regulation, water supply, and so on, and as *sinks* absorbing and processing society's waste (Coddington 1970). Although ecosystem services are as essential to our welfare as raw materials, they are dramatically harder to measure. Finally, we must also value the *site* (or location) of natural capital both in terms of its spatial relationship to human society and as a physical substrate for capturing solar energy and rainfall. Though site is neither a good nor a service per se, it is generally the single most important variable in determining the market value of land (the substrate on which all terrestrial restoration must occur), and nearly as important in determining the nonmarket values of the ecosystem services it does or could generate.

Most forms of natural capital simultaneously function as sources, services, and sinks, while site strongly influences the value of those functions. For example, a forest ecosystem may regulate and filter water flow (service), supply timber for building (source), and absorb carbon dioxide from the atmosphere (sink); its proximity to human populations (site) heavily influences the value of these other functions. A key question then is how to value all of these simultaneously for one geographic area and compare it to other areas, which may supply other important ecosystem goods and services. Note that some of these functions are primarily beneficial for local residents (filtration of water), whereas others provide global services (absorption of carbon dioxide).

Natural Capital as a Source of Raw Materials

The *source* component of natural capital consists of the stock of raw materials provided by nature that is essential for all economic production and includes renewable, replenishable, and nonrenewable resources. The restoration of natural capital can increase the stock of renewable and replenishable resources, but not nonrenewable ones; the extraction and use of all these stocks can seriously impair and degrade ecosystem function.

Sources can also be characterized as *stock-flow resources* (Georgescu-Roegen 1979). Stock-flow resources are physically transformed through production and embodied in whatever is produced. Use is equivalent to depletion, but stock-flow resources can be stockpiled, and humans can control the rate at which they are used. These resources are appropriately measured as physical quantities, such as cubic meters of timber, barrels of oil, or tons of fish. In addition to being essential to all economic production, stock-flow resources are vital elements of ecosystem structure, the building blocks of ecosystems. All stock-flow resources are *rival* in use, which means that one person's use of a given resource leaves less for others to use.

Fund-Services: Ecosystem Services and Waste Sink

The *service* component of natural capital consists of the ecosystem services that sustain all life on the planet and are essential inputs into many types of economic production. Ecosystem services are ecosystem functions of value to humans that arise as emergent phenomena when the various elements of ecosystem structure (the source component of natural capital) interact with one another to create a complex system (Costanza et al. 1997; Daily 1997). In a self-sustaining system, such services create the conditions that allow the biotic elements of ecosystem structure to reproduce. These services have been categorized in many different ways, but common categories include regulation services, production services, habitat functions, and information services (De Groot et al. 2002).

Ecosystem services can be characterized as *fund-services* (Georgescu-Roegen 1976). In contrast to stock-flows, fund-services are resources that are not physically transformed into what they produce, and hence are not depleted through use—human-made fund-services wear out, but natural ones are maintained by solar energy. A fund-service is the result of a particular configuration of stock-flow resources. Fund-services cannot be stockpiled, the rate at which they are provided cannot be directly controlled by humans, and they are appropriately measured as a quantity of service per unit of time. Most ecosystem services are nonrival in use, which means that one person's use of the resource does not leave less for someone else—for example, when one person benefits from the flood control services provided by a healthy, forested watershed, it does not diminish the amount of flood control left for anyone else.

The *sink* component of natural capital is the capacity of natural systems to absorb and process the waste products of economic production. Ecosystems act on waste in two distinct ways. Biologically active compounds can be transformed through processes such as cellular respiration, nitrification, and denitrificiation. If sufficiently dilute, waste products from the processing of biotic stocks can actually benefit ecosystems and restoration; for example, sewage is often used as a fertilizer in the restoration of forests. In excessive concentrations or in ratios inappropriate for biological cycling frequently produced by human activity, however, biotic wastes can seriously degrade ecosystems—for example, when too much raw sewage is dumped directly into aquatic systems. The same is true for some wastes resulting from the use of abiotic stocks, such as CO_2. In small concentrations CO_2 can benefit restoration, but in the concentrations currently emitted they are major factors in ecological degradation. For example, anthropogenic CO_2 is acidifying the ocean, reducing the viability and calcification of coral reefs (Royal Society 2005).

Ecosystems can also quite literally absorb metals and other persistent compounds by physically binding them to soil particles (adsorption) or absorbing them as molecules into biological tissue. Other wastes from abiotic stocks, such as persistent organic pollutants, take an extremely long time to break down, as their name implies, and they probably cause some ecological problems even in small concentrations. Wastes such as heavy metals and other elements can never be broken down. When the flow of any waste into the environment exceeds the capacity of the ecosystem to break it down, the waste will inevitably accumulate, resulting over time in concentrations that seriously degrade the receiving ecosystem. Although ecosystems can evolve to adapt to high nutrient or high metal environments, the time scale necessary for the natural system to respond is several orders of magnitude larger than the human lifetime, making the loss in ecosystem function permanent as relevant to society. Over very long time spans, these wastes may be transformed through natural processes into less toxic compounds, or buried where they do no harm.

Site

Finally, the *site* component of natural capital refers to land and water as physical substrates capable of capturing solar energy and rainfall. Biotic natural capital—source, service, and sink—requires this substrate, which is undergoing increasing conversion to economic activities. Site can be quantitatively measured in terms of surface area, solar radiation, rainfall, substrate, and other factors that affect its quality, but in a market economy its monetary value is determined almost solely by its relationship to human population centers. Thus, throughout most of the world the market value of land in urban areas is generally thousands of times more valuable than otherwise identical rural land. The value of the source, service, and sink functions of natural capital is determined by its proximity to population centers.

Critical Natural Capital

While natural capital sustains all life and all economic production, not all natural capital is equally important to human survival. It is therefore useful to define *critical natural capital* (CNC) both spatially and functionally, as those components of natural capital that are essential to human survival and for which there are no adequate substitutes (Ekins, Simon, et al. 2003). In most cases CNC cuts across source, service, sink, and site. This concept is important in addressing the question of when restoration is imperative and when decisions can potentially be left to marginal analysis of costs and benefits.

Many economists argue that manufactured capital is an adequate substitute for natural capital (for the extreme version of this argument, see Simon 1996), and therefore there is no such thing as CNC. From this perspective, sustainability requires nondiminishing quantities of capital as measured by value, though the specific type of capital does not matter. Ecological economists and ecologists, in contrast, generally assume that manufactured capital can substitute for natural capital only at the margin. For example, tractors and fertilizers can allow sustained yields on smaller and smaller plots of land, but only up to a point—we cannot feed the world from a flowerpot (Daly and Cobb 1994). From this perspective, sustainability requires nondiminishing quantities of CNC. The former position is commonly referred to as *weak sustainability* and the latter as *strong sustainability* (Neumayer 1999; see also chapters 2 and 4 for more on the distinction between strong and weak sustainability).

The existence of ecological thresholds and the complex nature of natural capital in which each component is related in some way to every other component complicate the precise identification of CNC. Individual species (source components of natural capital) exhibit thresholds in the form of *minimum viable populations* (MVP), and if populations fall below this level through harvest or habitat degradation, they become extinct. Unfortunately, we do not know what constitutes a MVP, which may range in number from many millions to just a few individuals. In a complex system, the loss of one species may trigger the loss of others in a chain reaction. Ecosystems may similarly be depleted below a *minimum viable size*. For example, studies suggest that the Amazon rain forest recycles rainfall, but if the forest falls below a certain unknown size there will be inadequate rainfall to sustain the system (Salati and Vose 1984). Another threshold results when waste emissions exceed absorption capacity resulting in reduced ecosystem function and an eventual accumulation of waste. Ecosystem thresholds may also be determined by the particular configuration and character of ecosystem structure, not just total quantity. For example, the same area of forest provides different and unequal services if it is fragmented or contiguous forest (for a case study in developing forest corridors, see chapter 8).

Multiple ecological thresholds are interconnected in a complex system, and what constitutes a viable level or configuration of a given element of CNC depends on the status of other elements. For example, climate change that results when we surpass the global waste absorption capacity for CO_2 may affect both the minimum viable size of an ecosystem and the MVP of a species. Restoring natural capital may affect the minimum viable size of species or ecosystem and can help them recover from otherwise nonviable states. Ecological systems science is critical to this approach as restoration of one type of ecosystem may be best accomplished by the restoration of other bordering systems. For example, aquatic restoration (streams, lakes, estuaries) must be connected to the restoration of at least some of the functions of upstream watersheds (forests, grasslands, and wetlands) and riparian zones to avoid a return to the degraded state.

Determining which and how much natural capital is critical has significant methodological challenges. However, this determination is of key importance in determining whether ecosystem restoration should be analyzed based on efficiency (marginal costs and benefits) or whether the value is infinite and should therefore be determined based on science and ethical attitudes toward uncertainty and toward future generations. In most cases CNC cuts across source, service, sink, and site. However, there are situations in which a particular piece of natural capital may be especially valued for only one or two of these attributes or become far more valuable in a particular place where natural capital is rare. Thus, comparing between source, service, sink, and site also requires appropriate valuation methodologies.

Units of Measure

Natural capital can be measured using monetary or nonmonetary methods (i.e., ecological thresholds, ecosystem health indicators, and physical measures) (see chapters 26, 32, and 33). Which approach we choose depends to some extent on our objectives and whether we embrace the notions of strong sustainability and CNC rather than weak sustainability. One objective is to compare the value of natural capital with that of the manufactured capital into which it can be converted. If the weak sustainability paradigm holds, then monetary valuation is perfectly appropriate for both manufactured and natural capital since the two can be substituted for each other. Though many ecosystem services are not exchanged in markets, economists have developed sophisticated (albeit controversial and costly) methods for estimating nonmarket values. For example, economists might ask people how much they would be willing to pay, hypothetically, for an additional unit of services provided by a healthy wetland (contingent valuation) or estimate the monetary damages from a marginal loss of wetland services such as flood control. Such values could be fed back into price signals via taxes or impact fees for wetland development, for example, theoretically leading to more efficient allocation of wetlands. We must remember however that monetary measures capture only exchange values, which are the value of one additional unit of service; they do not measure use values, which are the benefits from all units available. Monetary measures are simply not designed to measure any type of nonmarginal change. The distinction between exchange value and use value explains why diamonds, a mere adornment, have a far greater monetary value than water, which is absolutely essential to life.

Two major objectives of restoring natural capital are to ensure sustainable scale and just distribution. Many ecological economists argue that these objectives take precedence over efficiency and are incompatible with monetary valuation. As ecosystem services are created by nature, independent of individual effort, just distribution requires that decisions on their value and allocation be democratic. But most monetary values are derived from estimated demand curves. Since demand is preferences weighted by income, monetary valuation is based on plutocratic principles, not democratic ones. Monetary valuation also discounts the interests of future generations. Sustainable scale, which is the preservation of CNC, on the other hand protects the interests of the future. In addition, if the strong sustainability paradigm holds, there are no human-made substitutes for CNC, and in a complex system there are unknown thresholds beyond which CNC can collapse. Crossing the threshold from adequate to inadequate stocks of CNC is catastrophic, not marginal, and marginal valuation is inappropriate. Appropriate measures are physical, relying primarily on science.

Scale: How Much Restoration Is Required to Support Critical Ecosystem Functions?

When evidence suggests that natural capital is nearing a threshold of criticality, restoration (in addition to conservation) is imperative regardless of cost; though if there are several ways to achieve a goal, then cost effectiveness should be a criterion. Restoring vital function is the priority. Determining the physical size and configuration of CNC is inherently a question of measuring physical attributes and relies primarily on science. However, our ignorance concerning ecosystem function means that even physical estimates of ecological thresholds cannot be entirely objective. From a sample size of one (e.g., the planet Earth when considering restoration issues on a global scale), it is impossible to estimate irreversible ecological thresholds with any certainty, and how we choose to treat uncertainty is ultimately a normative decision.

The *sustainability gap*, or SGAP, has been proposed as a framework by which CNC can be compared to the current stocks of natural capital (Ekins, Simon, et al. 2003). SGAP is defined as "the difference between the current situation, the state of the natural capital stock or the pressure being put upon it and the sustainability standard" (Ekins, Simon, et al. 2003). SGAPs are expressed in physical units dependent on the function or service being considered by specific ecosystems. Determining this gap requires that we not only measure the current stock of natural capital but also calculate the natural capital required for sustainability. If the current stock of natural capital is below the minimum estimated to be required for sustainability (quantitatively or qualitatively), then CNC is being depleted. In such instances restoration is imperative to bring the physical stock of natural capital back to the minimum level, defined as that which is self-supporting and sustainable for both ecological and human systems. Ekins, Simon, et al. (2003) have developed a more specific framework, called *CRITINC*, that practitioners can use to evaluate specific examples of CNC by tracing ecosystem services back to specific ecosystem stocks. Again, we recognize that actual implementation of many of these measures is extremely challenging due to levels of uncertainty surrounding critical thresholds.

While the CRITINC approach emphasizes stock measures, Daly (1990) has defined three flow measures for sustainable use. First, stocks of renewable and replenishable resources cannot be used any faster than they are being renewed. Second, waste emissions must be less than waste absorption capacity. Third, essential nonrenewable resources cannot be depleted any faster than technology develops renewable substitutes. When we exceed the first two limits, we must restore natural capital sources and sinks or reduce resource extraction and waste emissions (Daly 1996). We cannot avoid uncertainty when measuring such biophysical constraints, and the possibility of crossing irreversible thresholds means we should always err on the side of caution.

Can monetary valuation provide useful information about CNC? As mentioned earlier, economists recognized two types of value: *use value* and *exchange value*. Use value measures the total contribution of something to our well-being. The use value of CNC is infinite. Monetary values in contrast are based on exchange value, which measures marginal benefits. The approach for measuring total value used in estimating gross national product (GNP) is to multiply the marginal value (or price) for each good or service by the total quantity produced. Of course, manufactured capital is bought and sold in markets, and its marginal value is created by market transactions. Marginal values of nonmarketed natural capital must be estimated using other much more complex methodologies.

Even setting aside the methodological problems, estimating total values as marginal value times (×) total quantity presents a serious problem when measuring natural capital. When a resource is absolutely essential and has no substitutes, such as CNC, marginal value essentially is infinite. For capital stocks near the criticality threshold, marginal values are exceptionally high and fluctuate dramatically with small changes in quantity supplied: in economic jargon, demand is price inelastic. A small decrease in supply will lead to an enormous increase in price, so that total value (price × quantity) paradoxically increases as total quantity declines.

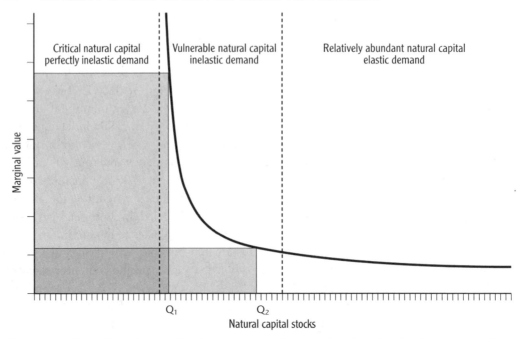

FIGURE 3.1. Hypothetical demand and supply curves for natural capital, showing increasing inelasticity of demand as capital stocks diminish, and the impact on total value as measured by price × quantity.

Figure 3.1 shows hypothetical demand and supply curves for natural capital stocks. When natural capital stocks are relatively abundant, we use them for nonessential goods and services. Marginal values are low and demand is elastic. As stocks diminish to vulnerable levels, we must dedicate what remains to meeting ever more essential needs. Demand becomes inelastic. This means that as natural capital stocks diminish from Q_2 to Q_1, the total value of the stock as measured by economists (indicated by the shaded areas in the figure) will increase, and the weak sustainability goal of maintaining a constant value of capital stocks becomes nonsensical. Even if we stick to marginal values, which are all that monetary valuation can legitimately be used for, the value of important natural capital will fluctuate dramatically with small changes in quantity, in which case meaningful values would need to be reestimated, frequently at high cost, and the cost of small estimation errors could be large.

Efficient Allocation: Which Restoration Initiatives Are Most Beneficial to Society?

When natural capital stocks fall below the criticality threshold, they must be restored. Efficiency is relevant only to how we restore the stocks, which should be done as cost effectively as possible. However, this marks only the minimum quantity of natural capital required. Once we are safely within the bounds of sustainability, we should instead consider desirable quantities of natural capital—not only whether but also how and where to restore natural capital. Desirability is a more subjective assessment than sustainability. In such analyses monetary values can be an effective tool in prioritizing restoration initiatives and in allocating limited funds for such endeavors.

The primary question concerning efficient allocation is whether the marginal benefits of a proposed restoration initiative exceed the marginal costs. As an ecosystem is restored from a degraded state toward full function, marginal benefits diminish and costs increase. Simply removing a disturbance to an ecosystem may allow the system to restore itself at a very low cost. Physical modification of a degraded system, for example, restoring hydrological flow patterns, is more expensive but may cost effectively reclaim key ecological services, including water filtration, waste absorption, and water regulation, despite the fact that the full former ecological structure has not been restored. As we approach true restoration of a system, that is, restoration not only of ecological function but also of the former ecosystem's "characteristic species and communities" (Harris and Van Diggelen 2006), costs may escalate dramatically. As long as natural capital stocks are relatively abundant, the marginal value of benefits is likely to change only slowly, as seen in figure 3.1, so that we can safely compare marginal costs and benefits. As we approach CNC however, marginal values begin to increase rapidly. While in theory it might still be possible to engage in restoration of vulnerable natural capital based on marginal analysis, the risks increase, as do the costs of continually reestimating marginal benefits. When valuation shows evidence of inelastic demand, it may be most efficient to simply reallocate resources from valuation to restoration as the costs of repeated valuations increase, and the costs of small errors could be catastrophic.

Distribution: Who Benefits from and Pays for the Restoration of Natural Capital?

In considering issues of distribution we must pay attention to the distribution of resources between generations as well as between different individuals within a generation. Intergenerational distribution is captured foremost in the decision to support strong sustainability, thereby ensuring a sustainable quantity of natural capital into the future. An intragenerational distribution requires that when weighing the marginal costs and benefits of restoration, we consider both who benefited from the degradation of the ecosystem and who would benefit from its restoration. Often these will be very different groups of people both geographically and in terms of socioeconomic class. In determining who should pay for the restoration of natural capital, monetary valuation should be used with caution. Whereas many economists assert that monetary values are an objective measure of scarcity and preferences as revealed by market decisions (see chapter 4), we argue that market demand is a function of individual preferences weighted by wealth and income; it accounts only for the preferences of those alive today, and gives greater weight to the preferences of the wealthy. The decision to discount the preferences of the poor and of future generations is normative.

Many financing mechanisms and policies are designed to balance the costs and benefits of restoration to different members of society. See chapters 26–33 for an extensive discussion of these issues.

Putting Theory into Practice

This chapter focuses on the theory behind a framework (figure 3.2) that can be used to support decisions regarding the restoration of natural capital. The framework is applicable regardless of whether we are considering the restoration of natural capital at the global scale, such as the sequestration of carbon dioxide, or at a local scale, such as water filtration. We

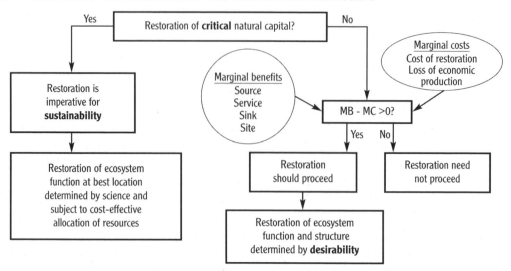

FIGURE 3.2. Applying ecological economics framework to questions of if and when to restore natural capital.

have been quite theoretical through most of this chapter, though much of this book focuses on specific examples of restoration to which this framework can be effectively applied.

Before applying this framework, the boundary within which the analysis is conducted and the goals of the proposed restoration must be clearly defined. In so doing, the identification of stakeholders becomes more transparent and the distributional issues of payment for restoration can be more carefully analyzed. The first step in applying the framework laid out in this chapter is to determine whether the ecosystem requiring restoration is on the brink of ecological thresholds that provide critical life-supporting functions. If the answer is yes, then restoration should be undertaken to close the gap between the natural capital required for sustainability and the current level of natural capital and should be done so as cost effectively as possible. In such cases, the primary goal of restoration should be to restore critical ecological function even if this means full ecosystem structure cannot be restored.

If the restoration initiative is determined to be safely below the critical ecological thresholds, then a marginal cost-benefit analysis can be used to determine which areas should be restored and to what extent. Any decision requires consultation with science and the communities that would benefit from the source, service, sink, and site resources provided by the degraded ecosystem. Keep in mind that the restoration of any natural capital is likely to have benefits at more than one scale even though the stated purpose of the restoration may be specific to the global or local scale (chapter 31). Table 3.1 shows the many ecosystem services provided by restoration of a forest at several different scales. The value of restoring the natural capital in a given section of forest may thus vary widely if it is primarily valued for global carbon sequestration versus local water filtration. Likewise the marginal value of wetlands for storm protection is likely to be much higher for local residents than its global marginal value for carbon sequestration. How the value of multiple benefits provided by natural capital at different scales is effectively used in prioritizing restoration projects is a developing area of policy and theory to which ecological economics theory can provide valuable insights. Once

TABLE 3.1

Ecosystem services provided by restored forest

	Local	Regional	Global
Service	Water regulation	Water filtration	Biodiversity
Source	Food, water	Timber: trade diminishes relevance of spatial scales	
Sink	Toxic waste	Nutrients	Carbon dioxide
Site		Key determinant of value of different services	

restoration is decided upon, efforts to assign the cost of the restoration to those who degraded the system in the first place should be made. In many cases governmental bodies will provide the funds of restoring natural capital due to the public nature for the goods and services rendered.

Contribution

In this chapter we have outlined a framework based on key principles in ecological economics that can be used to determine when restoration of natural capital is imperative and when it is appropriate to make decisions based on marginal costs and benefits. When natural capital, at any scale (local, regional, or global), is nearing an ecological threshold, as determined by science and measured in physical units, the restoration of natural capital is imperative to preserve life-supporting, self-sustaining ecological functions. In cases where existing stocks of natural capital are well within the sustainability thresholds, efficient allocation of resources, including the prioritization of restoration projects, becomes an issue of desirability and can be based on the marginal costs and benefits of the proposed restoration initiatives. In such cases, monetary valuation and market mechanisms may be useful in determining how much and which areas of natural capital it is optimal and desirable to restore. When monetary valuation provides evidence that demand for natural capital is inelastic, natural capital stocks may be vulnerable, and it becomes more efficient to spend scarce resources restoring natural capital than valuing it. Regardless of how a restoration decision is arrived upon, we stress the importance of distributing the benefits and costs of such projects justly among society and future generations.

Given the degree of scientific uncertainty involved in determining what is CNC, coupled with different ethical attitudes toward uncertainty, an objective decision-making rule for deciding when restoration is a question of necessity or of desirability is probably impossible. Agreement on the level at which we strive to measure CNC—locally, regionally, or globally—will be contentious. Which scientists or technocrats are entitled to make decisions will also be contentious. Such issues demand more discussion than space allows for here, but a framework for decision making is provided in chapter 27.

Chapter 4

Restoring Natural Capital: A Mainstream Economic Perspective

Eugenio Figueroa B.

Restoring natural capital is an important part of environmental projects, regulations, and policies in the developed world and is rapidly becoming so in the developing world as well. However, even though nature conservation has been shown to be less costly than transforming remaining wildlands to artificial uses (Balmford et al. 2002), the high restoration costs of degraded landscapes have yet to be properly analyzed from an economic perspective. This is especially relevant for developing countries faced with scarce resources and the needs of growing populations, many of whom live in poverty. For example, in many Latin American countries large financial investments have been used in restoring natural capital, especially in the mining sector prior to mine closures, and only a few studies have been carried out on the ultimate economic and social justification of these investments.

This chapter, in contrast to the previous one that discussed an ecological economics approach to restoration, presents a neoclassical (mainstream) economic approach in assessing natural capital restoration activities and recommends a multidisciplinary conceptual experiment to be carried out to enable and enrich its assessment, especially in developing countries

Assessing natural capital restoration projects, policies, and investments requires evaluation of environmental impacts, not only of their physical effects but also their relative social effects. Usually mainstream economists and ecologists disagree about the criteria to use to evaluate environmental services in general, and environmental degradation in particular. Many of their divergences arise from profound conceptual differences in dealing with some issues that are crucial in defining and determining environmental sustainability. It is expected that this chapter's pluralistic scope will increase interdisciplinary communication and reduce the broad conceptual gap separating economics and ecology regarding fundamental issues related to environmental valuation, such as substitutability, irreversibility, marginal effects, and scale dimensions.

Natural Capital and Its Restoration: An Economic Perspective

When analyzing the large investments and efforts made to restore natural capital in many countries during the 1990s, economists have been concerned with the costs and benefits of such undertakings, as well as how restoration should be implemented in the future. Resource scarcity forces societies to determine under which circumstances natural capital restoration is justified vis-à-vis satisfying everyday human needs. To analyze the type of conceptual issues involved, a formal economic model of social decision making will now be presented. Then a two tier approach is proposed to implement a multidisciplinary mechanism to assess restoration projects and policies in developing countries.

Optimal Restoration of Natural Capital

Here we examine a dynamic model of a society maximizing its welfare when it exploits its natural capital, Z, to produce consumption goods and services. The society's welfare function is defined as

$$(1) \qquad W = W(U_0, U_1, U_2 \ldots, U_t),$$

where W is a society's welfare function and U_t is the *aggregate utility* (or welfare) at time t, defined as a weighted sum of the utilities of the individuals in society. It is assumed that individuals satisfy their needs from consuming goods and services produced by the economy and harvested or otherwise utilized natural capital. Therefore, the aggregate utility in each moment t is assumed to depend on the consumption level, C_t, and the amount of natural capital at that moment, Z_t:

$$(2) \qquad U_t = U(C_t, Z_t)$$

and $U_c > 0$, $U_{cc} < 0$, $U_z > 0$, and $U_{zz} < 0$. The societal challenge is to maximize the present value of its welfare within an infinite time horizon:

$$(3) \qquad W = \int_0^\infty U(C_t, Z_t) e^{-rt}\, dt$$

At time 0, the stock of natural capital is assumed to be given and equal to Z_0. The economy produces at each moment different goods and services, which are represented by a composite commodity, Q_t. The production of this commodity is a function of the stock of manufactured capital, K_t, and the amount of natural capital used, E_t, since all goods are assumed to be produced using only manufactured and natural capital. As a result, E_t represents the total amount of natural capital used and therefore includes all the natural productive inputs (such as wood, minerals, water, air, soils, etc.), as well as forms of degradation (contamination, loss of ecosystem services and aesthetic properties, etc.) caused by human activities. Moreover, since some forms of natural capital have a capacity for self-restoration, it is assumed that they restore themselves according to function $g(\bullet)$, which has as an argument the stock of natural capital, Z_t. Additionally, natural capital can also be restored by human intervention through function $v(\bullet)$, which has as an argument the resources spent in restoring degraded natural capital, a_t.

Subsequently, the community's composite commodity is produced each moment through a production function involving the amount of manufactured and natural capital used (equation 7).

Using a standard optimal control theory problem, the society's welfare maximization problem can be represented as

$$(4) \qquad \max W = \int_0^\infty U(C_t, Z_t) e^{-rt}\, dt$$

s.t. (5)
$$\dot{K}_t = Q_t - C_t - a_t$$

(6)
$$\dot{Z}_t = -E_t + g(Z_t) + v(a_t)$$

(7)
$$Q_t = Q(K_t, E_t)$$

(8)
$$K(0) = K_0$$

(9)
$$Z(0) = Z_0.$$

In all equations $\dot{X} = \partial X/\partial t$, indicating a change or growth in a specific variable.

Equation 5 is the motion for the stock of manufactured capital of the economy, and it implies that the capital stock, K_t, changes at time t in the amount of investment generated at time t, $\dot{K}_t$, which is equal to what is produced at time t, Q_t, minus what is consumed at time t, C_t, minus what is spent in restoring natural capital, a_t.

The stock of natural capital, Z_t, in turn, changes at time t according to equation 6. It increases in the amount it is augmented by natural regeneration, $g(Z_t)$, minus the amount it is decreased by utilization and degradation E_t, plus the amount it is augmented by human-induced restoration, $v(a_t)$. Equations 7 to 9 restrict the maximization problem to the assumed production function and the given initial amounts of manufactured capital, K_0, and natural capital, Z_0.

Using a Hamiltonian function to solve the dynamic maximization problem in equations 4 to 9, and obtaining from it the canonical equations characterizing the optimal solution, it is possible to derive the following equations:

(10)
$$\frac{\dot{\lambda}}{\lambda} = r - Q_K$$

(11)
$$\frac{\dot{\phi}}{\phi} = r - g_Z - v_a$$

where λ and ϕ are the costate variables for manufactured and natural capital, respectively. Each of them measures the shadow price of its associated state variable at each point in time. The control variables are C_t, E_t, and a_t.

Equations 10 and 11 together state the dynamic efficiency condition that each asset or resource has to earn the same rate of return, and that this rate of return is the same at all points in time. The expression $\dot{\lambda}/\lambda$ in equation 10 represents the rate of change of the shadow price of manufactured capital for the economy, which in equilibrium has to be equal to the rate of change in its market price (capital gains). Markets for manufactured capital goods generally exist in modern economies and therefore there is a known value for $\dot{\lambda}/\lambda$. The expression Q_K in equation 10 is the marginal productivity of capital, showing how much product Q the community would obtain if, at the margin of production, it adds a unit of manufactured capital K to its production function $Q(K_t, E_t)$. Thus, equation 10 implies that the return to manufactured capital (i.e., capital gains plus marginal productivity, $\dot{\lambda}/\lambda + Q_K$) must be equal to the interest rate, r, which in equilibrium is also equal to the social rate of intertemporal preference (social discount rate).

The expression $\dot{\phi}/\phi$ in equation 11 represents the rate of change of the shadow price of natural capital. In equilibrium it should be equal to the rate of change of the market price of natural capital. However, natural capital will be in a steady-state equilibrium when its stock is constant over time, i.e., when $\dot{Z}_t = 0$. According to equation 6, this will occur when $E_t = g(Z_t) + v(a_t)$, or, in words, when at each moment of time the total utilization and degradation of natural capital (E_t) is equal to its augmentation produced by natural growth, $g(Z_t)$, plus its human-induced restoration, $v(a_t)$. This is true only for nonexhaustible resources. Moreover, if the demand for natural capital is constant over time, natural capital's price will remain also constant and $\dot{\phi}/\phi$ will similarly be equal to zero. Thus, equation 11 demonstrates that, in a steady-state equilibrium, society should spend resources in restoring natural capital until the point where the last dollar spent in restoration generates value equal to the rate of interest, r, minus the rate at which natural capital restores itself, g_z. Though, if the economy is out of the steady-state equilibrium, $\dot{\phi}/\phi$ need not be equal to zero.

For nonrenewable resources, such as minerals or fossil fuels, $g(Z_t)$ would be zero and $\dot{\phi}/\phi$ would be different from zero. In this case, equation 11 implies that optimal restoration is attained when the marginal resource spent in restoring natural capital generates a value equal to the rate of interest minus the price increase in the resource.

It is important to point out that in the real world it is not as easy for society to determine optimal solutions as those derived here. On the one hand, there are serious difficulties in defining the social preference or utility functions involved in the model. On the other, an explicit market for natural goods and services does not always exist and, therefore, there are no market prices for most of the goods and services provided by natural capital. This implies that it is quite complicated for society to determine either when its exploitation of renewable and exhaustible resources or its restoration of natural capital is at an optimal level.

Optimal Restoration of Natural Capital in the Real World

Two main messages arise from the previous section. First, even though restoring natural capital may sound like a reasonable aim in general, in a world of scarcity, decisions for using resources for this purpose need to follow certain rules. Second, the information needed to determine and follow such rules is not easy to obtain.

When determining optimal levels of natural capital restoration, developing countries should use the information relevant to their own conditions. This implies that, given their own social preferences, levels of development and greater abundance of natural resources relative to developed countries, developing countries are able to invest less in restoring natural capital when both confront similar natural capital degradation.

Currently, only a few studies have focused on the real needs of restoring natural capital in developing countries and their mining sectors (Ginocchio 2004; MMSD-AS 2002; Zolessi and Figueroa 2002), let alone the appropriate or optimal levels of investment for achieving this. In fact, most of these works deal with specific technical or scientific issues related to physical environmental damage, while very few studies have been carried out regarding the economic and social evaluation of restoring degraded natural capital.

Figueroa et al. (2002) propose general guidelines for improving how the Chilean judicial system deals with lawsuits related to environmental damage. Indeed, this example demonstrates the delayed response in developing countries for tackling environmental issues and could be applied to the present topic as well.

Attaining Optimal Natural Capital Restoration

Attaining optimal levels of restoration of natural capital is especially relevant and challenging for developing countries given their pressing social needs. Although abundant literature addresses nature restoration activities and analyzes different technical aspects (e.g., engineering procedures, pollution treatments, reforestation measures, etc.) of them, little documentation is available regarding the economic aspects of such activities. It is therefore difficult to assess whether the investments for restoring natural capital in developing countries are close or far from being optimal from an economic perspective. In Latin American countries, for example, where some mining activities are in remote areas, the social costs of natural capital deterioration could be small if managed properly. It might therefore be more profitable and appropriate for these countries to use their limited investment resources for urgent social needs in a social offset investment elsewhere.

Tradeoffs

When analyzing the cost of remediation and reclamation that have followed mine closure programs in various Latin American and other developing countries in the world, an economist might ask if the resources used in such activities could have been better invested, for example, in improving urban living conditions. This line of thought is familiar to economists but less so to ecologists and conservationists, whose priorities are often based on different perceptions, and hence a common ground needs to be sought. This book could make a relevant contribution toward establishing it.

Problems of Scale

Economists usually analyze human behavior and optimal decisions using the marginal approach of modern neoclassical economics. This is precisely the approach used in the economic model presented earlier, but as is usual in economic analyses little attention was paid to the scale of the economy as a whole and its relation to the natural resource base underlying it. It is obvious that the model establishes a relationship between natural resource endowments, that is, the stock of natural capital in the model (Z_t), and economic activities, that is, production (Q_t), consumption (C_t), and investment ($\dot{K}_t$); however, the scales of these activities affecting the ecosystems and their capacity to support and maintain them over time were not considered. This approach differs greatly from the ecologists' perspective for whom the notion of scale is generally central in the analysis, with thresholds of scale perceived to be unable to be crossed without major and irreversible losses in ecosystem functioning.

It is essential to merge these two approaches and to incorporate the type of thresholds and scale effects that concern ecologists in a model of the sort used here. This would enrich the economic analysis, though it seems that the theoretical and empirical knowledge of ecology is insufficiently developed to provide quantitative estimates of such thresholds and scale effects as to allow economists, by using them, to reduce the uncertainty of their own current economic models, which are also full of uncertainties. There is therefore an ongoing need for truly interdisciplinary work to identify available knowledge for understanding, defining, and quantifying natural capital's thresholds and scale effects that could be introduced into economic models.

Substitution and Irreversibility

The model presented earlier illustrates another area of disagreement between economists and ecologists that is at the heart of the discussion about the social appropriateness of using scarce resources for restoring natural capital rather than for tackling "more urgent" human needs. Economists generally consider that there is some degree of substitution between natural and manufactured capital. Moreover, for economists the concept of substitutability refers to the ability to alter production and/or consumption, without changing the desired overall flow of services, when the scarcity of some resource has increased. In the model presented, for the production function of the good produced by the economy, $Q_t = Q(K_t, E_t)$, it is assumed, that units of manufactured capital (K_t) and of natural capital (E_t) can be substituted to produce the good Q_t, since the function $Q(\bullet)$ is really an engineering type of relationship. Indeed, this kind of substitutability generally exists empirically (i.e., it is observed in the real world) and is similar to the physical properties of the substitutability concept used by ecologists. However, for economists, substitutability always refers to social (and/or individual) valuation of resources, goods, and services in the context of the economic problem, that is, the presence of scarcity.

To understand this in the context of our model, it is necessary to see that in the aggregate welfare function, $U_t = U(C_t, Z_t)$, that society maximizes in equation 4, manufactured capital (K_t), and natural capital (Z_t and E_t) enter indirectly, through the produced good (Q_t) that is consumed (C_t), or directly, through Z_t. It is through their relative contributions in value, in terms of the welfare they provide to society, that natural capital (in its Z_t or E_t form) and manufactured capital (K_t) are substitutable to generate the optimal level of the composite commodity (Q_t, the goods and services of the economy) and, ultimately, of society's welfare (W). Therefore, the economic notion of substitutability goes beyond the engineering concept of a production function, while the ecological notion of substitutability is closer, as ecologists generally use the terms *substitutability* and *reversibility* in reference to the biophysical properties of ecosystems themselves. In fact, for ecologists the reversibility of a condition is related to the resiliency of the ecosystem and its capacity to regain a former level of function after disturbance. Substitutability is in ecological terms a form of redundancy (Walker 1992) in the sense that it refers to the existence in an ecosystem of alternative sources of a given attribute when it has diminished.

Economic recipes to achieve optimal solutions are stated in terms of marginal conditions that largely disregard scale dimensions. This is because economics and economists are concerned with sustainability in the sense of maintaining human (individual and social) well-being over time. Basically, therefore, economists are concerned with the capacity of natural capital (Z_t), manufactured, and other forms of social capital (all included as K_t) to meet human requirements. Natural capital obviously satisfies part of these needs but does not encompass all the variables determining human well-being.

As Norton and Toman (1997) noted, if economic substitution possibilities are high enough, natural capital disruption is not a special cause for concern in the economic model, provided that society's total saving rate is high enough to compensate for natural capital reduction and thereby produces sustainable welfare paths. Even irreversible changes in the physical state of ecosystems are insignificant in this case, though the economic consequences of such changes need to be accounted for. However, as these authors explain, the converse also is true: if substitution in an economic sense is limited, then satisfying both current consumption demands and intergenerational equity concerns can lead to a greater need for safe-guarding natural capital. In this case, those types of biophysical irreversibility studied by ecologists could also raise worries about economic costs that are a significant social concern.

To summarize, an interdisciplinary approach directed toward the convergence of the concepts used by ecologists and economists, such as scale, marginal effects, substitutability, and reversibility, could improve the ability of current economic models to assure future human welfare levels with a reasonable degree of certainty, as well as the social relevance of the ecological models.

A Two-Tier Approach for Environmental Valuation

A multidisciplinary conceptual experiment is now proposed for assessing natural capital restoration projects and policies implemented in Latin American and other developing countries. The purpose is to determine how activities for natural capital restoration could be monitored, modified, and redesigned to improve their contribution to the social welfare of these countries.

The two-tier approach proposed by Norton and Toman (1997) for enriching current environmental valuation could lead to a more pluralistic and interdisciplinary system of decision making. This approach establishes a categorization of problems that determines the kind of decision rules that should be applied in different cases and identifies the decision rules themselves, implying that the context determines their application.

An example of a two-tier approach is proposed by Page (1977), which categorizes problems as intra- or intergenerational in their impacts and uses two criteria: efficiency and conservation. When, in a given context, intergenerational impacts predominate, the rule to apply is conservation.

To apply the two-tier approach, problems should be categorized according to the type of risk involved, followed by the identification of which criteria apply within different categories. The two parts of the process cannot be carried out in isolation. However, it is possible to improve the quality of the interdisciplinary sustainability discussion by asking the relevant questions separately, that is, by dividing the issues of which criteria (efficiency or conservation) are applicable, from the question of which specific policy should be implemented (Norton and Toman 1997).

This two-tier method has the further advantage of integrating the usual cost-benefit approach of welfare economics, as well as other options, while encouraging public discussion on which criteria to apply in sustainability calculations and measures. It can also foster a more adaptive, experimental process in which scientists, local communities, and policymakers collaborate on what to do in specific situations, and which criteria might be appropriate. This kind of discussion may lead to a process of value articulation, criticism, and experimentation with multiple schemes for valuing environmental goals. This is particularly important in developing countries, where public or community participation in policy decisions is rather weak and the democratic and political system in charge of channeling social preferences into concrete policies and projects operates with less efficiency than in developed countries.

To work together analyzing natural capital restoration projects, economists and ecologists could use this two-tier approach, incorporating previous attempts to integrate multiple criteria for action (Norton and Ulanowicz 1992). These attempts characterize situations in which the standard cost-benefit analysis can be applied, while in others they can be weighted using the *safe-minimum standard of conservation* (SMSC) criterion. This criterion asserts that a resource should be saved "provided the social costs are bearable" and, therefore, it places a larger burden of proof on developers to demonstrate that the costs of protecting an important resource are unacceptably high before undertaking the risk. Norton and Toman (1997) argue that the SMSC criterion is a concrete expression of a moral judgment and that large-scale, negative environmental effects may have unacceptable consequences for the intergenerational distribution of opportunities and well-being. In its favor, this criterion provides a larger scope for balancing benefits and costs than the precautionary principle, which argues for erring on the side of caution when uncertainty is high. For example, the environmental impact at mining sites, generally spatially confined and potentially reversible by natural succession in a number of years, could be incorporated into a cost-benefit analysis and be valued mostly on efficiency grounds. On the other hand, situations with widely dispersed impacts and/or impacts expected to persist for three or more generations, would be analyzed using the SMSC criterion, and valued mostly on intergenerational equity grounds.

Moreover, a scale dimension related to the ecological consequences of environmental impacts caused by mining activities could be incorporated into the first phase of the two-tier approach. In this way, situations in which environmental impacts affect features of ecosystems, whose destruction may threaten "fundamental ecological production functions," can be differentiated, analyzed, and treated differently from others where impacts are on a smaller scale. Different criteria to evaluate costs and benefits in each case could then be used; for example, valuation could use a conservation criterion or a production criterion, depending on the scale of negative ecological impact.

In agreement with Norton and Toman (1997), this pluralistic methodology for valuation within a two-tier approach can be combined well with the ecologists' concept of "adaptive management." This emphasizes that, in situations of high uncertainty, management plans should be formulated so as to improve knowledge and reduce uncertainty by approximation (Holling 1978; cf. chapter 11).

Combining the two-tier approach proposed here with ecological "adaptive management" may be the most balanced strategy, for example, during the implementation of closure plans in the mining industry, since these can occur in various phases over a protracted period. Moreover, environmental specialists in the mining sector are familiar with the concept of adaptive management.

Contribution

Investment in natural capital restoration is rapidly increasing and, therefore, it is important to assess the social convenience of this investment compared with other human needs. Only by bridging the deep conceptual differences between economists and ecologists will it be possible to reach the required social consensus to make more informed decisions about the exploitation and conservation of natural capital. Thus, there is a need for interdisciplinary work to enrich both economic and ecologic models for valuing natural capital. This chapter has proposed an economic model that could help in such a work. The next chapter reviews lessons learnt by the interdisciplinary Millennium Ecosystem Assessment team.

Chapter 5

Assessing and Restoring Natural Capital Across Scales: Lessons from the Millennium Ecosystem Assessment

RICHARD B. NORGAARD, PHOEBE BARNARD, AND PATRICK LAVELLE

Whereas the previous two chapters considered two alternative economic approaches on deciding when to restore natural capital, this chapter reflects on the lessons learnt by the *Millennium Ecosystem Assessment* (MA) for the assessment and restoration of natural capital, mostly from an ecological perspective. At the outset it should be stated that it is not easy to restore ecosystem services or biodiversity with our current paradigms and ecological tools. Though we understand intuitively that local natural capital is profoundly influenced by patterns and processes in the broader environment, it is difficult to translate abstract insights into practical actions for assessment and restoration. The MA therefore endeavors, vigorously, to improve our ability to gain insights about ecosystem services across spatial and temporal scales.

The MA is the most comprehensive and in-depth assessment undertaken of how people depend on and change ecosystems around the globe, involving about two thousand scientists over five years. With an assessment structured around ecosystem services and the physical and biological systems providing them, the MA scientists reviewed a vast literature, including how the relationship between people and the environment may change and respond to change in the future (MA 2005f).

The MA was based around a framework (figure 5.1) for understanding the interplay between humans and ecosystems (MA 2003). This recognizes three broad categories of ecosystem services of direct importance to people: provisioning, regulating, and cultural. It also notes the reliance of ecosystems themselves on supporting services, which sustain ecosystems and human livelihoods. Furthermore, the MA framework provides a way of thinking systematically about drivers of ecosystem change and helps identify intervention points to manage the way humans interact with ecosystems.

What can this process tell us about natural capital and the restoration of degraded ecosystems? The framework reminds us that we cannot understand socioecological systems and natural capital at a single spatial or temporal scale, an issue also highlighted in chapters 3 and 4. Many participating scientists conduct relatively local and short-term field research, while others work conceptually at global scales and for longer time periods. A major challenge has been to understand how processes at one scale affect those at another. By layering different spatial scales in figure 5.1, we are reminded to think locally, regionally, and globally, and in shorter and longer timeframes. However, while the diagram helps us consider interrelation-

36

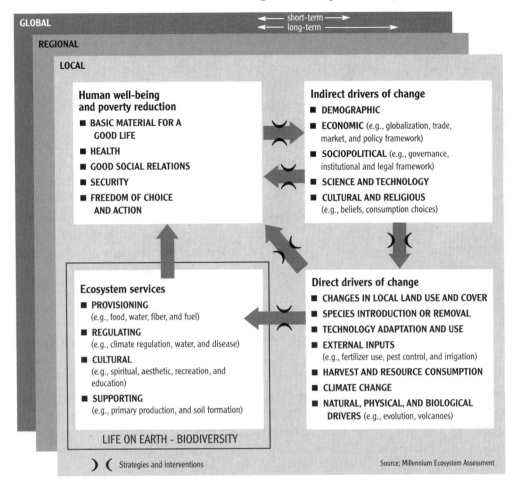

FIGURE 5.1. A framework for understanding ecosystem services and ecosystem change.

ships between social and ecological systems at a given scale, it does not allow us to comprehend the dynamics across scales. In fact, neither figure 5.1 nor any other diagram in the MA
abstracts any key method for making links across scales. This absence suggests the difficulties
of the process.

Nevertheless, restoring natural capital requires the assessment of the degradation drivers
and possible solutions across scales. On a small scale, restoration considers carefully the local
environmental conditions and how these might be improved and maintained, including convincing local people that their well-being has been improved (see the definition of *restoring
natural capital* in chapter 1). On a larger scale, the restoration and maintenance of natural
capital must be consistent with management of the broader landscape. For example, wetlands are influenced by upstream activities (such as inputs of nutrients, toxins, and seeds of
invasive alien plants) so that restoration can succeed only if planned within the landscape
context. In addition, to restore natural capital, it is necessary to identify who benefits from
ecosystem damage and/or who bears the costs of restoration.

As restoring natural capital becomes accepted practice, multiple interrelated sites, such
as conservancy farms or major watersheds, can increasingly be restored together. Perceiving

localities as dynamically connected requires a larger-scale analysis, while a broader social context may be essential for successful restoration and maintenance.

The Multiple Ecologies of Natural Capital

Natural capital is an economic metaphor for the ecosystem resources from which natural services—physical and biological—flow (see chapter 1 for a more comprehensive definition). How we understand ecosystems therefore influences our abstract notion of natural capital. Yet plants and animals are not intrinsically organized in well-bounded ecosystems. We organize them conceptually in different ways, depending on the ecological paradigm used.

Ecology consists of formal models, interpretive frameworks, and metaphors that help us understand ecosystems. We can think of an ecosystem in terms of interacting populations in an environment with a carrying capacity for each population, that is, a population biology model. Alternatively, an ecosystem can be viewed in terms of interacting biogeochemical cycles. Energetics models organize ecosystems according to energy flow among organisms, whereas food web models or hierarchy theory describe the interactions of these components differently again. Some of these models can be described with equations that have analytical solutions, while others can be simulated and run as computational models. In some cases, such as species-area curves, which predict that species numbers increase within a sampling area, a general rule has emerged from statistical analysis of empirical data. *Evolutionary ecology* (the study of how ecological contexts affect the evolution of organisms over time, and vice versa) and *landscape ecology* might thus be described as interpretive frameworks. Ecologists also use metaphors, such as *environmental or ecosystem engineer,* to remind themselves that species from microbes to elephants transform their physical environments. Assessment of the potential for ecosystem restoration uses insights from all these paradigms of ecology.

The multiple ways by which we understand ecosystems have three important collective traits. First, they do not all fit together within any overarching ecosystem metamodel. Ecologists do not first think broadly and then smoothly switch to the particulars. Nor can they always start with one particular issue and progress to another. Rather, ecologists see different bundles of particulars in subtly different ways, where different models of ecology do the bundling. Theoretical ecologists may have the luxury of working within one paradigm, but restoration ecologists must be able to jump between several. The absence of a metamodel makes these jumps necessary.

Second, because the frameworks of ecological analysis do not all cohere within a meta model, some insights gained from thinking in different frameworks may be complementary, and others contradictory. From a population biology perspective, one might imagine stable equilibria or bounded cycles. Such imagined equilibria are soon disturbed, however, on switching to a multispecies or assemblage perspective (Taylor 2005). Ecological understanding requires judging which model and insights are most applicable to particular situations. Having favorite models may blind a restoration ecologist to possible, or even fairly probable, outcomes. Alternatively, making assumptions about how an ecosystem will look in the future may lead one to select an inappropriate model that rationalizes a preconceived endpoint.

Third, some ecological paradigms apply best at a particular scale, while others apply at several. The scale at which ecologists think often depends on the organisms they study.

Studying wolves and caribou in a predator-prey relationship requires a different spatial scale than do meerkats or scorpions. Similarly, evolutionary timescales for pathogens are not the same as insects, fish, or albatrosses. In fact, a switch in mental frameworks is frequently required to accommodate the range of ways models relate to different spatial and temporal scales. Relying on one way of thinking about ecosystems can limit not only which parts and relationships are considered, but also the spatial and temporal scales of analysis.

The various approaches to understanding systems across scales, and their implications for policy and management, are recognized in an emerging interdisciplinary literature on adaptive social and ecosystem management (e.g., Berkes et al. 1998; Gunderson and Holling 2002). Ecology texts usually present the different models but say little about how to use the range of tools together. Authors of such texts frequently organize their chapters by scale, going from organism to community and landscape levels, without saying much about scale itself. Effective assessment and restoration of natural capital requires an understanding of interactions across scales as well as within scales.

Lessons from the Millennium Ecosystem Assessment for Restoration

The MA provides multiple lessons for thinking about spatial and temporal scales, while documenting the challenges to help subsequent researchers and practitioners set out from a realistic starting point. Most of these lessons are about analytical processes and paradigms for restoration, rather than recipe-book insights about methods.

1. Case studies cannot always be combined to understand the larger whole.

Many MA ecologists expected to be able to connect their own field knowledge with that of other natural scientists working in similar ecosystems around the globe. In reality, this has proven difficult as scientists investigating apparently similar ecosystems sometimes use different frameworks, or emphasize different key variables in their analysis, simply because they focus on distinct ecological problems. In addition, two ecosystems may appear alike yet are subject to different driving forces and are on unrelated trajectories. Similarly, the MA ecologists hoped to fit their understanding of natural systems with that of social systems, but they quickly discovered that social scientists use multiple frameworks and emphasize different variables. Most important, the links to ecological systems made by those stressing social systems were usually inappropriate for how ecologists modeled ecosystems, and vice versa. Furthermore, these conceptual mismatches were often compounded when looking at case studies across scales.

2. Data are poor or incompatible.

Ecologists rely on data collected by agencies with specific mandates. For example, in many (not all) parts of the globe, weather agencies have long historical records. However, weather data are often intensively recorded in populated areas but less so elsewhere, such as in mountainous regions where conditions change dramatically along transects. Similarly, data on population changes of various wild plants and animals are rare and often of poor quality, even for commercially exploited food species such as Brazil nut trees (Silvertown 2004) and fish (Smith et al. 2001). These kinds of data problems become particularly marked in developing countries. Nevertheless, even the United States has had considerable difficulty establishing its National Ecological Observatory Network (NEON) due to the different needs of contrasting ecological fields (NRC 2003).

The classification of ecosystems and terminology made it difficult in the MA to aggregate studies or even to confirm results from apparently similar investigations around the globe. In many cases, this was just a problem of incomplete data. In other cases, datasets were complete but not comparable; soils were classified differently, or terms such as "desertification" were defined so variably that data could not simply be compared. Hence, ecologists restoring natural capital should expect to be confronted with similar problems in any effort to assess natural capital across scales, or to understand how restoration at one site can inform that at another.

3. The middle is missing.

Most ecologists have developed their theories through fieldwork. The field of global ecology has emerged in the past quarter century (Rambler et al. 1989), leading ecologists to leap conceptually between small-scale patterns and global biogeochemical cycles. Of course ecologists have long known, for example, that migrating birds connect continents, and so on, but ecological models incorporating such linkages and case studies documenting their importance are few. While there have been many efforts to construct theories covering the middle scales for several decades (e.g., Brown 1995), applications are sparse. People interact on different scales, and globalization has extended and intensified large-scale linkages, not least through serious long-distance biotic invasions. Accelerating change also makes it problematic to compare studies from different time periods. Thus, it is not surprising that the participating MA scientists, especially those carrying out multiscale assessments, had difficulties extending local and regional analyses to larger scales. Nevertheless, the multiscale assessments of the MA were pioneering in at least partially overcoming these obstacles.

4. Scale is important.

As the MA's global analysis was initially conceived, scientists were expected to assess the conceptual and applied literature and evaluate the state of global ecosystems. The case studies, however, were typically local and could not simply be aggregated to understand larger phenomena—especially when ecosystem change in one region, such as montane deforestation, damaged an ecosystem nearby. Without a systematic way of considering both the mountains and the plains, little can be said in aggregate (Levin 1992). These types of complications led MA participants to introduce a whole new approach into the assessment process. Several scientists chose to "build up" from their own local field projects to the global scale. Adapting the MA framework and working in teams, they expanded their analysis of local drivers and effects to include these at the regional, national, and international level. Eventually eighteen approved and sixteen associated projects took this approach and contributed to the findings (Capistrano et al. 2005).

5. Teams can better leap the conceptual and empirical gaps.

While a single person may be aware of many paradigms, each individual knows and applies only a few approaches. By working together, MA researchers could quickly identify different patterns of thinking to help understand the essential dimensions of an issue. As a result, the complementary and contradictory knowledge in the teams forged a much stronger basis for judging the most important linkages and identifying practical examples. Similarly, the continual discussion and, hence, reinforcement of different conceptual arguments and data increased their perceived robustness of the conclusions.

The positive MA experience suggests that the assessment of natural capital and its restoration across scales requires shared team learning. Ecologists should undertake restoration planning and management with scientists and practitioners from different backgrounds and mindsets, on whom they can test their assumptions and arguments. The institutional setting for restoration should include scientists, managers, economists, and others with varied experiences, while interacting with landowners, planning agencies, economic stakeholders, environmental interests, and community groups. This will often generate difficult but ultimately usually productive discussions. The most important MA lesson is that restoration ecologists should interact with other stakeholders, and view this as an opportunity to strengthen systematic understanding rather than as a necessary chore. This becomes even more pertinent when assessing restoration across scales.

6. Working together across the natural and social sciences is difficult but essential.

Though ecologists think differently about ecosystems and approach restoration with varying experiences, they tend to speak the same way and share similar assumptions. This is often not the case when they work with social scientists (Lélé and Norgaard 2005). Economists tend to favor economic growth and seek economically efficient solutions (see chapter 4). Meanwhile, other social scientists may worry about social equity, cultural opportunities, or governance systems. Indeed, social scientists may seem especially anxious that ecologists understand them, but the reverse is also true. Both ecologists and social scientists come with their own vocabularies. It is worth acknowledging early on that both sides have weaknesses and strengths, and it is these that make working together essential but complicated. Nevertheless, it takes time to establish common ground, based on shared interests, approaches, and language.

7. Perspectives, policy issues, and players change along spatial and temporal scales.

MA scientists discovered that the political and policy agendas shifted from local projects to national or regional scales. New actors brought into the process at larger scales saw local projects on their own terms, in light of provincial or national politics. Regional and national politics have a life of their own rather than simply being the sum of local politics. It is important to accept this as "natural," rather than becoming frustrated and losing headway over the difficulty of holding local policy perspectives at larger scales. Rather, shifting political priorities and perspectives should be seen as another opportunity to see restoration through different lenses. Restoration ecologists need to shift modes and present the restoration project as best they can in the new policy context.

8. Economic approaches to the valuation of natural capital help frame discussions about value.

Other chapters of this book discuss economic analysis in general (chapters 2, 3, and 4) and market valuation of natural capital in particular (chapters 26 and 32). A few important lessons from the MA, however, are relevant.

Over the past few decades, conservation biologists have worked with economists, deriving economic values for species and ecosystem services to convince the public that these are valuable and deserve protection. Within the MA, many economists also worked on the policy end of the process by identifying how changes in economic incentives and valuation could improve ecosystem management and slow, or even reverse, degradation (Chopra et al. 2005). Furthermore, the MA also demonstrated that not only could economists contribute to the systematic understanding of the interaction of people and ecosystems, but they could be essential for changing how people impact the environment (figure 5.1). In this role, economists are at the beginning of the process, working with ecologists to sort out how economic institutions and cultural differences affect drivers of ecosystem change, rather than at the policy end, suggesting possible solutions after the damage has been done.

The valuation of ecosystem services, and thereby the value of natural capital itself, turned out to be problematic within the MA. One problem was that some values for ecosystem services in poor countries, such as ecotourism, depended on rich nations continuing to be rich and supplying ecotourists, and poor nations continuing to be poor and providing labor. Valuations of ecosystem services were fairly rare, so there was much interest in the possibilities of interpreting benefits from a particular study to other seemingly related situations. However, "transferring" economic valuations from one site to another proved as difficult as transferring the findings of ecological analyses, for many of the same reasons. These difficulties were accentuated by the extreme differences between rich and poor countries or between very different cultures. Nevertheless, the MA experience suggests that valuations of ecosystem services might best be used to frame discussions about value, rather than as the best way to assess value (see also chapter 3).

Dependence of Ecological Services on Multiscale Buffering Systems

The MA stimulated economists, sociologists, and ecologists to describe and analyze trends in ecosystem services and project these into different future scenarios. It soon became evident that we still cannot easily integrate and model the complex multiscale processes that produce ecosystem services (Holling 2001; Mattison and Norris 2005). Nevertheless, the MA showed the huge efficiency of any investments for improving the status and sustainability of ecosystem services, while also revealing the significance of buffering systems for regulating ecological processes at different scales (Lavelle et al. 2005).

Ecosystem services (e.g., flood prevention, climate regulation, or nutrient cycling) are delivered mostly at large scales, although they integrate complex processes at smaller scales. For example, Belnap et al. (2005) describe how lichens, mosses, and algae on the soil surface, together with soil-dwelling organisms, influence rates of rainwater infiltration and the water-holding capacity of the soil and thereby moderate water runoff from the soil surface. At a larger, spatial scale, plants and vegetation patches retard runoff and increase infiltration. In this way, organisms and processes at small and medium scales can control and buffer the rate of water and sediment discharge to rivers, and reduce flooding and siltation at larger scales and great distances from the rainfall event. Similarly, carbon storage in soils depends on a suite of ecological processes crossing many scales (figure 5.2), from tiny microbial assemblages to the landscape level, and ultimately global geological and climatic factors (Lavelle et al. 2004). Since scale is important for ecosystem function and, hence, human survival, restoration should occur at all scales and ideally be coordinated so as not to overlook essential links within the larger process.

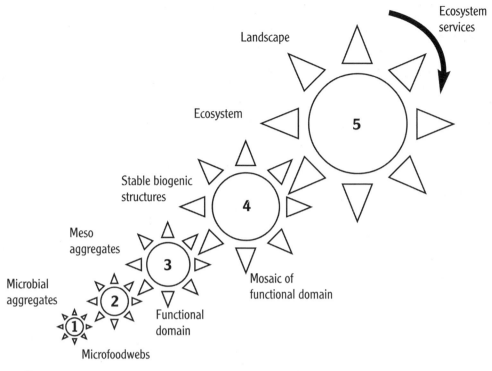

Figure 5.2. Self-organizing systems at different scales interact (almost like interlocking gears as indicated in the figure) across scales to deliver soil ecosystem services at landscape scale (Lavelle et al. 2004). The delivery of ecosystem services results from the integration across scales of processes that produce and regulate the service. See text for examples.

Contribution

The assessment of natural capital and efforts to restore it cannot be considered at a single scale. Rather, natural capital needs to be assessed and restored across scales. Ecology still has difficulties in tackling processes that involve different scales and different categories of processes. The many ecological models, interpretive frameworks, and ways of thinking about ecosystems do not cohere within a metamodel, and any particular approach works better at a particular scale or range of scales. The situation is similar in the social sciences. This means that the assessment and restoration of natural capital requires researchers and practitioners to *jump* between ways of understanding ecological and social systems. The process of jumping will be greatly facilitated by an interdisciplinary team approach, to help assure the quality and robustness of the arguments underlying the assessment and restoration of ecosystem processes.

Chapter 6

Assessing the Loss of Natural Capital: A Biodiversity Intactness Index

REINETTE BIGGS AND ROBERT J. SCHOLES

In the process of economic development, it is commonly observed that natural capital is partially converted to manufactured capital. We examine this conversion in the context of South Africa. If the decline in natural capital resulting from activities such as agriculture and mining is balanced by an increase in other forms of societal capital, it may be argued that the decline is simply a transformation of capital into a different form rather than a loss. Where the decline in natural capital is not accompanied by an increase in other forms of capital, it represents a clear loss to society in the long term. Understanding the relationship between natural capital and other forms of societal capital can assist in identifying conditions where restoration of natural capital is appropriate and required.

We suggest that a proxy for at least part of the natural capital decline in South Africa is the change in the population sizes of various forms of wild organisms (biodiversity) relative to the preindustrial period. These populations can be seen as an expression of natural capital since most ecosystem goods and services rely on specific organisms for their production, and the size of the flow depends on their abundance. Scholes and Biggs (2005) developed the *Biodiversity Intactness Index* (BII) to estimate changes in the mean population sizes of wild organisms when subjected to different land uses. The BII is conceptually equivalent to the *Natural Capital Index* (Ten Brink 2000) that has been applied in several European countries. Based on BII it is estimated that by the year 2000 the overall population sizes of plants and vertebrates in South Africa had declined by about 20% relative to their precolonial levels, with much higher declines in certain groups, such as large mammals (Scholes and Biggs 2005).

The objectives of this chapter are to apply the BII to changes in land use in South Africa between 1900 and 2000 to estimate the rate and pattern of loss of renewable natural capital. There is no general estimate of the reduction in renewable natural capital in South Africa from the time of European colonization in the seventeenth century to the present. We then compare the BII changes to variations in *gross domestic product* (GDP) and *fixed capital stocks* (FCS) over the same period to explore the relationship between natural and manufactured capital.

Economic Development and Changes in Society's Capital Assets

The wealth of a society can be thought of as the total worth of its capital assets. As defined in chapter 1, a capital asset is any stock that yields a flow of goods or services (Costanza and Daly 1992; Arrow et al. 2003). In this chapter, we define three broad categories of capital stocks: natural, manufactured, and human. From a theoretical perspective, it may be argued that, provided a society's capital stocks are properly valued and not declining, the flow of goods and services that sustain society can be maintained (Arrow et al. 2003). If a society's asset base is liquidated, flows may increase in the short term but will inevitably decline in the long term.

The valuation of natural and human capital has been attempted only in recent years (e.g., Costanza et al. 1997; Arrow et al. 2003). In contrast, the valuation of manufactured capital assets, given by the FCS in a country's national accounts, is relatively well advanced. GDP is a measure of the flows from manufactured capital (e.g., the value of factory products) and at least part of the flows from human capital (e.g., legal consultation fees, etc.) and natural capital (e.g., the value of timber extracted). While the GDP does not capture flows of all goods and services, and in particular does not account for many significant ecosystem services such as clean air and nutrient cycling, it is the best available integrative measure of the productivity of a society's capital base. Given the difficulty in valuing total natural, human, and manufactured capital stocks, GDP is often used as an indicator of development and the well-being of societies. Since GDP is a measure of flows rather than stocks, this can be misleading: Short-term increases in GDP may stem from an increased capital base with increased flows, or from the liquidation of capital stocks.

It is commonly observed that elements of natural capital are converted to manufactured capital during the process of economic development. Flows from natural capital tend to predominate initially as societies draw on their mineral wealth, forests, wildlife, and soil fertility to generate products for consumption and trade. The resulting revenues may be converted into human capital (through education and the development of institutions and services) and manufactured capital (value-adding industries and the accumulation of financial assets), which then begin to yield dividends of their own. In some cases, however, reinvestment of flows from natural capital into other forms of capital does not take place. Society then derives only short-term gains from the liquidation of the natural capital stocks and is ultimately left poorer. This is arguably the case with oil extraction in parts of central Africa.

Where investment in manufactured and human capital does occur, flows from these capital stocks may start to dominate in time. Once total wealth per capita rises substantially above the level needed to satisfy basic needs, some of the society's revenue is often reinvested in natural capital through protection and restoration of natural resources. This is the basis of the so-called inverse *Environmental Kuznets Curve* (Panayatou 1995), whereby some environmental quality indicators initially decline but then improve again in wealthier societies. In such cases, economic development can be thought of as taking a loan from natural capital to diversify the economic base and then later repaying part of it. In this sense, restoration can be seen as repayment of a loan.

Reconstructing Historical Changes in South Africa's Natural Capital

We use changes in the Biodiversity Intactness Index as a proxy for changes in natural capital. The method for calculating the BII is detailed in Scholes and Biggs (2005). Briefly, the BII measures the average change in population size of all plants and vertebrates, relative to the premodern state, as a consequence of different land uses. The reference for our study is nominally the precolonial period. In practice, contemporary abundances in large protected areas were used as a proxy for the precolonial state. Changes in population abundances relative to the precolonial period were estimated by a panel of sixteen experts for a range of land uses of increasing intensity (table 6.1). Estimates were made for approximately ten functional types (groups of species that respond similarly to human interference) per broad taxon (plants, mammals, birds, reptiles, and amphibians) for each of the six major biomes (forest, savanna, grassland, arid shrublands, fynbos, and wetland) in southern Africa. The estimates were aggregated to the broad taxon level by weighting by the species richness of each functional type. The BII is derived by weighting the relative population abundance estimates for each taxon in each land use by the area affected and the number of species (richness) in the taxon:

$$BII = \frac{\sum_i \sum_j \sum_k R_{ij} A_{jk} I_{ijk}}{\sum_i \sum_j \sum_k R_{ij} A_{jk}}$$

where
I_{ijk} is the population of taxon i, under land use activity k in ecosystem j, relative to a reference population in the same ecosystem type;
R_{ij} is the richness (number of species) of taxon i in ecosystem j; and
A_{jk} is the area of land use k in ecosystem j.

The estimated fractions of South Africa in various biodiversity impact classes were calculated per decade (table 6.2). Fluctuations in BII over the twentieth century were assumed to be only a function of changing land use, A_{jk}. Changes in the impacts of specific land uses on particular functional groups (I_{ijk}) and the potential species richness of different biomes (R_{ij}) were assumed to be negligible.

Reconstructing Land Use Changes Over the Twentieth Century

The spatial pattern of land use in South Africa between 1900 and 2000 was reconstructed per biome, at the magisterial district level. Data on the progression of cultivated area were obtained from agricultural statistics collated at the magisterial district level by Biggs and Scholes (2002). Where a district spanned more than one biome, the area under cultivation in each biome was allocated in proportion to the cultivated area in the 1995 Land Cover map of South Africa (Fairbanks et al. 2000; figure 6.1). The change in afforested area over the twentieth century was obtained in the same manner. The locations and dates of establishment for designated protected areas in the International Union for the Conservation of Nature and Natural Resources (IUCN), categories I to V, were obtained from the *World Database on Protected Areas* (IUCN and UNEP 2003). The urban land area in each magisterial district was calculated from the 1995 Land Cover map and decreased in proportion to the change in the percentage urban population in each province over the period. Land not in the protected, cultivated, plantation, urban, or degraded categories was assigned to the moderate use class.

The area of degraded land was reconstructed by building a narrative of the major land use–related changes and their causes, as well as their location and timing over the twentieth

TABLE 6.1

Land use classes and associated data sources informing the Biodiversity Intactness Index

Land use class	Description	Examples	Data source
Protected (BII = 100%)	Minimal recent human impact on structure, composition, or function of the ecosystem. Biotic populations inferred to be near their potential.	Large protected areas, "wilderness" areas	World Database on Protected Areas (IUCN and UNEP 2003). All designated protected areas of IUCN categories I–V
Moderate use (BII = 93.3%)	Some extractive use of populations and associated disturbance, but not enough to cause continuing or irreversible declines in populations. Processes, communities, and populations largely intact.	Grassland and savanna areas grazed within their sustainable carrying capacity	All remaining areas not classified into one of the other five categories
Degraded (BII = 56.7%)	High extractive use and widespread disturbance, typically associated with large human populations in rural areas. Productive capacity reduced to approximately 60% of "natural" state.	Areas subject to intense grazing, harvesting, hunting, or fishing; areas invaded by alien vegetation	1995 South African National Land Cover map
Cultivated (BII = 25.1%)	Land cover permanently replaced by planted crops. Most processes persist, but are significantly disrupted by plowing and harvesting activities. Residual biodiversity persists in the landscape, mainly in set-asides and in strips between fields (matrix), assumed to constitute approximately 20% of class.	Commercial and subsistence crop agriculture	1995 South African National Land Cover map
Plantation (BII = 27.2%)	Land cover permanently replaced by timber plantations. Matrix areas assumed to constitute approximately 25% of class.	Plantation forestry, typically pine and eucalyptus species	1995 South African National Land Cover map
Urban (BII = 12.7%)	Land cover replaced by hard surfaces such as roads and buildings. Dense populations of people. Most processes are highly modified. Matrix assumed to constitute 10% of class.	Dense urban and industrial areas, mines and quarries	1995 South African National Land Cover map

Source: 1995 South African National Land Cover map (Fairbanks et al. 2000); mean BII in 2000 (Biggs et al. 2006).

century. The formalization of land tenure restrictions, the increase in domestic livestock numbers, and the occurrence of severe droughts and periods of economic depression were the key considerations (Hoffman and Ashwell 2001). Based on the historical narrative, time-courses for the development of degradation were derived for each of seventeen regions of the country, with the extent of degradation in each decade expressed as a fraction of that in 1995

TABLE 6.2

Estimated percentages (%) of South Africa in six land use categories

Year	BII (%)	Protected	Moderate use	Degraded	Cultivated	Urban	Plantation
1900	90.65	0.00	95.94	0.41	3.37	0.26	0.02
1910	90.08	0.32	94.65	0.62	3.97	0.34	0.10
1920	89.71	0.32	94.02	0.85	4.29	0.37	0.16
1930	88.61	1.88	90.51	1.42	5.52	0.42	0.25
1940	87.57	2.70	87.87	2.06	6.56	0.50	0.31
1950	86.54	2.72	86.39	2.41	7.52	0.60	0.37
1960	83.91	2.79	82.80	2.68	10.37	0.67	0.69
1970	82.69	3.02	80.52	3.81	11.00	0.83	0.81
1980	82.41	4.03	78.76	4.26	11.14	0.87	0.94
1990	81.12	4.39	76.55	4.64	12.33	1.08	1.01
2000	80.35	5.30	74.76	4.95	12.23	1.29	1.47

Note: Total land area of the country is 1.2 million km².

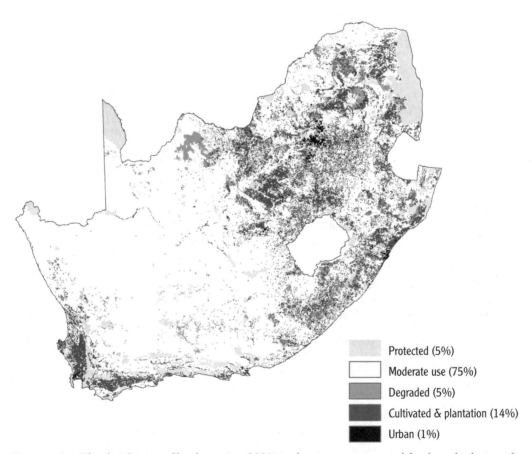

Protected (5%)

Moderate use (75%)

Degraded (5%)

Cultivated & plantation (14%)

Urban (1%)

FIGURE 6.1. The distribution of land uses (ca. 2000) in the six categories used for the calculation of Biodiversity Intactness Index (BII) (see table 6.1). The basic map is derived by lumping land-cover classes in the South African National Land Cover map (Fairbanks et al. 2000) and combining these with the protected areas of IUCN categories I to V from the World Database on Protected Areas (IUCN and UNEP 2003).

in the particular region. These "degradation time sequences" were applied retrospectively to the distribution of degraded land in the relevant regions as given by the 1995 South African Land Cover map to derive the total degraded area in each decade.

Confidence in the Historical Reconstruction

The level of uncertainty in the reconstruction of land use distribution over time determines our confidence in the estimates of the BII over the twentieth century. We estimate that the total error in the estimate of BII resulting from the uncertainty in the areas under different land uses is less than 2% absolute.

The largest impact on BII results from cultivation, which constituted 12% of South Africa in 2000, and has a mean impact score of 25% (i.e., a 75% reduction in biotic populations). By 2000, cultivation accounted for half the reduction in BII. The data on cultivated area (and plantations) are the most reliable in the set, with an error of no more than 1% absolute, which translates to a maximum absolute error of 0.7% in the BII score.

The second largest effect, accounting for a quarter of the reduction in BII, is due to moderate but extensive uses such as sheep and cattle grazing. Although the mean impact is low (93%), an extensive area is affected (75% of South Africa in 2000). Since the area under extensive "moderate use" is derived by difference (i.e., as the land surface not classified as protected, degraded, cultivated, plantation, or urban) (table 6.1), the uncertainty in the total area classified as such depends on the accuracy with which the other major classes are mapped. Of these classes, the cultivated and major protected areas have been well mapped.

The method used for mapping the development of urban areas was crude, but the total area is small (1.3%), so even with a mean impact score of 13%, the potential error is limited to less than a tenth of the observed BII reduction. The largest uncertainty is associated with the "degraded" class, which occupied about 5% of South Africa in 2000 and results in nearly a 50% reduction in wildlife population abundances where it occurs. Land degradation is responsible for about an eighth of the total biodiversity impact; if the uncertainty in the area classified as degraded is +2% absolute, then the absolute error in BII is 0.9% from this source.

The Conversion of Natural Capital into Manufactured Capital in South Africa

Taking the BII as a proxy for renewable natural capital, changes in BII, FCS, and GDP over the twentieth century support the conceptual model of a partial conversion of natural into manufactured capital during the economic development of South Africa (figure 6.2).

From Natural to Manufactured Capital Dominance

Economic activity in South Africa during the first two centuries following European colonization in the 1600s was based almost entirely on the use of renewable natural resources: grazing, cultivation, hunting, and timber extraction. The total population grew slowly, and the impact on the natural resource base was relatively limited spatially. Nevertheless, by 1900, the fertile valley soils of the Western Cape were mostly converted to crop agriculture, the limited extent of indigenous tall forest around Knysna was approximately halved by unsustainable logging and fires, and the large herds of antelope that grazed the interior plateau

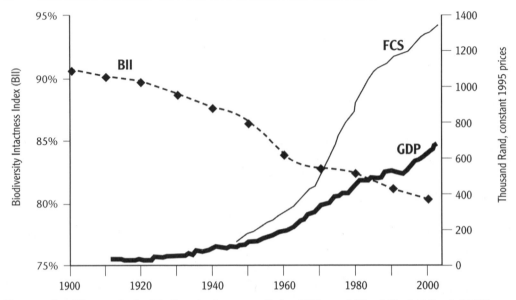

FIGURE 6.2. Changes in the Biodiversity Intactness Index (BII), total Fixed Capital Stock (FCS), and Gross Domestic Product (GDP) of South Africa between 1900 and 2000. Data on total FCS and GDP were obtained from the South African Reserve Bank. Data were not converted to U.S. dollars due to the absence of well-operating financial markets in the early 1900s. For indicative purposes only: 1 U.S. $ = R3.65 (mid-1995).

had been decimated (Hoffman 1997). We suggest that the concentration of these impacts on the highly plant diverse fynbos region and the lack of protection for indigenous mammal species (Carruthers 1995; see also chapter 7) accounts for the average 10% reduction in the populations of South Africa's flora and fauna that had occurred by the start of the twentieth century.

The discovery of diamonds in 1867, and gold in 1886 (Byrnes 1996), triggered the switch to the dominance of manufactured capital in South Africa that took place over a period of about a century. Mineral discoveries attracted European immigrants and African migrants and led to a rapidly expanding urban population. This necessitated the establishment of commercial agriculture and forest plantations to provide food and support timber for mining. A growing tax base and the strengthening of government institutions, coupled with marked technological advances in sanitation and medicine, led to a steady reduction in mortality rates so that the total population grew rapidly throughout the twentieth century (SSA 2004). The increased population, abetted by expanding markets and technologies, including newly imported livestock breeds, farm mechanization, irrigation, new maize cultivars, and agricultural chemicals, began to transform the landscape to an unprecedented degree and extent (Hoffman 1997). Between 1920 and 1960, aided by government loans and subsidies, there was an expansion in cultivation, particularly in the interior high altitude grassland region (Biggs and Scholes 2002), and a corresponding decrease in BII.

Land conflict and political domination by the Europeans led to the African population being restricted to about an eighth of South Africa by the Land Acts of 1913 and 1936.

African urbanization was initially blocked by legislation, resulting in excessive pressure on natural resources in the designated "African homelands" (Biggs and Scholes 2006), where population densities were five to ten times greater than those of climatically comparable rural areas in the rest of South Africa (SSA 2004). The average household income was also substantially lower, and lack of capital for agricultural inputs coupled with the small landholdings, insecure tenure, and restricted access to technology, advice, and markets led to deteriorating land productivity in many of these areas (Hoffman and Ashwell 2001). The greatest increase in degradation since 1900 occurred during the 1960s (table 6.2), when thousands of Africans were forced to settle in the homeland areas. Despite the end of apartheid, many of the homeland areas remain poor and heavily reliant on ecosystem products such as fuelwood (see chapter 18). Land degradation was also an issue of significant concern in white-owned commercial farming areas, notably during the 1930s, when widespread farming failures triggered by droughts and economic depression led to substantial increases in the degraded area.

Changes in BII and GDP During the Twentieth Century

We estimate that the BII declined from 90.7% to 80.3% in South Africa during the twentieth century; that is, by 2000 an average reduction of 20% had occurred in the population abundances of all indigenous plant and vertebrate species relative to precolonial levels. Over the same period, the cultivated area increased by 110,000 km^2 (from 3% to 12% of South Africa's territory), while degraded and protected lands each grew from very little to approximately 70,000 km^2 (5% of the land area). The vast majority of South Africa (95% of the land area in 1900, 75% in 2000) is under moderate use, mainly for livestock ranching, and increasingly for nature-based tourism.

The pattern of decline in BII over the twentieth century at a relative rate of 0.12% per year was unsteady and largely tracks the expansion of cultivated area. The major loss in BII occurred during the 1950s (absolute decrease of 2.6%) coupled primarily to the 3.2% absolute increase in the area under cultivation and afforestation (table 6.2). The 1970s, conversely, saw the lowest BII decline of any decade since 1900. This period corresponded to the leveling off of the area under cultivation, as favorable agricultural locations became limited and agricultural subsidies that encouraged the expansion of cultivation were withdrawn (Biggs and Scholes 2002). The 1970s was also a good rainfall decade (Preston-Whyte and Tyson 1988), thus reducing the rate of land degradation. The slow but continuing BII decline during the 1980s and 1990s was associated with marginal rises in the area under cultivation, as well as increasing degradation resulting from two protracted droughts and sustained high population pressure in many of the former homelands.

Increasing GDP broadly parallels the growing population. Inflation-corrected GDP increased seventeenfold over the twentieth century, with a growth rate of 4% per year and an average human population increase of 2.3% per year. The alternative livelihood opportunities presented by the burgeoning economy (Byrnes 1996) encouraged the urbanization of nearly 80% of the European population by 1950 (SSA 2004), mainly employed in the manufacturing and services sectors. The rapid growth in GDP between the mid-1940s and 1980 was derived mainly from these sectors, rather than from renewable natural capital, and reinvested in FCS, in the form of infrastructure (figure 6.2). The reduction in GDP growth during the 1980s and early 1990s were the result of severe droughts between 1981 and 1985 (Preston-Whyte and Tyson 1988) and between 1990 and 1993, as well as large fluctuations in the export price of gold (on which South Africa was heavily reliant) and rising security costs and labor disputes associated with the long-term effects of political suppression (Byrnes 1996). The end of apartheid in 1994 and improved macroeconomic management brought a return of foreign investment and rapid economic growth. By 2000, the majority of Africans were also urban dwellers employed in the secondary and tertiary sectors, which accounted for 90% of GDP (SARB 2004).

Plotting BII versus GDP for each decade (figure 6.3) suggests an inverse exponential relationship between natural capital and total flows of goods and services. Initially, the decline in BII per unit increase in GDP was rapid, as South Africa's revenue was accrued mainly from natural resources and particularly from the transformation of the landscape into agricultural land uses. These dividends were reinvested in the establishment of the manufacturing and service sectors, which in time began to yield dividends of their own. Further rises in GDP were accrued less from natural capital stocks and increasingly from manufactured and human capital. Therefore, the growth of GDP became increasingly decoupled from the reduction in natural capital. Although FCS data are lacking prior to 1946, given the high linear correlation between GDP and FCS ($r^2 = 0.981$, $p < 0.001$, $n = 58$), it is likely that a similar relationship holds between natural and manufactured capital, as indexed by BII and FCS, respectively.

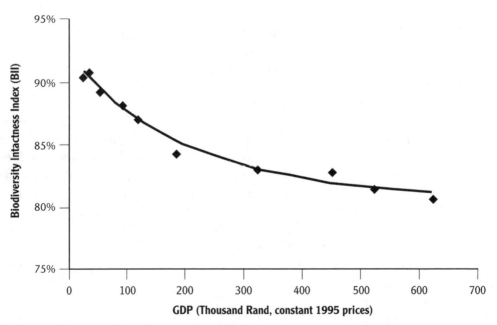

FIGURE 6.3. The relationship between Biodiversity Intactness Index (BII) and Gross Domestic Product (GDP) for South Africa. Data points are for the end of each decade between 1900 and 2000. The equation of the fitted regression line is BII = $0.804 + 0.110 \times 0.999994973^{GDP}$ ($r^2 = 0.976$, p < 0.001, n = 10). This form was empirically selected and should not be extrapolated.

Contribution

Taking the Biodiversity Intactness Index (BII) as a proxy for renewable natural capital, and fixed capital stock (FCS) and gross domestic product (GDP) as indices for manufactured capital, the changes in South Africa during the twentieth century reflect a partial conversion of natural into manufactured capital. Until the mid-twentieth century, the economy grew mainly by drawing on natural capital stocks, particularly through mining and land conversion to agriculture. From the 1950s, economic growth was increasingly based on the manufacturing and service sectors (SARB 2004), and the rate of BII reduction declined. The incommensurate units of the BII and GDP prevent us from directly quantifying the extent of reinvestment, but our findings suggest that the dividends from natural capital were largely reinvested in manufactured and social capital rather than liquidated. Our findings also suggest that over time economic growth in South Africa has become increasingly decoupled from a reduction in natural capital.

The BII is not an indicator of restoration requirement; we apply it simply as a measure of the remaining renewable natural capital. To establish a scientifically based restoration requirement, the BII score has to be related to the provision of critical ecosystem services. In practice, minimum requirements are likely to relate to populations of individual species or functional groups responsible for the provision of key services, in addition or in place of a minimum requirement on the overall BII score. In the absence of scientific knowledge about the link between BII and critical ecosystem services, the target BII level or restoration requirement is a political decision: it reflects the choices of society in terms of the desired balance between natural capital and other forms of societal wealth.

Using the historical perspective and current patterns to predict future trends, it is probable that the BII in South Africa will continue to decline slowly over the next few decades. The decoupling mechanism described above, together with the current reinvestment of manufactured capital into a rapidly expanding nature-based tourism sector, resulting in an increase in privately managed protected areas and indigenous mammal populations (Scholes and Biggs 2004), could conceivably lead to a small increase in BII over the first decade of the twenty-first century. However, this will probably be offset by other factors. While large areas of South Africa are not amenable to extreme land transformation, being too steep, too infertile, too dry, or too remote for intensive land uses such as cultivation or urbanization, they are vulnerable to degradation in the form of tree clearing, overgrazing, or alien plant invasion. Preventing the extensive areas (75% of South Africa's land area) under "moderate use" from becoming degraded has been identified as the principal challenge for biodiversity conservation in southern Africa in the medium term (Scholes and Biggs 2005). In comparison, restoration of currently degraded areas (which constitute 5% of South Africa's land area) will play a minor, albeit important, role in the conservation of natural capital in South Africa.

In addition to the threats posed by land use change, and accounted for in the BII approach, three significant concerns remain for biodiversity protection in South Africa. The first is climate change, which could potentially have a major effect on the region's biodiversity, particularly in the highly biodiverse fynbos and succulent karoo regions. Second, pollution resulting from increased industrialization and urbanization, especially in the form of nitrogen deposition on the land and acid drainage into rivers, could exceed the absorption capacity of ecosystems. This would lead to significantly higher environmental impacts, with a larger footprint, than currently associated with urban regions in South Africa. Third, the insidious effect of habitat fragmentation on population viability should be considered, particularly in light of projected changes in climate.

Acknowledgements

Thank you to the experts who contributed to the BII impact score estimation for their time and willingness to share their knowledge. Cultivated and plantation data were made available by Statistics SA.

PART II

Restoring Natural Capital: Experiences and Lessons

The first part of the book is concerned with definitions and concepts related to the restoration of natural capital, which include a search toward a synergy among social and environmental scientists, particularly ecological economists and restoration ecologists, for motivating, funding, and achieving activities that will restore natural capital. The second, central part builds upon this theoretical foundation by using nineteen case studies from around the world to illustrate challenges and achievements in (1) setting realistic, socially and ecologically appropriate targets, (2) refining approaches to funding and implementing restoration projects, and (3) using restoration of natural capital as an opportunity for social and economic upliftment.

Setting *targets* for restoring natural capital necessitates a clear definition of the social and ecological objectives of the proposed activities at scales that range from biomes and landscapes to those of populations and genes. Compromises may be essential for success because conflicts in goal setting waste the limited resources available for restoration. Much restoration takes place in utilized landscapes, so practical issues cannot be ignored. In many cases, historical-reference ecosystems are incompatible with modern life or do not offer sufficient benefits to be supported within rational, decision-making frameworks. Furthermore, they may no longer be realistic given irreversible local or global changes in the recent past. The envisaged flows from restored capital may range from the utilitarian to the aesthetic, from clean water to sightings of rare animals, but unless they contribute to the range of physical, cultural, and psychological factors that define human well-being, they are unlikely to be supported.

Case studies dealing with *approaches* to restoration make it clear that local support for restoration activities is as, if not more, important in achieving the sustained restoration of natural capital than is the technical design of the restoration intervention. Buy-in and participation of stakeholders is essential for success, particularly in cases where relatively few individuals pay in some direct way for restoration of services that benefit a wider public. Restoration of natural capital for future generations, or for national- or global-level benefits, may have costs for local land users who may lose access to certain resources. Such costs need to be recognized and local support encouraged through ensuring that those negatively affected in the short term can participate in goal setting and are compensated though incentives and processes that are tangible and immediate (economy as if nature matters).

The final group of case study chapters deals with social and economic *opportunities* presented by the process and outcomes of restoring natural capital. Ideally, restoration of natural capital should generate new lifestyles, livelihoods, and even employment opportunities (ecology as if people matter). Some inspiring case histories range from the social and ecological benefits provided by environmentally certified products to training and job creating through national-scale clearing of invasive, alien vegetation. Challenges to overcome include achieving buy-in at personal, corporate, and national levels; accessing funding; and adaptive management of long-term projects. The case studies highlight roles of individual entrepreneurs, nongovernmental organizations, government institutions, and collaborative initiatives in bringing about restoration initiatives, and the advantages of making them financially viable.

Local initiatives to restore natural capital tend to be disconnected—differing as they do in focus, from trees with medicinal bark to clean water, from employment opportunities to pride in natural heritage. Still, they all raise awareness of harmful effects that people have on their environment. By providing opportunities for stakeholders to participate mentally, physically, and/or financially in restoration, they all contribute in some way to social, as well as to environmental, restoration. In part 3, we formulate some strategies to promote the restoration of natural capital on all levels based on these case studies; in part 4, we offer a synthesis of the theoretical, applied, and policy issues in this book of relevance to the science, business, and practice of restoring natural capital.

Chapter 7

Setting Appropriate Restoration Targets for Changed Ecosystems in the Semiarid Karoo, South Africa

W. RICHARD J. DEAN AND CHRIS J. ROCHE

The restoration target is debatable in rangelands where westernization, commercialization, and cultural changes have had major impacts on the landscape, vegetation, and fauna (the natural capital), and where these changes are likely to persist. In this chapter we attempt to answer this question by focusing on the semiarid South African Karoo, an area of 323,900 km^2 in the western and central parts of the country.

When the European pioneer settlers moved into Karoo in the 1700s, the land they encountered was unlike anything in their experience (Christopher 1982). The land was dry and the vegetation sparse, but nevertheless it supported a diversity of herbivorous and predatory mammals and birds. Many of the large mammals were grazers (Skead 1980, 1987; Dean and Milton 2003) suggesting that grasses were widespread; but according to the travelers' records (summarized in Hoffman and Cowling 1990a), of varying abundance. The natural capital resources of the Karoo were the drought-adapted and resilient rangelands, the nomadic fauna (equids, antelope, ostrich) that moved with the unpredictable rain, and the dependable but widely scattered, perennial water sources. All three components were needed to sustain sparse populations of nomadic peoples and predators.

The present-day Karoo is very different in some ways to the land that the early settlers encountered. The region is still dry and the vegetation still sparse, but there have been major changes. The indigenous nomadic peoples, the /Xam San hunter-gatherers and Khoi pastoralists, and their languages (Bleek and Lloyd 1911) are extinct. Fences now divide the landscape, and great herds of antelope no longer trek across the land; large predatory birds and mammals have also disappeared (Skead 1980, 1987; Boshoff et al. 1983), and the vegetation has changed (Tidmarsh 1948; Downing 1978; Hoffman and Ashwell 2001). We have a relatively clear picture of why changes in large mammal diversity occurred. Since the ecology of large raptorial birds, including vultures, are inextricably linked to large mammals and predator-prey interactions (Boshoff et al. 1983; Macdonald 1992), we can infer that their demise in the Karoo is rooted in the loss of hunters and predators.

More difficult questions are related to land use. Central to the debate is whether the relatively heavy stocking rates that settlers imposed on the Karoo left a legacy of change that is apparent in the state of the present vegetation of the region. There have been changes in the amount of cover, and shifts in the dominant species of plants in parts of the Karoo (Acocks 1953; Downing 1978; Hoffman and Ashwell 2001; Dean and Milton 2003), so much so that many of the large herbivores that formerly occurred could no longer survive on farms of limited size without supplemental feeding. However, this remains largely untested, and anecdotal evidence suggests that in some cases replacing livestock with wild herbivores and removing artificial water points can lead to improved plant cover and productivity (Mark McAdam, Colesberg District, personal communication).

There is no doubt that environmental changes in parts of South Africa have diminished natural capital and quality of life (Milton et al. 2003). This is particularly evident in the Karoo where natural capital has been eroded away over time by high rates of stocking with sheep and other livestock (Dean and Macdonald 1994; Hoffman and Ashwell 2001).

Undoubtedly the region is in need of restoration, but what should the targets of such initiatives be and how should they be reached? A historical understanding of what has happened is essential to guiding restoration. Here, we summarize changes in land use, briefly touch on impacts on the rangelands, explain why restoration, sensu stricto, is not a practical target for the Karoo, and suggest interventions to partially restore natural capital in Karoo rangelands, incorporating human aspirations as well.

Land Use History and Impacts on Rangelands

Colonization and development of the arid Karoo by European settlers ousted the indigenous people, substituted domestic livestock for a diversity of wildlife, and completely changed grazing and disturbance regimes. Here we discuss the processes of change during European exploration and settlement of the Karoo.

The Early Years

The rural economy in the early European settlement years in South Africa was almost entirely based on rangelands and livestock (see chapter 6), with very little crop farming (Christopher 1982). As European settlers advanced across the semiarid Karoo, they displaced and exterminated indigenous peoples, for example, the /Xam San hunter-gatherers (Bleek and Lloyd 1911) and the nomadic Khoi pastoralists. Indigenous large herbivores were exterminated or shot for food (Talbot 1961) to reduce competition with domestic livestock (Acocks 1979). In general, the "hunting [of game] was used to clear land, reap income and provide meat so that slaughter of domestic animals [for provisions] could be avoided" (Beinart 2003). The drive to increase flocks of livestock was a primary reason for colonial expansion during the early period as trek boers (nomadic colonial graziers) engaged in grazing, hunting, and trading expeditions with the Khoi in the interior (Guelke 1979). The ready markets for provisioning ships and at the newly developed diamond and gold mines (Talbot 1961), and the export market for wool (Beinart 2003), not only increased pressure on pastoralists to increase flock sizes, but also increased hunting pressure on wild ungulates.

The early zoologists' records (Skead 1980, 1987; Rookmaker 1989) and farm names (Dean and Milton 2003) suggest that grazing and browsing wild herbivorous mammals, their major predators, and various scavenging birds were widespread in South Africa during the eighteenth and nineteenth centuries (figure 7.1). However, travelers' records do not consistently report large numbers of wild herbivorous mammals at the same places (Skead 1980, 1987), indicating that many of these species were not resident but nomadic within the Karoo, moving with the seasons and rains.

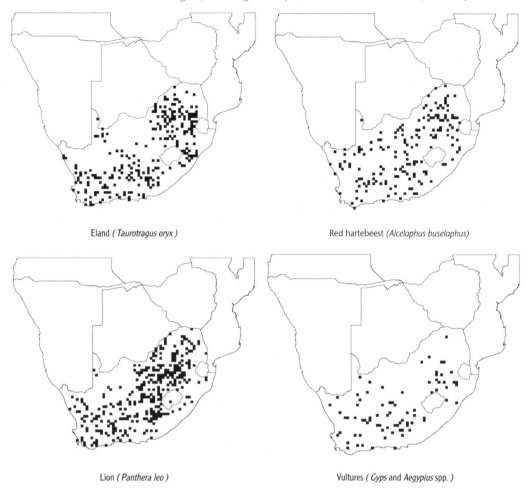

FIGURE 7.1. The distribution of two large herbivores, a large predator, and avian scavengers in South Africa, according to early zoologists' records (Rookmaker 1989). The distribution of red hartebeest may include records of Lichtenstein's hartebeest (*Alcelaphus lichtensteinii*) in extreme eastern South Africa.

The formerly vast herds of large herbivores on which precolonial San hunter-gatherers depended were replaced over the course of the nineteenth century with domesticated livestock. The original successor was the Cape fat-tailed sheep, estimated to number approximately 1.5 million in the Cape Colony by 1806 (Beinart 2003), replaced in turn by wool-bearing sheep (merinos) in the early 1800s (Talbot 1961). Wooled sheep doubled in numbers from around 5 million in 1855 to 10 million in 1875, and 12 million in 1891, and by 1920 the Cape Colony was the world's second biggest producer of wool (Beinart 2003). This position, ahead of much larger and better-watered countries, indicates the heavy, indeed excessive, extent to which it was stocked.

Regardless of whether or not these stock numbers were higher than those of the various wild ungulate numbers that existed prior to European colonialism, the grazing regime was changed and farming practices differed markedly from natural systems (Downing 1978). These natural systems were mimicked to some extent by nomadism or transhumance, but as the nineteenth century progressed, farming strategies changed. More formal systems of land ownership combined with fencing and boreholes resulted in a more intensive utilization of the land, which soon impacted natural systems that were adapted to sporadic periods of abundance and utilization followed by comparative droughts and ensuing rest (Talbot 1961).

The management of domestic livestock and rangelands by the early settlers was not based on any prescribed grazing system, but was simply ad hoc (Talbot 1961), driven by demand and supply—the undersupplied and ever ready market for both meat and wool drove up the number of livestock on the land, and these numbers were set by a (usually) overestimated carrying capacity based on what could be carried in the higher rainfall years (Talbot 1961). Other mismanagement practices had a marked impact on the rangelands—no fodder was reserved for times of drought, and the usual practice was to keep livestock on the land in the hope of rain, or during exceptionally dry spells to move livestock in search of water and grazing. The early settlers perceived drought years as unusual, whereas the reality is that they are equally as frequent as the wet years (Kokot 1948; Vogel 1994). Repeated overstocking of drought-stricken pastures impoverished plant cover and reduced primary and secondary production. This was exacerbated by changes in the ratio of grazers to browsers that affected the plant species composition of rangelands, and the removal of game and predators affecting ecological processes. This combination of anthropogenic effects on the arid ecosystem led to early concerns that Karoo rangelands could not carry large numbers of livestock (Downing 1978).

The Later Years

By the last decades of the nineteenth and beginning of the twentieth centuries, the herds of wild herbivores in the Karoo had all but disappeared (Macdonald 1992), and dramatic changes had occurred in stock farming practices, such as the introduction of fencing and the development of deep drilling technology (Talbot 1961). Additionally, the rinderpest epidemic of the late 1800s and early 1900s (Stevenson-Hamilton 1957) and the end of a major ecological process—the springbok (*Antidorcas marsupialis*) migrations or "treks" (Skinner 1993; Roche 2004) also wrought profound changes. While it is impossible to reconstruct the exact nature of the natural system of the Karoo under wild ungulate populations, it is clear that the phenomenon of springbok treks was a dominant feature of this natural system. An understanding of this poorly known process may shed light on the historical "pulses" of energy and nutrients and reaction to them by indigenous animals of the Karoo (see Roche 2004).

In the early part of the twentieth century there was an increase in the variety of domestic livestock in the Karoo (Talbot 1961) and a rise in stocking rates generally (Dean and Macdonald 1994). By 1930, the numbers of sheep in the Karoo had increased to nearly fifty million (Talbot 1961). This grazing pressure, coupled with devastating droughts, led to widespread degradation of rangelands (Hoffman and Ashwell 2001). The perception at this time was that there had been changes in the amount and timing of rainfall (Wilcox 1977), but careful investigations established that no such changes had taken place (Kokot 1948; Vogel 1988a, 1988b).

Recent studies suggest that there has been localized recovery of degraded rangelands rather than further deterioration (Bond et al. 1994). Recovery of karroid rangelands, whether localized or not, will be slow and as often as not will need to move from one stable state to another before any changes in the plant composition or cover take place (Milton et al. 1994). Such localized recovery is also highly weather dependent and, without rest from grazing following rainfall adequate for seeding, is unlikely to happen.

Approaches to Restore the Karoo's Natural Capital

Given that the natural capital of the Karoo is in need of restoration, should we aim at preventing soil loss in eroded drainage systems to regenerate perennial springs, or should we focus on repairing rangelands? Should rangeland capital be restored to support large wild herbivores, or to provide the maximum sustainable yield for people living on the land (Danckwerts and King 1984)? Which restoration targets are ecologically, economically, and socially realistic for the Karoo? Pertinent to setting restoration objectives is a decrease in the profitability of livestock ranching, which has led to diversification of rangeland use, including game farming, nature tourism, and hunting (Barnes 2001; Khuzwayo 2002; Goodman et al. 2002).

The Ideal of Natural Capital Restoration

Given the extremely large land sizes required to be able to revitalize ecological processes to some degree in the Karoo, it is clear that any landscape-level restoration plan would need to be the result of cooperation between the state, community, and private landowners and that multiple land uses would need to be entertained to ensure both profitability and sustainability. Lessons from international experiences in formulating approaches for restoration of natural capital, such as the proposed *Buffalo Commons* scheme for America's Great Plains (Popper and Popper 1987), can also be applied in a useful way.

However, the almost irreversible losses of processes such as large mammal–large predator–large scavenger relationships and disrupted water flow and flooding regimes (Acocks 1964) make restoration in the strict sense almost impossible. The most practical goal may be to aim at restoration of natural capital compatible with modern land use patterns, although restoration of some ecological processes may be entirely possible. First, it is essential to prevent further degradation of Karoo rangeland due to overgrazing, and to develop new initiatives for using the land profitably and in an environmentally sensitive way (Milton et al. 2003). Perverse incentives for landowners to plough or develop untransformed land (Botha 2001) have been created by the 2003 Property Rates Bill that obliges landowners to pay taxes on "unproductive land" (land not used to generate livelihoods) in South Africa.

Although in some parts of the Karoo, notably the broad transition zone of mixed shrubs and grasses in the eastern parts of the region that have become less grassy and more shrubby, attempts to rehabilitate these areas to their former grassiness may be fruitless (Bond et al. 1994). Similarly, attempts to rehabilitate vast areas of shrublands that have lost a large proportion of palatable plants are not practical due to the magnitude of the problem and the huge imbalance between cost and benefits. However, modified land use may be both practical and appropriate and can help open up opportunities for rehabilitation in the future. Modified land use options may include *sustainable use areas*, where there is off-take of secondary production mostly exported away from the area (i.e., pastoral farming for meat or wool production, including "game farming"), or *protected areas*, managed for conservation and tourism and from which domestic livestock are excluded.

Sustainable Use Areas

- *Conservation farming*: This aims to maintain the balance between utilization and conservation of farm-based resources (Donaldson 2002) to secure flows of ecosystem goods and services that are linked to biodiversity. General impressions and opinions not supported in the main by hard data are that as a result of conservation farming there has been an improvement in rangelands indicated by increased cover of usable forage plants (Donaldson 2002). Conservation farming requires little additional cost, and most important, draws the attention of the land user to key elements in the ecosystem and its processes.

- *Conservancies*: Recent conservation initiatives in the Karoo include consortiums of farms that are generally privately owned, but strongly linked to local nature conservation agencies. Conservancies are currently the most successful option for resident landowners that are affected by common resource issues (Botha 2001). Under the constitution of the conservancy, member farm owners pledge to protect natural resources as far as possible, while they continue to utilize rangelands. Conservancy landowners encourage research on ecological aspects of the conservancy and usually undertake to protect wildlife by controlling numbers of dogs and outlawing the use of traps and snares for hunting or poisoned carcasses for killing problem animals.

- *Game farming*: Farms that promote ecotourism and hunting have replaced traditional stock farming in many parts of southern Africa, often proving more profitable than stock farming in arid and semiarid areas (Barnes et al. 1999). The effects of wild mammalian herbivores at moderate densities are generally considered to be less deleterious to the shrubland vegetation of the Karoo than the effects of domestic livestock (Davies et al. 1986; Davies and Skinner 1986). Patch-selective grazing by wild herbivores (Novellie and Bezuidenhout 1994) may be instrumental in creating vegetation mosaics that benefit other animals, as has been suggested for certain mammals in Australia (Short and Turner 1994). It is outside the scope of this chapter to discuss the finer details of game farming except to note that poorly managed game farming can damage Karoo rangeland, especially where high densities of animals are confined by fences and supplied with water and supplementary nutrients throughout the year (Coetzee 2005). Expert opinion to evaluate habitat condition and advice on species and numbers of game are fundamental in setting up a game farming operation (Boshoff et al. 2001; Coetzee 2005).

Large Protected Areas

Given its size, the Karoo is remarkably poor in large protected areas (Siegfried 1989). The loss of the wild herbivores may have had effects on ecological processes that go far beyond the documented changes in the large raptor assemblages (Boshoff et al. 1983). Although these processes cannot be restored in their entirety, the creation of protected areas, particularly large protected areas intact with ecological processes, such as nomadic movement of game and patch selectiveness in grazing, should be important in rehabilitating some of the Karoo.

Very large protected areas are probably not attainable, even in the arid Nama-Karoo, because of sociological and financial constraints. However, ecotourism, based as it is on natural capital, could be a means of generating wealth in a sustainable way from areas that are relatively undisturbed (Cowling 1993) and could be a vital part of any conservation initiative in the Karoo.

Foreign ecotourists are willing to pay high prices for viewing African landscapes with indigenous plants and animals (Barnes et al. 1999). South Africa offers the ecotourist exceptional plant and animal species richness, beautiful scenery, and a good transportation infrastructure (Cowling 1993). But the ecotourism industry is largely dependent on unspoiled nature and the opportunity to view large, spectacular African mammals, particularly predators (see, e.g., Lindsey 2003). Eroded, or heavily grazed, "desertified" rangelands with run-down farm infrastructure do not fit the ecotourist ideal.

There is certainly a need to evaluate the ecological and economic effects of ecotourism in comparison with stock farming in the Karoo (Cowling and Hilton-Taylor 1994). However, land in national parks or used for ecotourism will generate a high income only if tourists are concentrated in a relatively small area, so ecotourism is probably not a viable alternative to stock farming throughout the entire Karoo. Furthermore, water use by large numbers of visitors to national parks or other protected areas in arid zones is high, and water supplies in the Karoo may not be constant enough to sustain large numbers of visitors. For this reason, ecotourism facilities in national parks or other protected areas in the arid zone should not be overdeveloped but should instead have a limited staff and infrastructure and should cater to the specialist ecotourist rather than the generalist.

Contribution

In this chapter we have explained why whole, naturally functioning Karoo ecosystems cannot be restored without vision and cooperation that extend across thousands of square kilometers. Protected areas are probably the best option for restoration, but unless they are set up to accommodate ecotourists, they are not financially viable. The cost of taking large areas out of agricultural production is high, and is not a cost a private landowner or the country can afford. At best, we can aim at restoring the natural rangeland capital to increase flows of secondary production in some areas through conservation farming, conservancies, and game farming. Priority should therefore be given to limiting further degradation, developing the restoration initiatives that are already in place, and developing long-term restoration and integrated development plans for the Karoo. In this respect, the *Integrated Development Program* (IDP) (Coetzee et al. 2002), run by local authorities, could be an appropriate vehicle for rehabilitation and restoration of degraded ecosystems at local scales.

Acknowledgements

We thank Mark McAdam of Hunter's Moon safari farm, Colesberg District, for helpful discussions.

Chapter 8

Targeting Sustainable Options for Restoring Natural Capital in Madagascar

Louise Holloway

Human well-being is wedded to ecosystem well-being, but we consistently attempt to divorce ourselves from the controlling factors that operate on other species. Although there is increasing evidence of the cumulative impacts and the unsustainability of many types of human–environmental interactions, it could be argued that the imperative to change our relationships with natural capital differs according to the degree to which we recognize how our well-being is bound to that of natural ecosystems. In Madagascar, the relationship is evident.

Much of the natural capital of Madagascar is its rain forests, rich in endemic plant and animal species, and the fertile but fragile soils that the forest feeds and protects. Here, direct human dependence on the local ecosystem results in more immediate feedback than is the case in many developed nations where people and markets draw upon natural capital from distant ecosystems. In fact, people in the project area can be perceived as part of the stocks and flows of the natural ecosystem.

This chapter evaluates a small-scale project to connect rain-forest fragments, restore soil fertility in food gardens, and raise awareness of natural capital dependence among local people, giving them access to options. The aim of the project—initiated and led by the author, with backing from the Wildlife Conservation Society—was to enhance the well-being of both human and nonhuman components of a rain-forest ecosystem by three interrelated approaches to restoring natural capital: facilitating processes that slow down degradation, repairing environmental damage, and generating ecologically sustainable cultivation systems.

The following provides an analysis of the successes and shortcomings of the project up to 2002 and indicates approaches that could result in improving efforts to restore natural capital in places like Madagascar.

Restoring Natural Capital in Madagascar

Madagascar is widely regarded as a world conservation priority area due to high levels of endemism. Indeed, Ganzhorn et al. (1997) convey clearly Madagascar's biodiversity value by stating that "a hectare of forest lost in Madagascar has a greater negative impact on global biodiversity than a hectare of forest lost virtually anywhere else on the planet."

Most of the 17.5 million inhabitants of Madagascar are directly dependent on flows from natural capital; the national economy is largely (80%) based on agricultural production, with as much as 70% of the population practicing subsistence farming and remaining reliant upon natural ecosystems for basic resources such as construction timber, fuelwood, and medicine. As a source of revenue, tourism (mainly wildlife tourism) ranks third, after fisheries and vanilla production (Carret and Loyer 2003). Hence, the quality and quantity of the natural capital is of paramount importance to Malagasy peoples' livelihoods.

With natural habitat loss estimated at >90% (Lowry 1997), habitat fragmentation is also a major driver of biodiversity decline. Small fragments preserve only a highly biased subset of the original flora and fauna, with widespread, generalist species mostly surviving at the expense of more rare ones (Gascon et al. 1999). In fact, most of Madagascar's forests are so fragmented that their long-term contribution to ecosystem functioning and species diversity is questionable (Ganzhorn et al. 2001). In addition, Madagascar is considered to be one of the world's poorest countries by traditional economic measures (such as gross national product [GNP] and gross domestic product [GDP]) and scores "poor" on both the *Human Wellbeing Index* (HWI) and the *Wellbeing Index* (WI) (Prescott-Allen 2001). Human–environmental interactions may be considered responsible for the poor state of human and ecosystem well-being. Though people colonized the island only 1,500 to 2,000 years ago, they introduced agronomic systems more suited to their Indonesian home environments. The intensity, frequency, and spatial scale of human impact, unlike natural disturbances, rapidly surpassed the self-repair mechanisms of the natural ecosystems. Biogeographic characteristics of Madagascar further impede repair processes since former rain-forest areas usually show arrested or deflected succession with low species diversity, dominated by nonnative plants (Gade 1996; Holloway 2000). A complex of cultural, socioeconomic, and political factors, operating over wide spatial and temporal scales, influences management. In particular, slash-and-burn, rain-fed, hill rice cultivation (*tavy*) directly contributes to land degradation, with increasing population pressure allowing insufficient fallow periods for soil fertility recovery (Conservation International 2004). Consequently, natural capital has been liquidated for low returns at a high cost by impeding attempts to improve living standards and undermining ecosystem integrity. Restoration is urgently needed.

A Strategy for Improving Human and Environmental Well-being: The Masoala Corridors Restoration Project (MCR)

There exists a paradoxical relationship between people and forests in Madagascar. Although forests are destroyed to provide fertile fields for rice cultivation, they also supply over 290 plant species used for foods, fuelwood, construction, and medicinal purposes. Even the environmental services provided by intact forests, such as water supply regulation, are recognized and valued, as eloquently captured by a Malagasy proverb: "Without the forest, there will be no more water; without water, there will be no more rice." Nevertheless, people perceive that the solution to land shortage and degradation is to carry out further forest clearance. To counter this paradox, a strategy has been developed that aims to improve ecosystem integrity and human well-being through catalyzing ecosystem restoration. The major objectives are as follows:

1. To raise community awareness of the unsustainability of their environmental impact so as to induce behavioral changes

2. To protect and reinforce linkages among forest fragments for facilitating natural migration in response to environmental changes
3. To reduce human pressure on forest resources and enhance livelihoods, by increasing local self-reliance and resilience to environmental disturbances through integrating sustainable gardens into land use patterns
4. To conduct field trials for improving the efficiency of ongoing forest restoration, while monitoring the project outcomes to inform future restoration initiatives
5. To stimulate, if successful, widespread application of the approach and its integration into national policy

Building Forest Corridors

This approach focuses on the rain-forest biome and restoring its connectivity, since fragmentation threatens the viability of endangered species populations. Though Madagascan rain forest has a poor capacity for self-repair after clearance and cultivation, in part due to a reliance on a few arboreal seed dispersers, the idea is to use the known ecological information for catalyzing natural forest regeneration processes. Forest corridors were established by planting local forest fruit trees, favored by frugivorous lemurs, in linked clusters between forest blocks. Lemurs, enticed by these species, will disperse seeds of other forest plants, via their feces, subsequently catalyzing natural forest development in these corridors.

Restoring Degraded Cropland

In conjunction with developing forest corridors, land degraded by long-term cultivation is being rehabilitated, while measures are introduced to prevent further environmental damage (Holloway 2003). One such approach has been to help local people establish permanent gardens that emulate aspects of natural ecosystems in structure and function. These are based on permaculture principles (Mollison 1988) and modeled on home gardens, as practiced for centuries in many parts of the humid tropics, but adapted to the specific characteristics of Madagascar's eastern rain forest. Yielding a reliable and continuous supply of foods and other goods for home consumption and for revenue, the gardens can be readily incorporated into local land use patterns, especially on near-exhausted *savoka* (hill rice-fallow lands). Another system, the savoka garden, involves the creation of an enhanced quality fallow area, which allows a shorter rotation between successful rice planting and increased use through restoration of degraded lands. Indeed, a key attribute of savoka gardens is the elimination of degraded land, thus reducing agricultural-driven deforestation. However, a prerequisite for establishing sustainable livelihoods is the behavioral change that arises and is maintained from comprehending the essential link between human and ecosystem well-being. Therefore, the full engagement of local stakeholders requires that they explore their cultural values in relation to their interactions with the natural environment, within the context of sustainability.

The combined development of natural forest corridors and rehabilitation of degraded land to create diverse, sustainable cultivation systems enhance ecosystem and human well-being. Key attributes of this approach include the following:

• Forest restoration and sustainable gardens are undertaken together.

- Working with ecosystem processes as well as with species (e.g., planting fruit trees that catalyze further regeneration by attracting seed dispersers).
- Simultaneously working at a range of spatial scales, from individual household needs to ecosystem viability, increases efficiency and sustainability.
- Different temporal scales to achieve immediate as well as medium- and long-term benefits (e.g., sustainable gardens can yield harvestable products within weeks, while restoring soil and ecosystem processes over a longer timescale).
- Being widely accessible by employing low technology at low cost, and using local resources and easily understood techniques.
- Being widely applicable but locally specific (e.g., by working with natural processes to improve nutrient cycling, while selecting species determined by the local bioclimate and human priorities).
- By catalyzing natural processes, this approach is efficient in terms of labor, resource, and financial inputs. As the local people become a driving force in the reconstruction process, this considerably aids the task of conservation and rural development organizations.

This approach was piloted in the Masoala Peninsula of northeastern Madagascar, which holds many locally endemic taxa (Kremen 2003) and, according to information held by the Missouri Botanical Garden (G. Schatz, personal communication), 50% of all Malagasy plant species occur in this region.

Masoala National Park

Masoala National Park covers 210,000 hectares, including three tenuous corridors linking the main forest blocks, upon which could depend the continued viability of many species. Park delimitation in the 1990s sought to avoid inclusion of human settlements and legitimate land claims in order to reduce potential conflict between conservation goals and local farmers. Unfortunately this virtually fragmented the park into several small blocks, undermining its value as the largest protected area of rain forest in Madagascar. As a result, the high perimeter-to-area ratio presents a special challenge for conservation management. Although opportunity costs to local stakeholders were inadvertently high because the delimited areas reduced potential cultivatable land availability, the park boundaries at least preempted the land shortage that would have eventually occurred through continued deforestation. Hence, the *Masoala Corridors Restoration Project* (MCR) was designed to mitigate these costs by addressing the restoration of degraded land as well as the restoration of forest corridors.

The initiation and evolution of the MCR should be understood within the context of its financial and institutional structures, as well as other social, economic, political, and environmental influences.

Internal Influences on the Course of MCR: Institutional Roles and Relationships

From 1992 until 2000, Masoala National Park management was overseen by the *Masoala Integrated Conservation and Development Program* (ICDP) comprising CARE International, the Wildlife Conservation Society (WCS), the National Parks Service (ANGAP), the Ministry of Water and Forests (MEF), and the Peregrine Fund. Within this institutional framework, WCS took responsibility for resourcing, coordination, and administration of MCR, which was initiated in 1997. In the context of Masoala, MCR was needed to maintain the integrity of the national park by providing ecological connectivity, as well as the potential for keeping a link with the remaining forests of eastern Madagascar (figure 8.1)

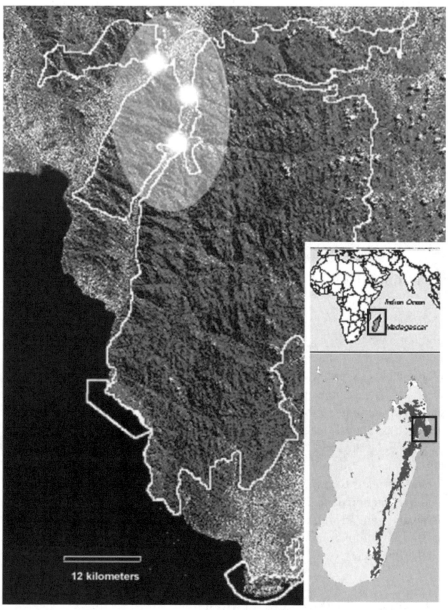

FIGURE 8.1. Masoala National Park (outlined in white) forms part of the most extensive and continuous remaining blocks of rain forest in Madagascar. Three tenuous corridors (highlighted with white circles) help maintain habitat continuity within the park. Thousands of people live by subsistence farming within the corridor zone (indicated by pale oval).

The aim of the ICDP was to achieve specific conservation and development objectives through a set of complementary activities conceived under the notional banner of sustainable development. WCS focused on park management, while CARE worked with human communities; MCR provided the interface. In theory, helping integrate the activities of the conservation and development agencies should have been advantageous, but instead MCR was perceived as "falling between two stools" because it included aspects that fitted each NGO's mission but also went beyond into the mission of the other. Before Cyclone Hudah in 2000, the project's more "development"-based activities overlapped projects run by CARE, and after the cyclone CARE changed its focus to infrastructural reconstruction (e.g., building schools and similar projects). As a result, WCS extended its responsibility to incorporate sustainable cultivation.

External Influences

Cyclone Hudah in April 2000 devastated much of the northern part of Masoala. Whole villages around the park were destroyed, and a section of the main restoration corridor, much of the project nursery and conservationists' homes, were swept away. Corridor-monitoring data and training notes were lost. Moreover, it has been estimated that the populations of seed-dispersing lemurs were reduced by as much as 50% (M. Hatchwell, personal communication).

In 2001, political conflict resulted in civil war following Madagascar's presidential election. As a result, MCR and virtually all park management activities ground to a halt because staff remained unpaid for several months. In addition, a global economic slump in 2001 and the events of 9/11 resulted in a cutback in WCS expenditure. Hence, potential MCR expansion, geographically and in operational scope, subsequently faded. However, in 2001 a New York charitable organization (that elected to remain anonymous) held a fund-raising event to support Masoala National Park activities, and it was decided to allocate the funds to MCR. The funds were channeled through CARE International in Madagascar who, in November 2002, collaborated with WCS and ANGAP in the sustainability awareness and livelihood aspects of the project.

Outcomes and Evaluation

As of 2005, a number of crucial elements of the envisaged project have not been implemented, including comprehensive monitoring of activities, due to inadequate financial and staff resourcing. Therefore, only some of the results can be tied causally to the project.

Local Ecosystem Well-being

The ecological aspect of the project has been successful, though on a smaller scale than originally anticipated. There is now a healthy young forest comprising a wide range of local native trees growing in one of the corridors. Birds are nesting in planted trees and the important, seed-dispersing, endangered red ruffed lemur (*Varecia variegata rubra*) traverses the planted area (Leon and Rabesodika, MCR staff, personal communication, 2004). Thus, the second objective of creating forest corridors has been partially met, though the project is still too young to evaluate its long-term success.

The fourth objective cited earlier concerning restoration trials and monitoring has been partly achieved with many useful outcomes. Trials revealed that germination and establishment occurs outside the forest, and that seeds germinate there earlier and in larger numbers than was previously suspected. In addition, seedling predation is lower, survival higher, and growth faster in savoka than within the forest. For example, in four years, vintanona (*Calophyllum laxiflorum*) reached a maximum height of only 64 cm inside the forest, yet 456 cm outside. In fact, in savoka, seven of the sixteen species monitored attained average heights over 3 m in four years, with some individuals reaching 6 m.

Local Human Well-being

In 2002, CARE and ANGAP agents held a weeklong "sustainability awareness" workshop with the local communities. Villagers were actively engaged in understanding their role in the ecosystem and how their potential environmental impacts could be reflected in the sustainability of their own precarious livelihoods. Their response was very positive with requests for technical assistance to create sustainable gardens and individual and group initiatives developed tree-based schemes on savokas for subsistence use. A dramatization based on the issues explored during the workshop was filmed for dissemination to the wider local community. Hence, the first objective of raising awareness about sustainability was met with many of the workshop participants stimulated to start sustainable gardens and savoka gardens immediately. Local stakeholders also voluntarily assisted with corridor tree planting, yet the project is unlikely to have been beneficial to them in measurable terms as of 2005. Despite uncertain outcomes, they proved receptive to ideas and willing to adapt through experimentation with new systems. A joint WCS/ANGAP/CARE–staffed "Viable Livelihoods Advisory Team" was established with the aim of supporting individual initiatives. This may have helped partially meet the third objective of reducing human impact by providing sustainable livelihoods, though this is too early to assess. Sadly, there has been little follow-through with this initiative, due to both insufficient resourcing and the nature of the institutional framework.

Some Institutional Outcomes

From the WCS perspective, the outcomes justified allocating resources to the untested approach of MCR (M. Hatchwell, personal communication). Although MCR offered an opportunity for close collaboration between conservation and the rural development NGOs at the outset, this in fact did not happen until funding was packaged in a way that necessitated cooperation. This raises the issue of how to devise appropriate institutional frameworks to facilitate the inherently holistic process of restoring natural capital (chapter 18).

Financial shortfalls and insecurities faced by MCR, with little assurance of project continuation from one year to the next, paralleled other sectors of park management. These contributed to some shortcomings in the meeting of MCR objectives. However, despite uncertainties, the establishment of the park was highly worthwhile, and this also applies to MCR, though further implementation is required. The final phase of Madagascar's environmental action plan (NEAP) is to develop sustainable biodiversity financing mechanisms by mainstreaming the environment into economic management, which is hoped to benefit MCR in the long term.

Wider Outcomes

Dissemination of information about MCR has generated sufficient interest on the part of other biodiversity conservation and development NGOs within Madagascar to integrate aspects of the MCR approach into regional plans (MCR objective 5). MCR is informing planning of the larger scale Andasibe-Mantadia Corridor Restoration project, which is integrating into its project design carbon sequestration and storage, as both a cobenefit and a funding mechanism (chapter 32). Finally, the MCR approach has been recognized by the scientific community as having high value for biodiversity conservation (Ganzhorn et al. 2003; Kremen 2003).

Issues Raised and Lessons Learned

MCR has illustrated that an integrated approach to the restoration of natural capital, involving prevention as well as cure, is highly appropriate in the socioecological context of the rainforest biome of Madagascar. MCR highlights three factors that also impinge on the restoration of natural capital in general: appropriate institutional frameworks (acceptance of responsibility), the continuity of enabling mechanisms, and the measurement of well-being.

Responsibility

How to equitably share responsibility for restoring natural capital is unresolved, since tackling the underlying causes of environmental degradation is essential if restoration is to be sustainable. Identifying the causes of degradation at a site can be difficult. Inappropriate agricultural practices in Madagascar are only proximate causes of degradation, driven by a complex of cultural, socioeconomic, and political factors stemming from historical, current, local, national, and international decisions, usually acting in synergy. Thus, responsibility for environmental degradation in Madagascar lies with both the Malagasy people and the international community.

Benefits of projects such as MCR span a range of temporal, spatial, qualitative, and quantitative scales, affecting beneficiaries in an equally complex manner. Table 8.1 illustrates the spatial and temporal distribution of some of the more obvious benefits of the project and shows that protecting and restoring biodiverse ecosystems is a service to the global and national, as well as local, communities.

Equitable sharing of responsibility for the restoration of natural capital among beneficiaries can be achieved by funding projects such as MCR through international agencies. Funds should be perceived, not as subsidies, but as payments for services rendered.

Enabling Mechanisms for the Restoration of Natural Capital

Madagascar is a signatory of relevant global conventions such as the United Nations Convention on Biological Diversity (UNCBD) and the United Nations Framework Convention

TABLE 8.1

Spatial and temporal scale encompassed by benefits and beneficiaries of the Masoala Corridors Restoration Project (MCR)

Benefits	Local (human/community)	Provincial	National	Global
	Human beneficiaries			
Biodiversity conservation	Healthier, more resilient local environment; increased biodiversity; natural forest products ▲▲▲▲▲|	Healthier natural environment; increased biodiversity; tourist revenue ▲▲▲▲▲	Healthier environment; increased biodiversity; tourist revenue; meeting obligations under international agreement ▲▲▲▲▲|	Healthier environment; retention of unique flora and fauna—existence value; global treaties ratified ▲▲▲▲▲|
Carbon sequestration and storage	Healthier local microclimate; potential carbon offset finance (not yet sought for MCR) ▲▲|	Amelioration of climatic oscillations; potentially strengthened economy ▲▲|	Strengthened economy; meeting obligations under international treaties ▲▲|	Mitigation of effects of climate change ▲|
Undeflected succession/refugia	Less labor input to fallows; improved natural seed dispersal ▲▲▲|	Improved ecology; possible tourism opportunities ▲▲|		
Genetic stock of potential medicines, crops, other useful products, etc.	Wider availability of herbal medicines—healthier people; availability of useful products for selective use; less hardship ▲▲▲|	Healthier populace—less pressure on local health system; possible strengthening/expansion of local economy ▲▲|	Healthier population; possibly strengthened economy; possible export trade ▲▲|	Possible new medicines ▲|
Watershed protection	Better water availability; reduced soil erosion; more reliable crops; healthier people ▲▲▲▲|	Improved water availability; reduced risk of landslides ▲▲|	More resilient economy ▲▲|	
Water regulation	Improved water availability ▲▲|	Regulation of hydrological flows ▲▲▲|		
Soil retention	Better, more reliable crops; healthier people ▲▲▲|	Reduced water runoff; reduced risk of flooding ▲▲▲▲|		

Nutrient cycling	Better, more reliable crops; healthier people; increased range of crop possibilities ▲ ▲▲ ▲▲ ▲▲	Possible increase in local trade; healthier local markets ▲▲ ▲▲	More sustainable economy ▲▲ ▲▲	Reduced risk of need for food aid ▲▲ ▲▲ ▲▲
Biological control	Reduced risk of pest problems; improved crops ▲▲ ▲▲	Possible increase in local trade; healthier local markets ▲▲ ▲▲	More sustainable economy ▲▲ ▲▲	Reduced risk of need for aid ▲▲
Food security	Healthier people; more time for other economic opportunities ▲ ▲▲ ▲▲	Increased local trade; healthier local markets ▲▲ ▲▲	More sustainable economy; reduced risk of need for food aid ▲▲ ▲▲	
Resilience to environmental disturbance	Reduction of risk of disturbance; buffering from cyclones ▲▲ ▲▲	Reduced risk of population displacements ▲▲ ▲▲	Reduced risk of emergency infrastructure needs ▲▲	
Improved pollination	Better, more reliable crops; possible increased crop range ▲▲ ▲▲	Possible increase in local produce and trade ▲▲ ▲▲		
Nontimber forest products/raw materials	More local needs met ▲▲ ▲▲	Possible strengthening of local economy ▲▲ ▲▲	More resilient, sustainable economy ▲▲ ▲▲	Possible trade opportunities ▲▲ ▲▲
Cultural	Renewed cultural opportunities ▲ ▲▲	Enhanced local culture ▲ ▲▲	Improved sustainability ▲▲ ▲▲	

Note: A selection of specific benefits is listed to illustrate benefits at different spatial scales. However, this does not illustrate the full qualitative benefits nor the synergistic effects manifested as a result of other activities. The temporal spread of benefits is indicated by the number of arrows (1 = 0–2 years; 2 = 3–10 years; 3 = longer-term benefits = 11+ years). The size of the arrow gives an indication of the scale of the benefit: the bigger the arrow, the greater the benefit.

on Climate Change (UNFCCC), which inform national policies such as those embodied in NEAP. Such policies attract and direct donor support, though they are not sufficient yet to facilitate large-scale restoration. To achieve this, appropriate institutional frameworks, combined with policymakers, donors, investors, and practitioners aware of the value of restoration, are also required.

The *Global Partnership in Forest Landscape Restoration* (GPFLR), a partnership of organizations, governments, communities, and individuals, is a positive example of a move toward coordinated restoration. By building upon existing structures, linking policy with practice, conservation with development, and recognizing the economic values of forests to people, the aims can be achieved (e.g., chapter 20).

Measuring Well-being

Economic indicators such as GDP are not measures of general well-being nor of sustainability. Until market mechanisms account internally for resource and environmental costs, the market will continue to drive degradation of natural capital (e.g., see chapters 6 and 19). Indices inclusive of sustainability in its widest sense, such as the Wellbeing Index (WI), which links human well-being with ecosystem well-being (Prescott-Allen 2001), may be a prerequisite for successful utilization of market-driven approaches to natural capital restoration. However, to accomplish this will require a paradigm shift in mainstream economic thinking. Alternative indices of well-being are starting to find their way into national accounting at the political level. If the restoration of natural capital project emphasizes the use of a variety of indices (ISEW, HDI, WI) alongside GDP and gives appropriate comparative weighting to the indices, the opportunity is there to ensure market uptake of restoration values.

A relevant, market-based development is that of land use carbon, because it recognizes payment for environmental services. A recent concerted move to generate carbon projects with strong environmental and social co-benefits presents a real opportunity to achieve sustainable funding for restoration programs. An additional bonus of integrated projects is their appeal to a wide range of investors, each perhaps interested in paying for different benefits (CCBA 2004). Markets can also be strongly influenced by legislation. For instance, 15,000 companies have been presented with quotas on their greenhouse gas emissions by the European Union, prompting an upsurge of interest in land-based carbon activities (along with technological emissions reduction measures) as a way of achieving their quotas within the timescale demanded (http://europa.eu.int/comm/environment/climat/emission.htm).

Contribution

Most livelihoods in Masoala are derived directly from local natural capital. MCR demonstrated that once people became aware of the link between their natural capital consumption and the resulting environmental damage, they were willing to modify their behavior to improve their well-being and that of local ecosystems. Hence, it appears that external resources in the form of technical and material support are required to catalyze this process. This contrasts with developed countries that extract much of their natural capital from distant ecosystems, while at societal and individual levels remaining largely unaffected by the negative effects of consumption. As a result, awareness may be insufficient to catalyze changes in behavior toward sustainable use of natural capital. Recognizing this is important in the design of restoration initiatives and for promoting sustainable use of natural capital globally.International agreements, institutional advocacy, and support from sectors of the international community, as well as from local people, have played a positive role in facilitating MCR. This reflects those who have engaged in and paid for MCR in a variety of ways, locally, nationally, and internationally. All have benefited, along with native biodiversity.

The key to breaking the paradox of people's relationship with rain forest in Masoala was the inclusion in the project design of cultivation systems mimicking natural ones, based on useful species. By emulating natural processes, people grew to understand them and to value the forest and its biodiversity. This is a lesson transferable to Madagascar as a whole and perhaps also to restoration projects in general.

Acknowledgements

I would like to thank Matthew Hatchwell of the Wildlife Conservation Society for useful discussions concerning MCR and surrounding issues, and Dr. Colin Tingle for providing valuable comments and improvements.

Chapter 9

Landscape Function as a Target for Restoring Natural Capital in Semiarid Australia

DAVID TONGWAY AND JOHN LUDWIG

Restoring natural capital is a "big-picture" concept that integrates the conceptual frameworks underlying both economics and ecology (see chapter 1). The objective of this chapter is to illustrate a case of restoring natural capital in semiarid parts of Australia so that it produces a landscape that is stable and retains water and nutrients that support economically productive rangeland. In this way we try to integrate both economic and ecological principles, but central to this effort is the restoration of landscape function.

Restoring landscape function is intrinsically a complex, interactive, and long-term process, requiring the participation of, for example, land managers, ecologists, economists, sociologists, and engineers. Conceptual frameworks can build understanding and enhance communication between participants working to solve complex problems (Low et al. 1999), including the restoration of degraded land (Walker et al. 2002). One such framework, labeled *trigger-transfer-reserve-pulse* (TTRP), views landscapes as dynamic, interacting systems in time and space (Ludwig and Tongway 2000) and has proven useful for addressing many land management issues and environmental problems in Australia (e.g., Ludwig and Tongway 1997; Tongway et al. 1997; Tongway and Hindley 2003).

Only by considering landscapes within a temporal perspective can progress toward restoration be monitored and hypotheses generated for what causes natural capital to be augmented or lost. Here we present a reliable (well-tested) and robust (precise and repeatable by different users) restoration assessment procedure called *Landscape Function Analysis* (LFA), which has been used successfully to track recovery of landscape processes in damaged rangelands and mining sites of Australia and elsewhere (Tongway and Hindley 2004).

Harvesting of Natural Capital

The capacity of a landscape to provide extractable natural capital in the form of goods and services is an assessable property. However, the historic or current provision rate of goods and services is not necessarily a reliable indicator of natural capital abundance or a guarantee of sustained supply. For example, the wool extracted from Australia's rangelands can be quantified in terms of bales produced; yet these data cannot be used to formulate long-term projections. This is because merino sheep wool grows only marginally less well even when pasture is extremely limited, so that starvation occurs suddenly, interrupting wool production unpredictably (Freudenberger and Noble 1997). Hence, slow-moving variables, such as minor reductions in wool growth, can act as indicators of sudden "flips" in ecosystem functioning; they signify that major thresholds have been crossed, affecting a decline in landscape production and function (Scheffer and Carpenter 2003).

Manufactured capital is human made and subject to a set of rules to meet societal and corporate needs. It may appear intuitively more easily understandable than natural capital because of the human mindset responsible for its design and structure, as well as its linearity and perceptible impact. Nevertheless, as the TTRP conceptual framework proposed here is based on resource availability in space and time, it facilitates a close correspondence between manufactured and natural capital.

Landscape Function

Restoration of natural capital metaphorically expresses, in economic terms, landscape rehabilitation to a high level of biophysical functioning. However, it has a more restricted meaning, as natural capital tends to be conceptually "the bottom line" in an accounting procedure, whereas landscape functioning embraces the spatial and temporal dynamics leading to natural capital accumulation. In effect, many interacting "currencies" in natural ecosystems contribute to natural capital accumulation, which may be continuous, serial, and/or periodical, and involve both negative and synergistic effects. For example, soil sediments eroded from rangelands may flow on to pollute and damage Australia's Great Barrier Reef (Prosser et al. 2001). Soil erosion demonstrates flow-on effects and synergistic interactions between neighboring landscapes and may be perceived as negative or as a loss of natural capital. These are easy to observe but difficult to measure. Simple indicators are needed to rapidly assess soil erosion and deliver this information to land managers for any required remedial action.

Lavelle (1997) and Herrick and Whitford (1999) have summarized factors and processes affecting the physical, chemical, and biological natural capital in soils at a range of scales. They document the intimate, sequential interdependency of many organisms within the soil and their respective roles in acquiring, utilizing, storing, and transforming natural capital. In addition, Lavelle and Spain (2001) describe how different processes assume importance as scale increases from clay particle size (10^{-6} m) to catchments (10^4 m), hence providing an integrated, qualitative articulation of the nested hierarchies of processes from within soils to landscapes.

A Framework for Understanding Dynamic Natural Capital

The conceptual framework trigger-transfer-reserve-pulse (TTRP) describes how natural landscapes function over space and time to retain and use vital resources (Ludwig and Tongway 1997, 2000) or, in this context, what might be called the "economics of vital resources." This framework was originally developed to understand the interacting processes within a time perspective relevant to Australia's semiarid pastoral landscapes, which have low and highly unpredictable rainfall. The TTRP framework and many of its underlying assumptions are currently being evaluated in other semiarid landscapes globally (e.g., Wilcox et al. 2003; Ludwig et al. 2005).

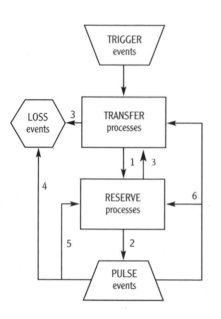

Ref. No.	Processes
1	• Run-on • Capture/storage • Deposition • Saltation capture
2	• Plant germination, growth • Nutrient mineralization • Uptake processes
3	• Runoff into streams • Rill flow and erosion • Sheet erosion out of system • Wind erosion out of system
4	• Herbivory • Fire • Harvesting • Deep drainage
5	• Seed pool replenishment • Organic matter cycling • Concentration by soil fauna
6	• Physical obstruction • Absorption processes

FIGURE 9.1. The Trigger-Transfer-Reserve-Pulse framework. Numbered arrows represent processes where a triggering event causes resources to be acquired, spatially transferred, transformed by biota, and cycled or lost from the landscape (after Ludwig and Tongway 1997).

The TTRP framework specifically examines the processes by which physical and biological resources may be acquired, used, cycled, and lost from a landscape (figure 9.1). A trigger event, such as rainfall, initiates processes including runoff/run-on (1) where some water becomes stored in the soils of vegetation patches (the reserves). If soil water reserves are adequate, a pulse of plant growth is initiated (2) accompanied by animal production and microbial mineralization, all of which contribute to building biomass or natural capital. However, other processes such as runoff and erosion (3) can cause loss of soil and water resources from the landscape (i.e., negative natural capital flow). A feedback loop (5) represents a myriad of largely biologically mediated processes that are the "engine-room" of natural capital accession, transformation, and cycling. These vital processes include seed-pool replenishment, organic matter processing, nitrogen fixation, soil carbon sequestration, soil macrofaunal and microbial activities, and soil nutrient transformation (e.g., mineralization of organic nitrogen to available forms: ammonium and nitrate ions). Furthermore, soil macrofaunal (e.g., earthworms, termites) activities create pores and galleries, resulting in higher levels of "soil health" due to increased water infiltration and availability and root and microbial respiration. An additional feedback loop (6) represents other biophysical processes, including how plant growth pulses build denser vegetation patches, which in the next trigger event reduce runoff and enhance water infiltration and retention (Ludwig et al. 2005). Denser vegetation cover also prevents physical crust formation (Moss and Watson 1991).

Natural capital can be assessed at any time by measuring the content in the Reserve and Pulse boxes (figure 9.1). For example, the Reserve box could be examined for the size of its soil, seed, or mineralizable nutrient pools, amount of water stored in the root zone, plant population size, or biomass of soil fauna. Concurrently, the biomass or size of the Pulse box, represented by plant and animal populations, can be determined to quantify natural capital. In the TTRP framework the dynamics and efficiency of the processes shifting and transforming the Reserve and Pulse box contents are more important than content sizes at any one time.

Landscape Function Analysis (LFA)

The TTRP framework has facilitated assessment and monitoring procedures that rapidly examine the status of the processes by which natural capital is acquired, used, and retained. These procedures are encompassed within *Landscape Function Analysis* (LFA), which is described in detail in a series of manuals (Tongway 1995; Tongway and Hindley 1995; Tongway and Hindley 2004). Briefly, LFA collects data at two scales. At a broader scale, the locations of patches and interpatches are mapped; patches tend to accumulate natural capital, whereas interpatches tend to shed it. At a finer scale, nested within the patch and interpatch pattern, eleven simple, rapidly collected, soil surface indicators are assessed that estimate the effectiveness of a range of processes. These indicators are then combined into three general indices reflecting the landscape's surface stability, infiltration capacity, and nutrient-cycling potential. In conjunction with other measures, such as vegetation patch structure, these three landscape surface indicators are interpreted to assess whether natural capital is being lost, maintained, or enhanced over time, as illustrated by a mine site rehabilitation example (table 9.1).

As LFA procedures focus on landscape processes and not on any particular form of soil, vegetation, or biota, they can be implemented across a range of landscape types, uses, and managements. For example, Tongway and Hindley (2003) applied and verified the methodology to nine mines in Australia and Indonesia, with landscapes varying from sandy deserts to tropical rain forest, and in different geological settings from which were extracted gold, nickel, bauxite, coal, uranium, and mineral sands. In addition, LFA procedures have been widely used to assess landscape processes and attributes, reflecting natural capital across Australia's rangelands (Tongway and Smith 1989; Tongway et al. 1989; Tongway 1993; Ludwig and Tongway 1995; Karfs 2002).

TABLE 9.1

Indices of stability, infiltration, and nutrient cycling

Rehabilitation period (years)	Stability Index	Infiltration Index	Nutrient Cycling Index
Zero: freshly prepared, unseeded land	40.6	34.2	14.1
1	43.9 (2.1)	25.1 (1.7)	12.1 (0.9)
2	50.9 (4.2)	29.6 (1.5)	16.7 (2.6)
3	61.6 (2.6)	30.1 (1.3)	22.8 (2.2)
4	60.0 (4.9)	30.4 (4.7)	25.8 (5.3)
8	61.5 (4.1)	37.2 (2.4)	29.3 (2.9)
13	82.5 (1.2)	50.2 (4.1)	45.6 (5.2)
20	81.5 (1.4)	65.9 (2,5)	63.4 (2.5)
26	86.7 (0.9)	66.9 (2.0)	71.3 (4.2)
Reference site	75.5 (3.7)	48.4 (2.9)	44.3 (4.2)

Note: The scale, from 0 to 100, is derived from eleven measurements obtained using the Landscape Function Analysis (LFA) monitoring procedure. All the indices increase over time, implying that landscape function is improving, as is the accession of natural capital.

Perspectives on the TTRP Framework

Prior to development of the TTRP framework, rangeland degradation was described mainly in terms of vegetation composition and structure. Soil erosion status was reported in vague terms and not connected to the vegetation assessment by an explicit framework. Processes mediated by various biota were implicit in the monitoring information but not quantitatively assessed. Hence, these descriptive and compositional assessments were unable per se to specify degradation levels or the means for designing successful rehabilitation. The TTRP framework facilitates a much more "econometric" examination of landscape function, as it is based on the availability and use of limited vital resources by biota in space and time. More recently, the loss of native species and other issues of biodiversity have been included in the definition of landscape function (Ludwig et al. 2004).

The TTRP framework is more directed to the processes by which natural capital is acquired than its quantification. The former is perhaps of greater interest to ecologists, whereas the latter is more the focus of economists, though within the framework, natural capital valuation is entirely compatible across both disciplines. For example, the accession of "new" exogenous natural capital and the loss of existing natural capital are an integral part of the framework, and as such it is well suited for use in a participative approach (e.g., adaptive learning workshops) to better understand the issues of restoring natural capital.

Knowledge of the multiple "currencies" in the ecological world (such as organic matter, mineralizable nitrogen, soil-stored water, etc.) and the timescales and processes affecting their interactions is still incomplete. Because of the need to deal with management issues "today," the TTRP framework is an inclusive concept, which while explicitly acknowledging ecological complexity, measures only net outcomes of intimate interactions, rather than waiting for a complete knowledge. Nevertheless, the temporal and spatial sequence of processes represented in the framework has been observed to be appropriate for assessing ecosystem functioning across a range of landscape types and management systems at different scales (Ludwig et al. 1999, 2000, 2002; Ludwig and Tongway 2002; Tongway and Hindley 2003).

A Continuum of Landscape Functionality

The TTRP framework recognizes a continuum of functionality in every landscape, ranging from highly functional to highly dysfunctional (figure 9.2). Highly functional, semiarid woodlands have been shown to possess high levels of natural capital, in terms of topsoil retention, nutrient pool size and cycling, and aboveground biomass (Tongway and Ludwig 1990; Ludwig and Tongway 1995). Moreover, in TTRP terms, landscape analysis indicates that the biophysical mechanisms for natural capital retention are active: mobile resources flowing off bare slopes are effectively captured in grassy and woody vegetation patches (figure 9.2a), while the biological feedbacks from Pulse to Reserve and Pulse to Transfer are both complex and efficient. This also indicates that functional biodiversity is high and structurally complex (Ludwig et al. 2004; McIntyre and Tongway 2005).

Conversely, a dysfunctional landscape has fewer surface obstructions (figure 9.2b), resulting in a lower capacity to intercept and retain resource inputs such as water, soil, and seeds in runoff. Thus, stored natural capital is at a greater risk of being rapidly transported from the local landscape, such as rangeland hill slopes. Depletion of natural capital to low levels may transform the landscape system into a different state (Gunderson and Holling 2002).

Continuum of landscape functionality

Functional   Dysfunctional

←——————————————————————————————————————→

(a)   (b)

FIGURE 9.2. A continuum of landscape functionality in the semiarid woodlands of eastern Australia from (a) highly functional, where natural capital is acquired and stored (soil enrichment in patches of grass and trees), to (b) totally dysfunctional, where natural capital is lost (through death of plants, soil erosion).

Responses to Stress and Disturbance

The functionality of landscapes can differ in terms of their response to stress and disturbance (Tongway and Ludwig 2002). For example, robust landscapes are able to maintain a high delivery rate of goods and services as stress and disturbance increase (figure 9.3a), although they will eventually drop to a lower capacity. In contrast, fragile landscapes rapidly lose functionality (figure 9.3b), that is, they rapidly lose accumulated vital resources and the capacity to acquire fresh resources, and hence the capacity to deliver goods and services. The resilience of the landscape will determine its response, for example, to human-driven disturbance with a rapid fall in functionality being viewed as a critical threshold (Tongway and Ludwig 2002). Above this threshold, natural capital storage and accumulation processes are sufficiently effective for a self-sustaining landscape. Below this threshold (figure 9.3c), stored capital is too low and the processes for retention are ineffective (i.e., the landscape is dysfunctional). Certain goods and services may still be extracted from dysfunctional landscapes, but their continuity in space and time is liable to disruption.

In nature, there are typically parallel subsystems leading to similar outcomes. This structural complexity is sometimes called *redundancy* (Walker 1992). Indeed, nature is typically endowed with multiple pathways and processes to achieve similar ends or outputs, depending on which mechanism is more active at a particular time. It is this complexity that confers a landscape-buffering capacity to oppose stress and disturbance and that restores the system after a natural or induced perturbation.

Trajectories of Natural Capital Restoration

There are four principal questions when restoring natural capital:

1. Is natural capital accumulating?

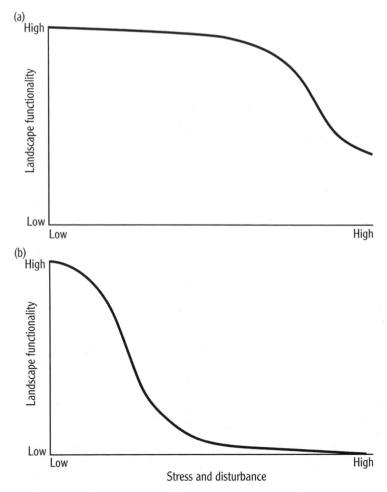

FIGURE 9.3. The response of landscape functionality to stress and disturbance for (a) robust and (b) fragile landscapes. The landscape functionality axes could also be labeled as low to high natural capital.

2. If so, at what rate?
3. Is the level of accumulation sufficient for self-sustainability of the rehabilitating landscape?
4. Have the mechanisms for natural capital accumulation become sufficiently complex to confer buffering capacity on the landscape, enabling it to survive stress and disturbance?

In our work on thirty-five mine sites across Australia (Tongway et al. 1997; Tongway and Hindley 2003), three main types of rehabilitation trajectories were observed (figure 9.4), which indicate how landscape functionality changes over time. Trajectory A represents the accumulation of natural capital and landscape function, so that after a reasonable time, the landscape passes through a conceptual, critical threshold for self-sustainability, and at longer timescales continues to improve. Trajectory B illustrates a slowly responding treatment, where, although there is a detectable increase in landscape function, the rate is so

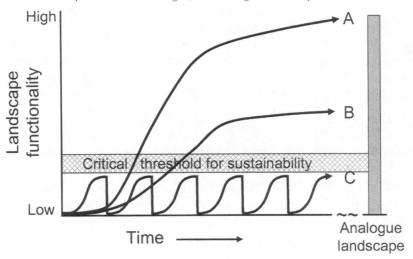

FIGURE 9.4. Three trajectories of landscape functionality for rehabilitating mine sites toward that of nearby reference sites: A = successful, B = moderately successful, and C = unsuccessful. The landscape functionality axis can be equated with restored natural capital.

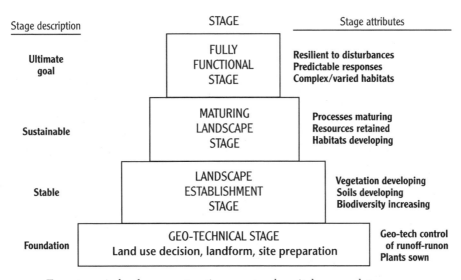

FIGURE 9.5. Four stages in landscape restoration as natural capital accumulates.

slow that the critical threshold is not exceeded for many years. During this time, the rehabilitation may be subjected to severe perturbations such as fire, drought, or storms that could threaten its success. At the extreme, trajectory C includes settings where site preparation and species selection are inappropriate, to the extent that disturbances result in no net natural capital accumulation.

Contribution

Tongway et al. (1997) proposed a stepped pyramid for conceptualizing and assessing the rehabilitation of mine sites (figure 9.5), in which recovery proceeds through four stages to a stable and fully functional landscape. Success particularly depends on applying ecological principles in the initial landform design and site preparation stage. A complex landscape will emerge that possesses a multiplicity of life-forms (biodiversity) and regulatory processes (functional diversity). Such landscapes will be buffered against environmental and management disturbances both by their accumulated natural capital and by the complex diversity of the processes responsible for new natural capital accession.

In assessing the return of natural capital in terms of landscape functionality, it is important to distinguish between those indicators of landscape health that are explicitly involved in fundamental biophysical functioning and those that simply reflect reaction to change. The key test would relate to their relative contribution to resource retention, use, and transformation. The Landscape Function Analysis approach can be used to track the health of a recovering landscape and to demonstrate to the benefits of restoring natural capital. Such evidence is important for achieving social acceptance of the rehabilitated landscape that is part of the final evaluation of whether the restoration targets have been achieved.

Chapter 10

Genetic Integrity as a Target for Natural Capital Restoration: Weighing the Costs and Benefits

CATHY WATERS, ANDREW G. YOUNG, AND JIM CROSTHWAITE

Introductions of plants and seeds may be essential for the restoration of natural capital in sites damaged by mining or overexploitation (see, for example, chapters 8–14 and 20–23). The goal of such reintroductions is to replenish stocks to improve and sustain the flows of ecosystem goods and services, which could contribute to human benefits. However, there is considerable debate about how to assess the risks that translocated plants and seeds may pose for the ecological integrity and utility of reestablished plant populations. In this chapter, we use Australian case studies to develop a decision framework that weighs up benefits of using local versus other sources of plant genetic material, or natural capital, against the ecological risks and economic costs.

The native seed industry is expanding in response to increased demand for larger scale revegetation in Australia (Waters et al. 1997), especially within the mining industry that uses 70%–80% of all the commercial native seeds collected (Mortlock 1999). While legislative requirements for revegetation after mining vary between states, most aim to restore native communities compatible with the surrounding ecosystem (Coates and van Leeuwen 1997). The mining company and commercial wildland seed collectors usually collect seeds. Almost two-thirds of seed and seedlings are sourced from wildlands, a reliance that is likely to be maintained (Waters et al. 1997; Mortlock 1999; Mortlock 2000).

Areas being revegetated by the mining industry are small compared to required native-tree plantings to reduce salinity caused by forest clearance and land degradation, or in response to the market for carbon sequestration. In low rainfall areas of Australia, new, woody, perennial crops such as the Western Australian oil mallee industry (*Eucalyptus* spp.) are being successfully incorporated into agroforestry systems (Bartel 2001). However, this farm and plantation forestry potentially poses a threat of genetic pollution to native forests, comprising over eight hundred species of eucalypts endemic to Australia (Potts et al. 2003).

Voluntary conservation groups, in particular Landcare, have upscaled Australian revegetation programs. Between 1996 and 2002, Landcare received a US$170 million federal investment through the Natural Heritage Trust Fund (NHT 2002). A key Natural Heritage Trust activity has been reversing the decline in extent and quality of native vegetation, while restoring habitat for threatened species. In Australia, it is estimated that by 2050 seventeen million hectares of agricultural land, some 52,000 km of roads, and two hundred towns could be affected by dryland salinity. Large-scale revegetation is required to reduce the amount of groundwater recharge and to remove nutrients draining into rivers and streams (CRC 2003).

Restoration projects should lead to net environmental improvement and should not cause persistent adverse environmental impacts on-site and within the surrounding landscape (chapter 1). These goals are usually thought to require local seed—a widely held perception supported by funding bodies in Australia and elsewhere. For example, the Society for Ecological Restoration (SER) suggests that "under normal circumstances, the reintroduction of *local ecotypes* is sufficient to maintain genetic fitness" (2002). It is further assumed that, in a world undergoing rapid climate change, reduction of genetic fitness may hasten losses of species and the services they provide. The underlying assumptions are that local provenance seed is (1) the best adapted, (2) has the highest genetic quality, and (3) will not contaminate the resident population or lead to a loss in genetic diversity (Knapp and Rice 1994).

Here we present case studies from the Australian flora to demonstrate how these assumptions may be flawed. As an alternative to unquestioned use of local seed, we suggest use of a risk assessment framework that applies natural capital concepts yet allows for concern, uncertainty, and irreversibility considerations to be taken into account (O'Riordan and Cameron 1994).

Testing Assumptions Using Case Studies

Assumption 1: Local provenance seed is the best adapted, and nonlocal seed will result in reduced fitness.

This assumption appears logical where gene flow between populations is limited and a high degree of adaptation can be expected. In these situations, ecotypes can occur within a species holding a distinct set of morphological and/or physiological characteristics (Dunster and Dunster 1996). However, as the scale of such variation remains largely unknown, describing seed sources as "local" provides little useful information to the restorationist.

There are about 1,300 native grass species in Australia, yet few have been studied in detail. Ten common grass species of the semiarid rangelands showed morphological differences between populations that could be attributed to place of origin (Waters et al. 2003). One species, *Austrodanthonia caespitosa* (wallaby grass), had five distinct site groupings; taller, larger plants tended to come from cooler, higher rainfall areas, and smaller, shorter plants from warmer, drier areas. Average growth rates of different populations of this species varied with temperature and rainfall (Hodgkinson and Quinn 1976), and flowering was cued by changes in day-length at southern, moister environments, but by rainfall in semiarid environments (Hodgkinson and Quinn 1978). It is thus likely that the southern populations of *A. caespitosa* would perform poorly when seeded into these semiarid environments.

Conversely, in the widespread native grass *Themeda australis* (kangaroo grass), control of flowering shows a broad range of adaptive responses (Evans and Knox 1969). Although the appearance of the plants differs regionally, the germination ecology of *T. australis* was similar over its climatic range (Groves et al. 1982). For some native grasses, the magnitude of variation is more difficult to recognize. For example, *Microlaena stipoides* (microlaena) cv Griffith occurred originally in Canberra (southern Australia), yet reintroduced populations grow well some eight hundred km north (Whalley and Jones 1995), suggesting a large adaptive range. However, populations of *M. stipoides* growing in a patch of *Lolium perenne* (perennial rye grass) were found to be genetically distinct from *M. stipoides* obtained from a patch of *Poa pratensis* (Kentucky bluegrass) in the same paddock (Magale-Macandog 1994), indicating that microevolutionary differentiation has occurred. Both fine- (Waters et al. 2004) and large-scale (Waters et al 2005) intraspecific variation has been observed for *Austrodanthonia*. Collectively, these studies illustrate that different scales of ecotypic variation occur and that it can be difficult to distinguish populations diverging as a result of genetic drift (variation due to chance) or natural selection (adaptive variation).

Assumption 2: Local provenance seed has the highest genetic quality.

Genetic constraints to plant performance, such as inbreeding, can compromise the genotype fitness of local seed sources. The primary concern with inbreeding, and the associated decline in genetic diversity, is reduced population viability. Inbreeding can negatively affect a wide range of fitness traits ranging from seed weight to germination, growth characteristics, and reproductive output (Fenster and Dudash 1994). Fitness loss is more marked when conditions are harsh (Dudash 1990). Inbreeding depression is partly determined by life history, so that obligate outcrossers are more affected than species with other breeding systems.

The issue of inbreeding influences revegetation and restoration when seed is sourced from small or isolated populations in which the likelihood of self-pollination (selfing) is high. Such situations are typical of most agricultural landscapes in southern Australia, where native vegetation is highly fragmented. Here, sourcing seed locally to maintain environmental adaptation means obtaining seed from disturbed remnants, often several kilometers or more from the next native bush sites. If inbreeding increases, the reduction in fitness and subsequent poor performance of the reestablished plants may outweigh any advantages gained from using locally adapted genotypes.

Where only small populations are available, the tradeoffs may be unbalanced, as in the case of *Swainsona recta*, a small pea plant endemic to grassland and grassy woodland environments in southeastern Australia. Its fate over the last hundred years has paralleled that of its habitat, as grasslands have been reduced to about 0.5% of their original pre-European extent (Kirkpatrick et al. 1995). It is now known only in seventeen populations, ranging in size from three hundred to four hundred plants. Recovery strategies are likely to rely on reestablishment of new populations and augmentation of remaining small grassland fragments, which will require seed sourcing for restoration plantings. However, this pea species varies in appearance across its range, reflecting possible environmental adaptation and suggesting that local sourcing may be wise. Unfortunately, this limits collection from fairly small, isolated populations of fewer than two hundred plants.

Analysis of the population size effect on inbreeding in *S. recta* showed that as population size drops, inbreeding increases (Buza et al. 2000). Growth trials revealed that seed from small populations had reduced fitness (including slow growth and low seedling survival) compared with seed from large populations. For *S. recta*, it may be worth sacrificing some local adaptation by sourcing seed from larger populations to avoid the deleterious effects of inbreeding.

Assumption 3: The use of nonlocal seed will contaminate the resident population and lead to loss in genetic diversity.

There is a perception that nonlocal seed sources may spread foreign genetic material to resident populations through cross-fertilization causing reduced progeny fitness. This may man-
ifest itself in either reduced reproductive performance in resident populations or reduced hy-
brid fitness in subsequent generations. However, there is limited evidence for this occurring,
and, in fact, importing new genetic material may be necessary if a small local population is to
remain viable. In particular, this can be the case where a remnant population, being used in
revegetation as a seed source, has little genetic variation for genes of major effects such as disease resistance or self-incompatibility. In large populations, with high genetic variation, self-
incompatibility systems (such as rejection of own pollen) limit inbreeding and the associated
negative effects. However, in small populations, this can cause significant reproductive limitation. Thus, sourcing seed from a limited number of local populations could be a mistake if
these are small and contain little genetic diversity at the self-compatibility locus.

In the threatened grassland daisy *Rutidosis leptorrhynchoides* in southeastern Australia, local populations with less than two hundred plants exhibit 20%–90% reductions in mate availability due to self-incompatibility and an associated decline in seed set (Young et al. 2000).
Sourcing seed from only one or two small local populations would not be advisable for genetic
variability, whereas broader seed collections from multiple populations will maximize possible mate availability. There is a limit though to this strategy, because within its geographical
range, this daisy varies genetically (Young and Murray 2000). Thus, a little mixing of nonlocal
gene pools would be beneficial, but sourcing from distant populations could be deleterious.

These case studies illustrate that it is overly simplistic to suggest that seed sourced from
distant locations is "poorly adapted" and will "contaminate the local genetic material"
(Whisenant 1999), while reducing the vigor and competitive ability of the restoration species
(Knapp and Rice 1994). In Australia and elsewhere, species-specific studies and a synthesis of
this knowledge to understand the risks associated with revegetation, in particular genetic pollution, are lacking. In the absence of such scientifically derived information, debating the relative merits of using local or exotic provenance material is of limited practical application to
the restoration practitioner. Moreover, application of the precautionary principle outside the
context of a risk framework will make restoration impracticable where there are insufficient
quantities of seed.

Risk Assessment Framework for Seed Selection

While the choice of plant material is fundamental to any successful restoration program,
there is little guidance offered in the literature for land managers or restoration practitioners
to balance the desired seeding objectives with available genetic or economic resources. Although Jones and Johnston (1998) describe an integrated approach that embeds genetic concepts into a seeding recommendation, they fail to provide a practical framework that incorporates an assessment of the associated risks.

We now examine the framework components of a general risk assessment for maintaining
genetic integrity within native plant restoration. A risk framework also highlights where further
research is needed before a particular revegetation proposal can proceed. In the risk framework proposed (figure 10.1), seeding recommendations are based on consideration of seeding
objectives, known and unknown ecological adaptations, risks, and net economic benefits.

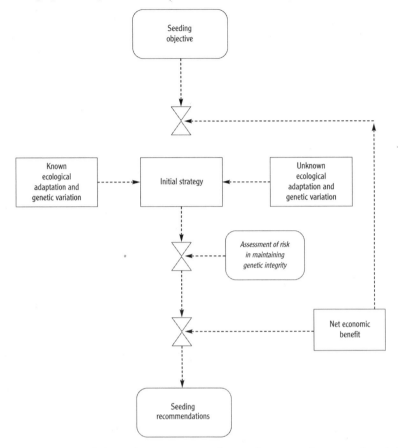

FIGURE 10.1. Risk framework for making recommendations for native plant reseeding (adapted from Jones and Johnston 1998).

Seeding Objectives

The site potential and the desired landscape will influence the seeding objectives. We have assumed that natural regeneration is not possible and that reseeding is required.

Known Ecological Adaptations

Some prior knowledge of ecological adaptations or genetic variation can be used in the development of the initial seeding strategy. For example, *Themeda australis* is one of Australia's most widespread grass species and known to vary across its range (Hayman 1960; Woodland 1964). Inland populations are indifferent to day length in their flowering behavior, while flowering in northern and southern populations is triggered by day length (Evans and Knox 1969).

Unknown Ecological Adaptations and Genetic Variation

Past attempts to guide seed collection, when little information exists on the variability within a species or its adaptive significance, have suggested that where a species remains common, seed should be collected nearby (Knapp and Rice 1994; Mortlock 2000). Where a species no longer occurs close to the revegetation site, collection should take place within a defined radius around the site (Millar and Libby 1989; Jones and Johnson 1998), which assumes that genetic dissimilarity increases with distance. However, geographic distance per se appears to be an unreliable indicator of genetic integrity, and collections should be supplemented from sites of similar composition and/or physical attributes (Millar and Libby 1989).

Risks in Maintaining Genetic Integrity

The method suggested here for assessing the risks of genetic integrity loss also factors in the potential economic benefits. Consideration of genetic integrity is based on (1) natural capital benefits of revegetation, (2) condition of the seed collection site, (3) required seed collection range, and (4) individual species characteristics.

DETERMINATION OF NATURAL CAPITAL BENEFITS

These are based on enhancement of biodiversity and amelioration of environmental problems. In allocating one of the four benefit values (low, medium, high, very high), the restorationist needs to consider the conservation significance of the restoration site, irreversibility issues if wrong seeding choices are made, and potential impacts on surrounding ecosystems.

DETERMINING THE CONDITION OF THE SEED COLLECTION SITE

This needs to be considered using a number of locations, landscapes, populations, and individual plant attributes (table 10.1).

TABLE 10.1

Determination of the seed collection site value based on landscape and plant characteristics

Collection site value	Site attributes	Landscape attributes	Population attributes	Individual plant attributes
High	Soil characteristics closely match the restoration site; habitat quality high; minimal disturbance; accessible	High degree of connectivity; climatic site characteristics closely match those of the restoration site	High number of reproductive individuals; diverse age structure; dense	Disease free, healthy plants; high seed set
Medium		Medium degree of connectivity		
Low	Soil characteristics grossly different to the restoration site; habitat quality low; high level of disturbance; accessibility difficult	Low degree of connectivity; climatic site characteristics grossly different to those of the restoration site	Low number of reproductive individuals; limited variation in age structure; sparse	Unhealthy, diseased plants; low seed set

TABLE 10.2

Determination of the required collection range for native plant seed

Quality of the seed collection site	Natural capital benefit of the restoration			
	Very high	High	Medium	Low
High	Narrow	Narrow	Intermediate	Intermediate
Medium	Narrow	Intermediate	Intermediate	Regional
Low	Intermediate	Intermediate	Regional	Regional

Note: Based on both expected natural capital benefits at the restoration site and the quality of the seed collection site.

REQUIRED SEED COLLECTION RANGE

Assessment of the environmental benefits of the revegetation and the condition of the seed collection site are then combined to determine the required collection range for native seed (table 10.2). The collection range should be narrow where both the value of the seed collection and revegetation sites is high. Conversely, a regional seed collection might be acceptable if the collection sites are of low quality and the selected restoration localities have low genetic integrity.

INDIVIDUAL SPECIES CHARACTERISTICS

These are important in determining an acceptable collection range (table 10.3). They are not intended to provide prescriptive boundaries, rather that local knowledge should add detail where appropriate. For example, pollination characteristics or flowering times may be important where local fauna and flora have specific requirements.

DETERMINATION OF RISKS

Risk scores can be derived from the combined assessment of the revegetation and collection sites (table 10.2) and the assessment of species characteristics (table 10.3). Where the predicted collection range (table 10.2) matches the assessment of species characteristics (table 10.3), a risk score (range 0–1) can be obtained. An identical match will result in a score of 0, a moderate match a value of 0.5, and a complete mismatch with a value of 1.

Net Economic Benefits of Reseeding

Techniques for valuing ecosystem services that flow from natural capital are still in their infancy, yet, as in ecology, the principles of economics can be applied in a context of uncertainty and lack of information. Knowing what benefits and costs are relevant, having information about specific costs associated with the reseeding, but limited quantification of benefits, may be sufficient to make decisions. Here, a threshold test, which asks how high the benefits would have to be in order to justify the costs of reseeding, would be appropriate. Judgment can then be applied as to whether the benefits will exceed the costs.

TABLE 10.3

Determination of acceptable seed collection ranges based on species characteristics

Acceptable collection range	Longevity	Reproductive system	Morphological variation	Genetic variation	Species range
Narrow	Long	Selfing	High, unknown	High, unknown	Discrete
	Long	Selfing	High, unknown	High, unknown	Discrete
Intermediate	Long–medium	Selfing or mixed mating	Unknown	Unknown	Discrete or continuous
	Short–medium	Selfing or mixed mating	Unknown	Unknown	Discrete or continuous
Regional	Short	Outcrossing	Limited	Limited	Continuous
	Short	Outcrossing	Limited	Limited	Continuous

Source: Adapted from Mortlock 2000.
Note: A narrow collection range reflects a high degree of required precaution; a low degree is required for regional collections.

The relevant categories of benefits and costs, which should also account for the goods and services generated by natural capital, go beyond a traditional economic analysis (Clewell and Rieger 1997). The natural capital lost is routinely ignored when native ecosystems are developed or managed to meet primary production expectations. These lost products can include genetic material, timber, pasture, wildlife habitat, and recreation, while the associated eroded services comprise maintenance of soil hydrology (e.g., combating salinity), preventing desertification, as well as storage of excess carbon dioxide. Restoration of these values can help determine the viability of restoration efforts to planners and policymakers. The revegetation costs, to compensate the natural capital loss for low or moderate risk scores, can be deducted from the development profits.

Applying the Risk Framework

If the risk score is high, then the restoration proposal should proceed only if it involves seed from within the acceptable collection range and it can be justified on economic grounds. Seed from outside the limit is unacceptable as it is likely to result in a loss of biodiversity assets that are critical natural capital. Their loss may be irreversible or considerable uncertainty could exist about the effects of introducing new genetic material (Ekins, Simon, et al. 2003). In this case, the decision to protect the restoration site from possible genetic pollution or maladapted seed sources is consistent with strong sustainability (chapter 2), and the onus is on the restorationist to prove that using seed from outside the collection range is acceptable in terms of the restoration project goals and risks to neighboring natural capital stocks.

Where the risk is moderate, the net economic benefits need to be significant to use seed from outside the acceptable collection range. For example, use of such seed should enhance the restoration site without causing irreversible damage and biodiversity losses. Although some loss in noncritical natural capital may occur, this should be compensated by the restoration effort. If the risk is low (no loss in natural capital), revegetation can take place with seed from any source, provided that this is soundly based on economic grounds.

In the proposed framework, the onus is placed on the restorationist to ensure that reseeding will not damage critical natural capital. Development bonds are inadequate in these circumstances, because the costs associated with loss of critical natural capital involve much more than potential damage repair. Assurance bonds, as proposed by Costanza and Perrings (1990), may be more relevant. Hence, the bond needs to be high enough to induce the restorationist to undertake or commission the necessary research and development. Only then will this give confidence that irreversible losses are unlikely and that reseeding can go ahead without jeopardizing the bond.

Contribution

The decision-making tool presented here weighs the benefits of using local versus other sources of plant genetic material against the ecological risks and economic costs. It considers the environmental benefit of the revegetation activity, the condition of the seed collection site, and individual species characteristics. Consideration of these factors can aid the determination of an appropriate collection range within which native seed can be "safely" sourced, enabling the restorationist to modify seeding proposals to obtain the best outcome. This decision-making process is for a single reseeding proposal, but it can be adapted for multiple potential revegetation projects, which should always compare risk scores, net economic benefits, and other relevant decision criteria.

A key feature of the risk framework is defining an acceptable collection range, comparable to the concept of safe-minimum standard (Bishop 1978). The SMS rule allows decisions in favor of development if the social protection costs were judged to be very high relative to the costs for endangered species. A practical risk framework for guiding the assessment of the genetic integrity of seed sources is imperative given the current and forecasted expansion of efforts to restore natural capital in an increasingly human-dominated world.

Chapter 11

Restoring and Maintaining Natural Capital in the Pacific Northwest, USA

Andrew Carey

The Pacific Northwest is renowned for its magnificent forests, the biological diversity they contain, and the Pacific salmon that breed in their streams. These forests supply essential ecological services of cooling, clean air and water, flood amelioration, and waste assimilation while offering recreational and spiritual opportunities. In the recent past, they provided important regional economic activity from timber, fish, and game, as well as related employment that supported rural communities. Overexploitation of these resources and consequent reductions in environmental quality caused a precipitous diminution in economic activity. Conflicts among stakeholders regarding access to the dwindling resource base became acrimonious to the point of intractability for responsible public entities. With the intent of avoiding impasse, the state of Washington and the federal government jointly issued a request to develop a regional management plan that would restore lost biodiversity, recover threatened species, provide a sustainable flow of wood products and ecological services, reinvigorate declining industries and local communities, and thereby satisfy all stakeholders. In other words, was there a way to restore and maintain the natural capital associated with forests? In response to this request, a group of forest ecologists, myself included, and scientists from other critical disciplines was assembled and commissioned as the Washington Forest Landscape Management Project.

Only the collective goals of sustainability and intergenerational equity seemed to bring all stakeholders together around the concept that the forest landscapes would be restored and thereafter managed to retain desirable levels of natural capital indefinitely for the benefit of this and future generations. The project developed a novel approach that expanded the traditional bounds of forest management and incorporated multiple social, economic, and environmental values (Carey, Lippke, et al. 1999). We called this approach *Active Intentional Management* (AIM) for multiple values, a title that emphasized social dimensions such as collaborative learning and management. As it turned out, the approach of AIM was really a prototypical application of the concept of restoring natural capital, as described in this volume. In this chapter, the AIM approach and experience are described as exemplary of restoring natural capital.

The AIM Approach

AIM addresses multiple spatial scales, including ecological reserves at regional and smaller scales, riparian protection at landscape and smaller scales, and active management of forest vegetation at both landscape and smaller scales (from hectares to square meters) (Carey, Lippke, et al. 1999; Carey 2003). Vegetation management includes timber harvesting, seedling planting, thinning to maintain biodiversity while obtaining wood products, long rotation times between major felling, and other techniques to promote healthy, adaptive forests (Carey 2003). The strategic approach of AIM invokes deliberate attempts to do the following:

1. Address multiple values, including wood, water, wildlife, and naturalness.
2. Incorporate multidisciplinary science in a systems approach to ecosystem and landscape management.
3. Involve people from various sections of the community in collaborative learning and management.
4. Address issues of environmental, economic, and social sustainability in an equitable manner.

This approach is fully consistent with restoring natural capital with respect to the augmentation of flows of natural goods and services for use and benefit of all stakeholders and to an increase in the social awareness of—and respect for—the importance of natural capital, as discussed in chapter 2.

Consensus Building

Achieving consensus among stakeholders is essential for AIM, which can be attained only with the assurance that management decisions reflect the breadth of current ecological, sociological, and economic knowledge. Adoption of an AIM program by stakeholders requires improvement in public awareness about ecosystems. Four propositions were developed in this regard:

1. To change the public perception of an ecosystem and its development from that of outward appearance of the vegetation to that of ecological processes and an appreciation of the biological complexity that contributes to an understanding of ecological processes (Carey, Kershner, et al. 1999)
2. To develop an appreciation of the need for ecosystems to be resilient in the face of ever-changing environmental conditions and for ecosystems to be adaptive and ever evolving if they are to persist (Holling 2001)
3. To develop an understanding of how local forest ecosystems are adapted for resilience
4. To develop awareness regarding the processes of self-organization that lead to biocomplexity, adaptiveness, and stability in a dynamic sense. This awareness should be coupled with an understanding that biocomplexity is fundamental to the capacity of an ecosystem to satisfy diverse values and user requirements by modern pluralistic societies.

Validation and Evaluation of AIM

AIM techniques were implemented on state and federal lands and proved instrumental in giving stakeholders insights into these four propositions. This led to reconciliation among stakeholders in collaborative management groups, who were formerly strongly in opposition. In order to broaden this awareness and validate the importance of AIM as an effective management program for restoring natural capital, the following procedures were proposed:

1. Develop illustrative models that can be used to contrast management outcomes of the AIM approach with (a) no management (or benign neglect), (b) conventional timber management (the most convenient benchmark), (c) alternative ecosystem management approaches (additional benchmarks), and (d) highly resilient, biologically diverse, natural forests (i.e., an old-growth baseline).
2. Simulate alternatives and compare results.
3. Conduct field experiments.
4. Evaluate results with third-party criteria, such as those presented in the SER *Primer* (2002).
5. Allow independent analyses by other organizations that confirm the validity of modeling.

Develop Models

With respect to the first item, three illustrative models of Pacific Northwest forest ecology were prepared. The first model concerns the role of birds as agents in the biological control of insect pests, the second considers roles of soil fungi in ecosystem functioning, and the third incorporates a "keystone complex," or ecosystem framework, that includes Douglas-fir, spotted owls, three squirrel species, and ectomycorrhizal fungi (see Carey, Lippke, et al. 1999; Carey 2003).

Simulate Alternatives

Our group assembled an interagency, multiuniversity, multidisciplinary team to build computer models that simulate potential landscape management alternatives for the Olympic Peninsula, Washington (Carey, Lippke, et al. 1999). We were fortunate to have access to detailed landscape data, including empirical growth and yield models for timber, timber prices, market distances, costs of alternative methods of silviculture and harvest, and road construction expenditure. Additional information included quantitative descriptions of forests and streams, published, expert-based, wildlife habitat–relationship tables, and data-based models for selected wildlife populations and communities. After much debate, our team of experts selected five standard criteria that could be quantified as measures of economic output to make comparative evaluations of forest management alternatives. The selection of these criteria was crucial if a compelling case for the AIM alternative was to be made. As it turned out, the criteria provided a quantitative basis for assessing the value of natural capital for each alternative:

1. Ability of a forest ecosystem to support wide-ranging old-growth species, based on area estimates of older, complex forests required to support one pair of spotted owls

2. Capacity of a forest ecosystem to support vertebrate diversity based on published accounts of the habitat requirements of 130 species, evaluated as the percentage of the maximum possible
3. Forest-floor function, defined as the biotic integrity of the forest-floor, small-mammal community, which represented the top of the forest-floor food web and part of the prey base for weasels, foxes, coyotes, bobcats, owls, and hawks
4. Ecological productivity of a forest ecosystem, as indicated by the biomass (kg/ha) of three squirrel species and as representing the system's production of truffles, mushrooms, fleshy fruits, and tree seeds (consumed by squirrels) and the capacity to support their medium-sized predators (weasels, owls, and hawks)
5. Production of black-tailed deer and Roosevelt elk, taken to represent the system's capacity to support large predators, such as wolves and mountain lions, as well as subsistence hunting by indigenous peoples and sport hunting

The simulations showed that simply protecting second-growth forest (the benign neglect option) caused the landscape to undergo successive stages of forest development (Carey, Lippke, et al. 1999). Measures representing the five standard criteria yielded unsatisfactory results in the first stages, which reflected degraded watershed conditions and oversimplified forests. A much longer time (ca. two hundred years) would be required for these forests to achieve biocomplexity, even with the assumption that this would happen without the inclusion of legacies (some large, old trees, alive, dead, and fallen) and in an impoverished landscape. On the other hand, timber management with minimum constraints (the conventional timber management option) produced a landscape inhospitable to >20 vertebrate species and allowed no recovery of degraded streams. The sustainability of this landscape was uncertain, but its net present value was maximal. Timber management with wide, no-entry riparian buffers, drawn to comply with federal guidelines, produced narrow, well-separated strips of older forest in the long term, which were unlikely to function fully because of their continued adjacency to clear-cut and young forests. In fact, clear-cutting was intensified in nearby uplands due to removal of streamside forest in accord with normal management practices. In contrast, a forest that received AIM produced significant ecological benefits, including supporting a pair of spotted owls, maintaining the capacity to sustain vertebrate diversity, achieving near-potential, forest-floor function and ecological productivity, while promoting deer and elk numbers comparable to the timber management regime (table 11.1). Surprisingly, costs of the AIM alternative were relatively low—only a 15% loss in net present value compared to maximizing the net present value of timber extraction (table 11.2).

Assuming increased riparian protection was mandatory (and it later became mandatory) and eliminating costs of improved management of riparian and landslide-prone areas, AIM caused only a 6% decrease in net present value whereas other economic values increased: decadal revenues rose by 150%, forest-based employment quadrupled, and the wood products manufacturing sector diversified with greater reliance on high-quality wood products and value-added manufacturing. The final landscape mosaic maintained >50% older, complex forests and <15% in recently harvested areas in any decade, resulting in a landscape fully permeable to dispersing old-forest species. Two recent analyses, one of state trust lands in western Washington and one carried out by a timber investment organization, confirmed the economic feasibility of AIM.

TABLE 11.1

Measures of ecological performance and landscape health

Ecological measure	Timber (NPV)	AIM
Habitat for spotted owls	No	Yes
Numbers of cervids (deer/elk)	423/134	401/200
Vertebrate diversity (% of maximum possible)	64	100
Forest-floor function (% of maximum possible)	12	100
Ecological productivity (% of maximum possible)	19	94
Landscape health (mean of the above %)	32	98

Source: Adapted from Carey, Lippke, et al. 1999.
Note: From the last 100 years of 300 years of simulated management of a 6,828-hectare forest area in western Washington, USA, in the last 100 years using maximizing *net present value* (NPV) of timber and *active, intentional management* (AIM) to produce ecological services and economic goods. Landscape health is defined as the mean of the capacity to support vertebrate diversity, forest-floor function, and ecological productivity. The last 100 years represents steady-state sustainable outputs; the first 100 years constituted a conversion to steady state.

TABLE 11.2

Wood production and values

Economic measure	NPV	AIM
Cumulative wood volume (10^6 m^3/ha)	1.6	1.4
Tree quality (cm)	36.0	76.0
Net present value (10^6 $)	70.4	57.9
Decadal harvest (10^3 m^3/ha)	50.0	48.7
Decadal revenues (10^6 $)	26.0	42.5

Source: Carey, Lippke, et al. 1999.
Note: Landscape management for maximizing *net present value* (NPV) and *active, intentional management* (AIM) for multiple values for a 6,828-hectare landscape in western Washington, USA. Decadal averages are for the last 200 years of a 300-year simulation. Tree quality is defined as diameter at 1.5 m above ground at rotation age.

Conduct Field Experiments

An experiment near Olympia, Washington, evaluated an essential AIM technique to induce heterogeneity in secondary-growth forest canopies (creating a fine-scale mosaic) by variable-density thinning to increase biocomplexity. In addition, the canopies of secondary-growth forests were manipulated in two ways: the first employed a conventional, intensive forest management practice that consisted of multiple thinnings to induce equal spacing between trees of the same size and species, with defective trees removed. The second involved the retention of legacies from the preceding old growth, followed by benign neglect, as described further by Carey (2003).

Both conventional thinning and benign neglect–legacy management produced imbalanced small-mammal communities, with some species being low or absent that are common in natural forests. Canopy mosaics, as prescribed by AIM, had immediate positive impacts on forest-floor mammals. In particular, planting of shade-tolerant midstories apparently restored the biotic integrity of the small-mammal community. Flying squirrels, a key species in the ecology of Pacific Northwest forests, remained rare in the previously thinned stands, perhaps due to dense understories promoting excessive chipmunk abundance. These dense understories apparently impeded foraging for truffles by flying squirrels. Decreased canopy connectivity following equal-spaced thinnings may inhibit travel through the canopy, while large foliage-free gaps between the shrub layer and canopy, in accord with AIM recommendations, likely increased exposure of squirrels to predation.

Canopy mosaics also increased numbers of wintering birds. However, cavity-excavating birds remained low in abundance, as is the norm in younger forests where decay in live and standing dead conifers is rare compared to old forests. Nevertheless, it appears that promotion of deciduous trees early in stand development can offset this deficiency by providing short-lived trees that will decay from within and expedite cavity excavation. Consequently, attention should be paid to intentional management for decadent deciduous trees and conifer trees (decay allows cavity excavation by birds) to stimulate the pest-controlling function of these birds.

Conventional thinning produced species-rich understories, but these contained numerous exotic plant species that were frequented by only a few highly abundant native bird species. Legacy management with substantial regeneration of conifers and benign neglect consistently produced depauperate understories. Induced canopy mosaics produced under AIM techniques markedly increased diversity and the abundance of native species in both conditions, yet they also encouraged numerous exotics, with some persisting for ten years. It may not be possible to promote native diversity without fostering concomitant exotic diversity. The exotics that persisted, however, did not occur in large enough numbers to displace native species and may disappear with time. Importantly, canopy mosaics produced by AIM techniques, and the associated spontaneous establishment of native species among under-planted tree seedlings, are leading to increased spatial heterogeneity.

Healthy forest soils in the Pacific Northwest are dominated by fungi, rather than bacteria. Near-surficial fungal mats in all experimental plots were apparently destroyed by mechanical disturbance, which caused the replacement of originally occurring fungi (*Hysterangium* and *Gautieria*) by another fungus (*Melanogaster*). Loss of mats that reduced fungal diversity is putatively important to soil fertility. In addition, *Gautieria* is a favored squirrel food. By creating canopy mosaics through AIM, truffle diversity increased to a degree rivaling that in natural old-growth forests. Although the negative impacts of experimental thinning on truffle production (as opposed to diversity) was noticeable, it was of brief duration. Other observations revealed an increase in mushroom diversity and abundance and a reduction of some uncommon plant species in response to the development of dense understory. Hence, the retention of both thinned and unthinned patches in mosaics would be advisable to conserve fungal mats and rare plants.

In summary, inducing heterogeneity into homogeneous, closed canopies had positive effects on diverse biotic communities even in the short term (<5 years) in stands managed with conventional thinning or solely legacy retention. Therefore, managerially induced disturbance at the proper scale and intensity can function much the same way as small- to intermediate-scale natural disturbances in promoting biological diversity. Simulation and experimental results both provide support for AIM. However, both used criteria that were chosen by the modelers and scientists involved in the project, and the objectivity of the outcomes needs independent verification to fully validate these apparent benefits of the AIM approach.

Validate Results with Third-Party Criteria

The results of AIM were evaluated for compliance with ten attributes of restored ecosystems that were developed by the Science and Policy Working Group of the Society for Ecological Restoration International (SER) and presented in the SER *Primer* (2002). Ecological restoration of degraded ecosystems, as conceived in the *Primer*, is one way to assure the restoration of natural capital, according to its definition in chapter 1 of this volume. Here we compare AIM's results with each of the SER attributes:

1. *The ecosystem contains a characteristic assemblage of species that occurs in reference ecosystems.* Many second-growth forests lack the characteristic species and assemblages found in old-growth and other naturally complex forests (Carey, Kershner, et al. 1999; Carey, Lippke, et al. 1999; Carey 2003). However, it appears that the AIM technique of inducing heterogeneity into homogeneous canopies also promotes the establishment of a characteristic assemblage of species in the reference ecosystem that consists of old-growth forests. Computer simulations suggest that AIM has the capacity to restore not only the presence of key species but also the structure of biotic communities in forest ecosystems that have been degraded by past mismanagement.

2. *The ecosystem consists of indigenous species to the greatest extent practicable.* Experiments suggest that AIM can promote native species diversity with minor additions of exotic plant species, which commonly are generalists adapted to open sites and none of which are expected to persist. These same exotic species colonize natural forests as well.

3. *All functional groups necessary for continued development and stability are represented.* Experimental results suggest that AIM can maintain diverse functional groups (including truffles, mushrooms, soil bacteria, soil nematodes, and litter arthropods). In addition, restoration of small-mammal communities indicates that the establishment of diverse, functional food webs, at least when variable-density thinnings incorporate small patches of undisturbed soil to ensure the continued persistence of certain rare plants and matt-forming fungi.

4. *The physical environment is capable of sustaining reproducing populations of the species necessary for continued stability or development along a desirable trajectory.* Simulations support AIM as a strategy to promote development of complex forests that contain all keystone (species especially important to ecosystem functioning) and flagship (species emblematic of an ecosystem type) species and keystone complexes (Carey, Lippke, et al. 1999; Carey 2003). However, experiments and managerial implementation of AIM have been in place for too short a time to evaluate population viability, and this uncertainty emphasizes the need for continuing monitoring and, if necessary, adaptive management.

5. *The ecosystem functions normally for its ecological stage of development; signs of dysfunction are absent.* The AIM experiment began with two historic conditions showing marked signs of dysfunction, including root rot, lack of shade-tolerant regeneration, absence of large-cavity trees, and low populations of cavity-excavating birds. In addition, there were incomplete and imbalanced small-mammal communities, in-

cluding low densities of arboreal rodents as well as either reduced plant-species diversity or high plant-species diversity with abundant exotic species. Much of this dysfunction has been ameliorated, within a remarkably short time, using a single AIM technique.

6. *The ecosystem is suitably integrated into a larger ecological landscape.* A major component of AIM is integrated landscape management; AIM integrates healthy forests into natural-cultural mosaics (Carey, Lippke, et al. 1999).

7. *Potential threats to the health and integrity of the ecosystem from the surrounding landscape are reduced as much as possible.* AIM strives to maintain sustainable natural-cultural mosaics.

8. *The ecosystem is sufficiently resilient to endure normal periodic stress events.* Two test sites for AIM techniques have shown early resiliency to windstorms and ice storms, which provide experimental evidence that AIM produces resiliency. In spite of these early test results, additional observations will be needed over a longer period of time to determine if AIM ecosystems can endure normal or unusual stress events. AIM draws upon historical forest management, forest ecology, and disturbance ecology to apply treatments promoting resiliency to normal periodic stresses and future "surprises" (Holling 2001).

9. *The ecosystem is self-sufficient to the same degree as reference ecosystems—and it may evolve.* AIM restores to second-growth forests the biocomplexity characteristic of long-lived, resilient, natural forests. In fact, modeling suggests that the endpoints for AIM forest ecosystems should be relatively stable for long periods. The use of shifting, steady-state mosaics maintained through "creative destruction" provides opportunities for adaptation to changing environments and suggests AIM has potential to satisfy this criterion.

10. *The ecosystem provides specified natural goods and services for society in a sustainable manner, including aesthetic amenities and accommodation of activities of social consequence.* This is a major component of AIM.

These ten comparisons with attributes of restored ecosystems do not offer conclusive evidence that AIM produces restored ecosystems and thus restored natural capital, but they are highly suggestive of that possibility.

All evidence leads to the conclusion that AIM successfully leads to ecological restoration and ultimately to restoration of natural capital. Economic modeling data present a compelling case for adopting AIM in the Pacific Northwest in order to restore natural capital and to resolve acrimonious stakeholder disputes over resources lost to exploitation. Less than two decades ago, conservation biologists argued the merits of single large reserves versus multiple small reserves and of the need for conserving genetic diversity and restricting active management. Simultaneously, forest managers focused their attention on plantation management, transportation networks, and watershed restoration. Now it is recognized by both groups that active management is required to restore degraded ecosystems and to produce fully functional forests outside of reserves. Research on AIM techniques has shown that reserves can become self-fulfilling prophecies of highly isolated "islands" of diverse forests within depauperate second-growth forests and developed areas, while conventional timber management can oversimplify forest stands to the detriment of ecosystem health and landscape function. As human demands grow, intentional systems management will be necessary to conserve the biodiversity of natural-cultural landscape mosaics and the ecological services and goods they provide (Carey, Lippke, et al. 1999).

AIM does not seek to restore any particular pre-Columbian ecosystem state. Rather it strives to restore ecosystem function, resilience, adaptiveness, biotic integrity of vertebrate communities, and diversity of vascular plant communities and other functional groups. In this sense, restoration trajectories initiated by AIM are intended to produce adaptive ecosystems of the future rather than to reconstruct the past. The approach is dynamic and allows for self-organization, "creative destruction," and ecological innovation, and it absorbs ecological surprises.

Contribution

AIM is an approach for the restoration of natural capital that satisfies the conflicting desires of stakeholders. Carefully conceived computer models demonstrate its economic worth in terms of sustainable, high-quality natural capital and the capacity to generate social capital in terms of steady employment and stakeholder satisfaction, with only a slight (6%) decrease in profits that would be expected from current, short-term, forest exploitation. AIM unabashedly seeks to serve human needs, such as providing clean air and water, recreational and spiritual experiences, wood and other forest products, economic activity, and employment. AIM works within the knowledge that we are attempting to be just and moral to one another, to future generations, and to other species, as we restore ecosystems and their natural capital.

Chapter 12

Restoring Natural Capital Reconnects People to Their Natural Heritage: Tiritiri Matangi Island, New Zealand

John Craig and Éva-Terézia Vesely

The most irreplaceable natural capital of New Zealand is biodiversity, namely its unique plant and animal species and their habitats. However, like much of the New World, New Zealand has an ecosystem management approach influenced by colonial values (Pawson and Brooking 2003), with ecosystems divided into either productive or protected areas. The productive landscapes are dominated by introduced (alien) species, managed mostly for short-term societal returns (see also chapter 31). In contrast, protected landscapes have been established to preclude common forms of extractive development and are largely held in public ownership, often managed with "benign neglect" (chapter 11). There are no indigenous mammals (except bats) in New Zealand; however, the universal presence of introduced mammals (such as rats, deer, cats, and stoats) is continuously eroding natural values in all ecosystems (DoC/MfE 2000), making the so-called management appear more akin to "wanton neglect" (Craig et al. 2000).

New Zealand's first State of the Environment Report (MfE 1997) describes the ongoing decline of indigenous biodiversity—loss of native species, genetic diversity, and the supporting habitats and ecosystems—as the country's most pervasive environmental issue. The public realization of this trend has led to an increasing commitment to restoration, with government agencies concentrating efforts on secluded areas and islands, where control and even eradication of introduced pests is more easily achieved. Many of these efforts have grown out of programs focused on rare species. The consequence is that urban New Zealanders, who make up more than 80% of the population (PCE 2002), are increasingly separated from their natural heritage. This means that biodiversity and functioning ecosystems, the building blocks of natural capital, are not recognized by most people in their everyday lives (Pyle 1993; Stewart and Craig 2001). In a country where biodiversity management is dependent on political largess, the problem becomes self-perpetuating (Craig 2006); each generation of urban dwellers sees less native species and expects and demands less (Kahn 2002).

To counteract this trend, there has been an increasing effort by some individuals and community groups to restore habitat for native species close to their homes. For many people, this has meant moving to peri-urban areas, where remaining fragments of native ecosystems still exist; others have adopted nearby public lands and initiated restoration programs in partnership with government agencies. Similarly, schools and other groups have turned their attention to their own properties.

Funding these restoration programs is problematic. Costs of controlling or, in the few situations where it is possible, eradicating pests are large. As access and entry to government-controlled areas is free, it is difficult to obtain ongoing revenue from any restoration area. This is because New Zealand societal values hold that the benefits of indigenous biodiversity restoration should be free to everyone. Where monetary value is not associated with native species and public lands are available at no charge, private investment is rare. Market pressures arise only where international buyers are seen to put a premium on the products of sustainably managed lands. This is beginning in the dairy industry and is leading to the fencing of waterways (PCE 2004).

The disconnection of the majority of New Zealanders from native biodiversity, in association with the ongoing decline in native birds, reptiles, and invertebrates, makes the restoration of this form of natural capital difficult. The social attitude of apportioning value and potential income to introduced species in production landscapes, while not assigning value to native species, along with the failure to recognize the importance of ecosystem services, means that restoration and conservation remain peripheral activities that function largely through welfare (PCE 2001). The focus of government action on rare species and more secluded areas further reduces the chances of building public support for wider scale restoration.

Restoring natural capital requires the concomitant construction of social capital through enhancing the lives of, especially, urban New Zealanders, by providing tangible experiences of their own natural heritage. Hence, restoration is needed where urban people can connect with their natural heritage and where the benefits can be easily demonstrated, while additional work continues in more distant areas. In this chapter we describe just such a restoration approach for Tiritiri Matangi Island. Here a public-led and cofunded program is restoring the biodiversity capital and generating educational, recreational, and other biodiversity-based products that people want.

The Natural Capital of Tiritiri Matangi Island

This island was an ideal candidate for restoration, having escaped invasion by most introduced animal and plant pests. With an area of 220 ha and located in the sheltered waters of the Hauraki Gulf, Tiritiri Matangi Island is only 4.5 km from the nearby Whangaparaoa Peninsula and only 25 km from Auckland, the largest city in New Zealand with 1.2 million people. Settled by indigenous Maori people after AD 1500, who introduced the Pacific rat or kiore (*Rattus exulans*), the island was partly cleared for living areas and agriculture. When British colonists arrived, it became government owned and was farmed for over a century. It was also the site of an important lighthouse. Despite almost total forest clearance, some native species survived, including the bellbird (*Anthornis melanura*), a pollinator and seed disperser that became locally extinct on the nearby mainland in the 1860s. In 1971, on the advice of the botanist Allan Esler, the government terminated the grazing lease and left the island to regenerate "naturally." Prior to the restoration program, Tiritiri Matangi Island received about three hundred visitors a year, who were mainly private boat owners landing for summer picnics.

University research began on the island in 1974, when it was largely covered with grass, with only small forest patches surviving in gullies, the largest of which was 4 ha (Mitchell

1985). Seedling and seed dispersal research was undertaken to investigate the likelihood that unassisted forest regeneration would occur. This work rapidly showed that when tree and shrub seeds fell into the grass, lack of light and high grass density prevented germination (West 1980). In addition, some of the tree and shrub species had been reduced to very small numbers, while data revealed that the two avian honeyeaters, the tui (*Prosthemadera novae-seelandiae*) and the bellbird were severely food limited for many months of the year (Stewart and Craig 1985; Craig and Douglas 1986). The partial loss of a major winter-food tree, *Vitex lucens*, during a cyclone demonstrated the precarious status of the resident bellbird. Clearly "natural" or unassisted regeneration would not be fast enough to save these bird populations or restore forests within the lifetime of current human generations.

A restoration plan was developed that would result in planting much of the island in local trees and shrubs, which would provide food, especially for the honeyeaters. The focus on birds was an acknowledgement that the presence of the kiore would preclude the introduction of reptiles or large invertebrates. After considerable consultation the restoration plan, which had been developed with three key goals, received mixed support (Craig et al. 1995). Public conservation groups were supportive, government agencies were either supportive or cautious, while scientists were strongly negative. The scientists consulted by the government argued that the replanting would not work, that long trials were needed first, and that it would not be possible to mix the general public with rarer species (Craig et al. 1995). The goals were, as follows: (1) that the island would be replanted in forest starting with species supporting selected, rare bird populations; (2) that selected bird species would be reintroduced; and (3) that the public, especially Aucklanders, would have access to the island to experience their natural heritage. The overall vision was to recreate the forest of the coastal Auckland region as a functioning ecosystem, fill it with the birds and reptiles living there prior to human arrival, and ensure that people could experience this. Plant species chosen were those known to occur or thought to have been on the island. Hence, linking the building of both social and natural capital was seen as the key to success.

Restoring the Natural Capital of Tiritiri Matangi

The restoration plan matched the strong preservationist ideals generally held by New Zealanders at the time. The nation had a social climate where the government subsidized clear felling of indigenous forests for timber, as well as native vegetation clearance, including wetlands drainage, to allow "more productive" use as agriculture. In addition, biodiversity management and conservation was spread across multiple government agencies. In fact, rare species programs for some critically endangered species, such as the kakapo, were shelved for lack of finance. Against this background, a plan to restore an island to a functioning forest ecosystem and simultaneously increasing rare bird populations, while ensuring public access, was unusual.

In contrast to its other attitudes, the government offered a 2:1 subsidy for any donation toward conservation, which was clearly only a small financial liability in the social climate of the time. In 1982, a grant from World Wide Fund for nature (WWF), along with the government subsidy, provided the necessary start-up funding for the restoration of Tiritiri, which in the intervening decade of unassisted restoration had produced few changes, affirming the value of intervention. One and a half full-time staff initiated on-site planning, established a nursery, and prepared approximately 25,000 plants for each of the ten years of public planting that began in 1984. Initially progress was slow, but within four years planting by community groups and schools became a regular feature through the winter and early spring months. In fact, people paid to go to the island and plant, prepare hiking trails and access roads, and work in the nursery. Local commitment grew into an organization, Supporters of Tiritiri Matangi (SoTM), which is now the major player in the ongoing management and funding of the island restoration.

The deliberate attempt to restore to a past ecosystem required concept changes for rare-species management. The government attitude at this time (and currently) was dominated by ensuring human exclusion from rare-species locations, as anthropogenic activities were deemed the greatest threat. An ambitious plan to reintroduce ten nationally and locally rare birds was part of the restoration program. The first, the tieke or saddleback (*Philesternus carunculatus*), an endemic wattle bird previously restricted to a single, closed island, was introduced in 1984 at the start of the planting program. Thus, the rewards for the volunteers included immediate access to observe species not seen by the public elsewhere. In addition, resident species such as the bellbird were found only here in the Auckland region. Annual releases of other species, including the critically endangered takahe (*Porphyrio hochstetteri*), greatly enhanced the rewards. Indeed, volunteer numbers increased to such an extent that it was necessary to ration the number of trees available for each individual to plant.

The volunteers' group, SoTM, began in 1987 and raised additional funding for equipment and materials. They also organized building of trails, upgrading accommodation, and any other assistance needed by the resident government staff. SoTM is now the largest funder of the island and has a formal cooperative management agreement with the government agency (see also chapter 32 for voluntary financial mechanisms to restoration).

In addition to these volunteers, scientific research has been an ongoing activity, which, initially driven by Auckland University, included staff and graduate students (Craig et al. 1995). Besides exploratory research, the dynamics of most of the newly introduced bird populations and of the replantings became the focus for graduate students (e.g., Dawson 1994; Cashmore 1995; Armstrong 1995; Baber 1996; Girardet 2000; Jones 2000). Subsequently, Massey University has taken on an increasing role, with research questions broadening from an ecological focus to investigate people's perceptions, their effects on the wildlife, and the economics.

Representatives of present and potential stakeholders in the restoration project—SoTM, the Department of Conservation, the Hauraki Maori Trust Board, the University of Auckland, and Massey University—are aware of the range of strengths, weaknesses, opportunities, and threats (SWOT) associated with the project, as shown in table 12.1.

Measuring the Success of Restoring the Natural Capital

Restoration success can be measured in a range of ways, though the most logical approach would be to evaluate the outcomes against the original three goals.

The original ten-year program was completed on time with over 260,000 plants being raised and planted. Subsequent planning required the planting of some other small areas, but the nursery has now been dismantled and over 60% of the island is forested. Twelve locally or

TABLE 12.1

Strengths, weaknesses, opportunities, and threats associated with the Tiritiri Matangi restoration project

Strengths	Weaknesses
• Inspiring vision • Personalities of the rangers present • Specific local focus • Good opportunities for people to get involved and realize passions • Networking capacities of some of the people involved • Establishment of the Supporters of Tiritiri Matangi • Early volunteer rewards from endangered species translocations • Credibility from scientific research • Links with educational institutions • Management trust	• No financial planning for a long time • General approach initially was ad hoc • Bureaucratic slowness • Occasional management rigidity

Opportunities	Threats
• Future species translocations, such as native frogs, giant weta, and snails • Improved visitor management • Addition of a marine reserve in surrounding waters • The development of a historical perspective • Enhanced education and research opportunities	• Invasion of predators • Fire • Overexposure and/or overshooting the carrying capacity • Other islands copying the model without differentiation • Personality and vision incompatibility between different stakeholders

Source: J. Brown, J. L. Craig, M. Galbraith, C. Hayson, B. Walter, and R. Walter, 2002, personal communication.

nationally rare birds, and a nationally rare reptile, the tuatara (*Sphenodon punctata*), have been reintroduced to provide the most faunally diverse forest ecosystem in the region (see table 12.2). Visitor numbers have reached 35,000 per year and continue to grow, with the majority coming from Auckland. Hence, the original goals have been met.

Another approach to measure the success of the restoration project is from a natural capital perspective. The term refers to the capacity of the stock, that is, the ecosystem, to provide goods and/or services (Ekins, Simon, et al. 2003). The natural capital perspective encompasses the multifaceted nature of an ecosystem's functions, such as regulation, production, habitat, and information, and the various dimensions involved—ecological, economic, and sociocultural (Chiesura and de Groot 2003). The impact of natural capital restoration can be measured by an assessment of the changes in the natural capital stock, the associated flows, and the values created.

Using Tiritiri Matangi Island, an island ecosystem, as a form of natural capital, it can be shown that an extensive transformation has occurred due to the restoration effort. The forest cover has increased from 5% to 60% (and pasture decreased from 95% to 40%). Approximately 140 ha are now covered in forest, with the remainder maintained as grass, either as bird habitat, or to provide views for visitors, or to allow unassisted regeneration. With the exception of one formerly grass-covered valley system, which has become colonized by New Zealand flax (*Phormium tenax*), the majority of these areas have changed little in the last thirty years. In addition, populations of twelve rare species have been translocated to the is-

TABLE 12.2

*Wildlife reintroduced to Tiritiri Matangi, their ecological service,
and initial and current numbers, as of 2006.*

Species reintroduced	Ecoservice	Number released	Current numbers
Birds			
Red-crowned kakariki	Seed control	60+	200+
Saddleback	Seed disperser, pest control	24	300+
Brown teal	Pest control	19	8+
Whitehead	Pest control	30	1000+
Takahe	Pest control	8	18
Robin	Pest control	24	190
Little spotted kiwi	Pest control, soil formation	16	30+
Stitchbird	Pollination, seed dispersal, pest control	37	188
Kokako	Seed dispersal, pest control	7	14
Fernbird	Pest control	19	25+
Tomtit	Pest control	32	20+
Reptiles			
Tuatara	Pest control	60	50+

TABLE 12.3

*Restoration-induced changes in service flows and their associated functions and values,
from Tiritiri Matangi Island*

Natural capital	Flows of services	Change in flows after restoration	Associated environmental functions	Associated values
		Functions of		
	Conservation of native species	increased	Habitat	Ecologic, social, and economic (nonmarket)
	Water purification	increased	Regulation	Ecologic and economic (nonmarket)
Restored Tiritiri Matangi Island	Carbon sequestration	increased	Regulation	Ecologic and economic (potentially market)
		Functions for		
	Recreation	increased	Information	Social and economic (nomarket)
	Education	increased	Information	Social (potentially economic)
	Research opportunities	increased	Information	Ecologic, economic, and social
	Grazing	decreased	Production	Economic (market)

Note: See O'Connor (2000) and Chiesura and de Groot (2003) for detailed explanations of terminology.

land and are increasing in size. These changes to the island translate into new flows and services and their associated values being provided (see table 12.3). These include the recreational experience of a native island ecosystem, educational uses, research opportunities,
rare-species conservation, and other ecological services. Decreased flows include the grazing
service provided for cattle and sheep, which have been removed from the island.

Flows of Natural Capital and Their Associated Values

The involvement of local people in the restoration of natural capital, including the reestablishment of indigenous species, removal of alien species, and return of aspects of a functioning pre-settlement landscape, has generated a range of new values for the island.

Conservation of Native Species

The translocation of species has created flows of seed dispersal, pollination, seed control, pest control, and soil formation services, with a contribution to endangered species' conservation rated as 0.15 in terms of Conservation Output Protection Years (COPY) (Cullen et al. 2005). This is the sum of gains in the threatened species' conservation status on a quadratic scale from 0.00 (extinct) to 1.00 (not threatened). While the overall population size of many of the translocated species is small, they are important when considered as part of a networked metapopulation. Consequently, the insurance value of newly established island populations against loss of other wild populations is highly appreciated by ecologists (Girardet 2000; Jones 2000).

Furthermore, the public has perceived the restoration of the island as very valuable. In fact, the benefits that the Auckland region population derive from maintaining conservation activities on the island have been estimated by contingent valuation at NZ$7.7 million per year (1 NZ$ = 0.5397 US$ [1993]) (Mortimer 1993).

Water Purification and Carbon Sequestration

The restored forest provides a range of ecosystem services, including water purification and carbon sequestration. The extensive forest cover ensures that the water flowing into the surrounding sea has minimal nutrient and pollution loads. In addition, the newly planted forest has a carbon sink capacity potential of fifteen tons per hectare (Whitehead et al. 2001). Besides the ecological value, this latter ecosystem service has a potential market value, as almost half of the forested areas were established post-1990 and would qualify for "Kyoto" carbon credits.

Recreation

The flux of recreational services provided have increased significantly after restoration, with visitor numbers rising from several hundred a year to 35,000 in 2004 (Lindsay 2004). In particular, the island has been largely set up for day visits to see birds, with two walks available through patches of more mature forest, as well as replanted areas. So the SoTM provide trained volunteer guides to explain the history, as well as plant and bird information, and approximately 80% of the visitors pay for this service. On a day trip, the majority of visitors would have little difficulty in seeing at least six of the reintroduced rare species, along with the more common birds. Basic overnight accommodation is available for up to fifteen people, and these visitors have the opportunity to see nocturnal species, including the iconic kiwi (*Apteryx oweni*). With the development of better quality trails and an informed guiding option, 80% of visitors in 2003 were "extremely satisfied" with their experience (Lindsay 2004).

By using average consumer surplus values from nonmarket, recreation-valuation studies carried out in New Zealand and the United States (Kaval 2004), the value of recreation enjoyed by the 35,000 Tiritiri visitors in 2004 is estimated to be between NZ$0.9 and NZ$2.3 million (1 NZ$ = 0.6294 US$ [2004]).

Education

As Cessford (1995) explains, one of the main justifications for allowing public access to specially protected islands is that the visitors can understand the conservation issues associated with the sites. This may foster reevaluation of their attitudes and promote greater involvement.

Many visitors indicated that their island experiences had changed the way they thought about conservation, with almost 50% previously nonvolunteers showing an interest in participating in some volunteer work (Cessford 1995).

Increased voluntary participation has ramifications for conservation and restoration projects, as this labor source is often a significant share of the total investments. In addition, Stewart and Craig (2001) found that frequent visitors to the island held stronger environmental views than first-time visitors. The former were more proactive in promoting conservation and participation, while believing taxes should provide a major funding source.

Research

Scientific information has been gained from different aspects of the restoration project, including ecology and conservation management. From investigating the revegetation ecology on Tiritiri (Cashmore 1995), these findings have been subsequently applied to a restoration project on the nearby Motutapu Island. In addition, research associated with native bird translocations—Dawson (1994) and Baber (1996) on the takahe; Armstrong (1995) on robins (*Petroica australis*); Girardet (2000) on the little spotted kiwi (*Apteryx oweni*); and Jones (2000) on the kokako (*Callaeas cinerea wilsoni*)—has advanced the understanding of these previously little-studied species. There has also been considerable research on other species, as well as on visitor issues

Contribution

Colonial attitudes combined with current short-term economic reasoning and a public attitude not sufficiently valuing native species have ensured the decline of indigenous biodiversity and their habitats and ecosystems in New Zealand. There is increasing awareness of the need to restore native ecosystems, but there remains little perception that such systems are a form of capital adding to human health and well-being. The restoration of Tiritiri Matangi Island is an example of a public-led and -funded program that has restored an island ecosystem, and which produces associated flows that are in increasing demand.

The project manager from the Department of Conservation ranked the success of Tiritiri out of a possible 100 as, restoration of ecosystem (99); endangered species conservation (90); and advocacy and education (100) (Cullen et al. 2005). Conversely, when comparing six New Zealand projects from a threatened species conservation-outcome perspective using the

Conservation Output Protection Year (COPY) indicator, Cullen et al. (2005) found Tiritiri to have performed poorly. Having the smallest project area, Tiritiri cost thirty times more per hectare per year than did the project with the largest area, and it produced one of the lowest COPY values for one of the largest investments. This is because by managing low though increasing percentages of the total populations of threatened species, only minor contributions were made to the species' conservation status. However, Tiritiri was envisioned as generating a much wider range of benefits, and, as shown, this successful restoration project has generated flows with associated social, economic (nonmarket), and ecologic values. Consequently, decisionmakers need to be aware of the range of outcomes and how they balance when making performance assessments in relation to restoration programs.

Increasingly, other restoration programs are in progress as New Zealanders demand greater access to their natural heritage. While there is a general belief that these restoration programs deliver value, especially through recreation and tourism, the quantification and balance among the created ecologic, economic, and social values remain fertile areas for future enquiry.

Chapter 13

Restoring Forage Grass to Support the Pastoral Economy of Arid Patagonia

Martín R. Aguiar and Marcela E. Román

Arid Patagonia, in southern South America, includes grass- and shrub-dominated ecosystems (steppe), with precipitation occurring mainly in autumn and winter (May–September), and totaling generally less than 350 mm per year. Despite the short history of pastoralism (ca.100 years) compared to other ecosystems in temperate South America, the impact has been highly damaging and extensive, with no relics of precolonial rangelands remaining by the 1950s (Soriano 1956).

Today there is no reference site for the original structure, composition, and functioning of these ecosystems. The most common restoration recommendation is reduced domestic-herbivore stocking rates. This management approach erroneously assumes the spontaneous natural recovery of threatened species, when in fact there is an urgent need to experiment with different restoration procedures that actively facilitate their recovery.

In this chapter we present a plan to restore the forage-grass populations that are the natural capital in grazing-degraded Patagonian rangelands. In particular we propose to recover *Bromus pictus*, a native perennial grass that is heavily grazed by both domestic and native herbivores. This plan is based on information gathered since 1999, relating to the establishment of new plants and forage production. It has four elements: first, the current grazing system has been investigated for its role in natural capital erosion of the Patagonian steppe; second, the biophysical constraints are identified for natural capital restoration in grazing rangelands; third, a "low-input" grazing management intervention is described for restoring natural capital; finally, the model is evaluated from a combined biophysical and economic perspective.

Utilizing the Natural Capital on the Patagonian Steppe

Agroecosystems need to be analyzed from three different perspectives: biophysical, productive-technical, and socioeconomical. Although these three perspectives are complementary, most productive-technical decisions are based on the socioeconomical perspective. Management decisions are often based on the biophysical attributes of the animals (i.e., tolerances and requirements) and rarely on the biophysics of the resource base (i.e., plant population dynamics and soil characteristics). The historical management of the Patagonian arid steppe provides a good example of this lack of complementarity, while shedding light on its trajectory to the current state.

The Patagonian steppe evolved as an arid ecosystem since the Tertiary, when the Andes formed (Soriano et al.1983), creating a rain shadow, with westerly Pacific winds losing their moisture as they climbed inland over the mountains. The vegetation is dominated by perennial tussock grasses and low shrubs (Soriano et al. 1983; Ares et al. 1990) with aboveground primary production limited in winter by low temperatures (<2°C in July) and after October by water availability (Jobbágy and Sala 2000). Many large herbivores grazed on the steppe until the Pleistocene extinction, about 10,000 years ago (Markgraf 1985), after which the only large herbivores remaining were the guanaco (*Lama guanicoe*) and the lesser rhea (*Pterocnemia pennata*), a large flightless bird (Bucher 1987; Baldi et al. 2001). Two major features in recreating the functioning of these original grazing systems hinges on understanding (both extant and extinct) herbivore plant selectivity and migration patterns. Native herbivores would have selected their diet at spatial scales ranging from regional to individual plants (figure 13.1a), while the grazing regime (i.e., frequency, intensity, forage composition) was probably controlled by biophysical attributes, such as macroclimate, local climate, landscape topography, plant community composition, and individual plant status (Bailey et al. 1996). In addition, vegetation composition and the quantity of biomass consumed at different times of the year would have been controlled through a network of feedbacks, such as increased herbivore consumption causing plant population decreases followed by grazer population declines. Reduced forage availability would also have forced animals to move to other grazing areas or regions. The arrival into this ecosystem of hunter-gatherers started around 12,000 years ago (Markgraf 1985). Even in low numbers and with or without climate change they may have played a major role in causing rapid extinction of most of the megafauna (Markgraf 1985). Unfortunately, few archaeological deposits or early colonial documents are available for analyzing the productive-technical and socioeconomical perspectives of this precolonial ecosystem, although pollen records and preserved plant macroremains provide useful data for reconstructing the vegetation from these early periods (Markgraf 1985).

With the advent of sheep husbandry, early in the twentieth century, a new ecosystem developed. Humans increased in importance as consumers and land managers and also developed an economic market (figure 13.1b). As there are only a few historical records of management in the region (e.g., Texeira and Paruelo 2006), the current understanding of grazing impact is derived from vegetation studies under different herbivore densities. Although the nature of precolonial grazing pressure is difficult to estimate, the significant difference between the two systems was most likely the impact of animal distribution in space and time. With the introduction of domestic herding, fences and water points became the main controllers of grazing distribution over the landscape, impacting greatly on patch and plant selectivity (figure 13.1b). In some cases, natural pastures are allocated either to winter or to summer grazing, though generally grazing occurs throughout the year (Soriano 1956; Golluscio et al. 1998). The consequence of the continual selection by livestock for palatable grasses is a decrease in reproduction and an increase in mortality of these grasses resulting in their reduced abundance. This in turn causes unwelcome changes to the vegetation of the arid Patagonian steppes (León and Aguiar1985; Ares et al. 1990; Perelman et al. 1997). A modeling exercise indicates cascading effects on ecosystem function, as decreases in the grass/shrub ratio promote changes in water dynamics that ultimately reduce primary production and herbivore biomass (Aguiar et al. 1996). The decline in sheep stocks during the last fifty years is therefore the result of a continuous demographic process rather than a change in

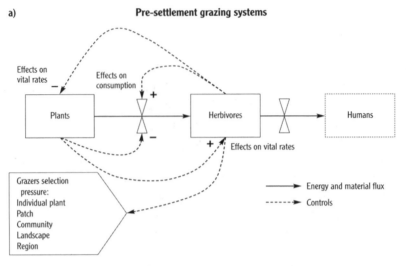

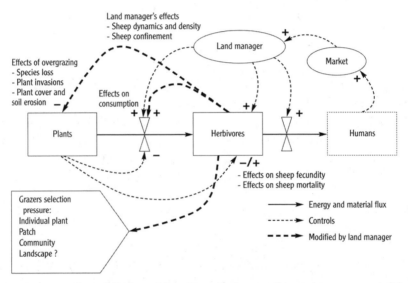

FIGURE 13.1. Conceptual model of pre- (a) and post-settlement (b) grazing systems on Patagonian steppe. In pre-settlement grazing systems, native herbivores were able to select plants freely (block arrow). Post-settlement grazing is dominated by sheep in constrained areas and controlled by humans. Humans are an active component of the system as consumer, land manager, and a part of the market. Land managers determine animal densities, fences, and waterwheels, which in turn control animal distribution.

management decisions (Texeira and Paruelo 2006). In other words, the biophysical characteristics of the system in conjunction with the grazing (technical) management are constraining its productive output, with little possibility of sustaining yields over time. Although credits, subsidies, and/or tax exceptions are utilized to restore sheep stocks, generally after major climatic events, this financial buffering will only enhance the continual degradation and the biophysical constraints limiting the stock of palatable plants and species.

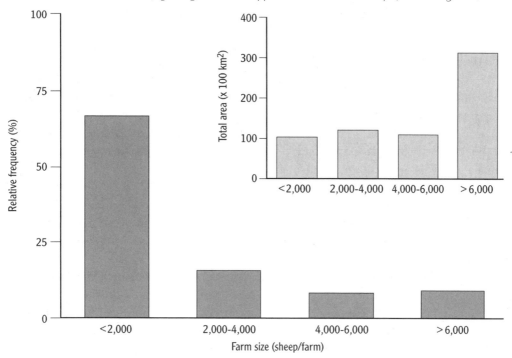

FIGURE 13.2. Size class distribution of sheep farming enterprises in the Chubut province of Argentina (n=815), expressed in terms of the number of sheep in the farm. Inset: total area occupied by each farm size class.

Curiously, compared to plant-grazing management, animal management is more developed (Golluscio et al. 1998). For example, for sheep this includes sanitary considerations and a husbandry calendar advising optimal dates for breeding, marking, weaning, and shearing. From the socioeconomical perspective, the Patagonian region investigated here, the western portion of Chubut province of Argentina (from 42° to 46° S latitude), includes four counties. According to the most recent available National Agricultural Census (INDEC 1988), ~67% of the farms possess <2,000 sheep (figure 13.2), whereas ~9% of the farms have >6,000 sheep, which accounts for almost half of the area (figure 13.2 inset) and regional sheep population. Included in the category "<2000 sheep" is the social group with <500 sheep that has the most precarious living; this category includes the communal lands occupied by indigenous people. From the biophysical and social perspectives, these units represent extreme poverty (Alvarez et al. 1992; Golluscio et al. 2000).

Restoration of palatable grasses appears to be essential for the overall sustainability of these agroecosystems. From the biophysical perspective, diversity at the species (i.e., genetic) and plant community level needs to be maintained. From the productive-technical perspective, restoration has the potential to increase wool production and make sheep husbandry economically sustainable. This is an important goal also from the socioeconomical perspective since it is desirable to enhance livelihoods in rural areas rather than facilitate the human migration to shantytowns.

The Biology of *Bromus pictus*

A perennial tussock grass native to Patagonia, *Bromus pictus* reproduces by large seeds (14.4±0.4 mm in length and 7.2±0.4 mg in weight), as well as by tillers. Seed viability is >90% (Rotundo and Aguiar 2004), with seedling emergence occurring in the cool wet season (April through September). It is economically important because it is a highly palatable species that sheep graze preferentially. On average, plant density in exclosures occurs at 6 plants m^{-2} and decreases to between 1 and ~0.6 plants m^{-2} in grazed and heavily grazed rangeland. In heavily grazed rangeland, *Bromus* finds refuge beneath shrubs or is mixed with nonpalatable grasses (Oesterheld and Oyarzábal 2004).

Seed production during a normal year is directly related to the size of the plant, though, on average, adult plants produce 128 seeds/plant/year (Aguiar and Rotundo, unpublished data). Flowering and fruiting occurs as the season progresses, and dispersal starts in midsummer and extends for a month (Aguiar and Sala 1997). Most seeds are wind dispersed and accumulate close to natural barriers, such as other plants or litter (Aguiar and Sala 1997). In natural conditions, emergence takes place as soon as rain season starts (May) and is controlled by burial depth rather than horizontal seed distribution (Rotundo and Aguiar 2004); hence grazing can increase emergence rates because sheep promote seed burial (Rotundo and Aguiar 2004). It has been demonstrated in grazing exclosures that the emergence rate is 10% of available seeds in a microsite, whereas in grazing paddocks this is increased to 20%. Nevertheless, 7% of emerged seedlings survive (Aguiar and Sala 1997), and survival is not significantly affected by grazing.

Restoring Patagonia's Heavily Grazed Natural Capital

The target envisaged was to increase the density of *Bromus* tenfold from 0.6 to 6 plants m^{-2}, yet still manage sheep grazing. This was considered achievable by eliminating grazing during most of the year, but especially during the spring and summer growing season (August to December) to allow plant growth, accumulation of plant reserves, and flowering and seed production. In early January, grazing is resumed until the forage accumulated in the growing season is consumed. Meanwhile, grazing at the end of the growing season fosters seed dispersal and burial (Rotundo and Aguiar 2004), and as summer is a critical period for ewes and lactating lambs, better nutrition increases lamb survival and growth after weaning. Since *Bromus* forage is of high quality, this grazing management should lead to recovery of the grass populations and growth of the sheep flocks. Presently, the growth rate (λ) of flocks is around 0.9, which is sustainable only with the addition of new sheep from elsewhere (Texeira and Paruelo 2006).

Grazing experiments take many years to establish and complete. For this reason we used the best available ecological data to simulate recovery times for forage-grass populations and then analyzed the economic feasibility of restoration for various types of pastoral enterprises.

Simulated Restoration of Grass Populations

To estimate the grass seedling establishment through time for rangeland under this management, the following equation was used:

Seedlings/m^2 = (plant density × seed production plant^{-1}) × (emergence × survival),

where plant density is expressed in plants m^{-2}; seed production is the number of viable seeds produced and dispersed per plant; emergence is expressed as a proportion of dispersed seeds, and survival as a proportion of emerged seedlings. Three assumptions were made: first, a starting plant density of 0.6 plants m^{-2}; second, those seedlings that survived one year are considered established; last, new plants begin to produce seeds in the third year after emergence.

We simulated the restoration dynamics for a range of initial plant densities (from 3 to 0.0001 plants m^{-2}). Recovery time (target density of 6 plants m^{-2}) varied between one and sixteen years, respectively (figure 13.3a). Experimental data indicate that *Bromus* population dynamics are controlled mostly by intraspecific rather than interspecific competition. As *Bromus* density in old exclosures is ~6 plants m^{-2}; it is considered that this is the species-carrying capacity in this ecosystem, and population growth should slow down as density approaches or exceeds this threshold. However, this particular restoration density dependency was not simulated.

Other simulations indicated that recovery dynamics had different sensitivities to interannual variability in seed production. Reducing seed production and emergence rates by 20% each had the greatest combined effect at the lowest population density but also slowed down the recovery dynamics for 0.1 and 0.6 plants m^{-2} (figure 13.3b).

Increments in forage production and consumption during the restoration were also estimated. An average value of green and recently dead biomass of 1.5 g/plant was used (Rotundo and Aguiar, unpublished data), while assuming a requirement of 1 kg dry forage biomass/sheep/day (Rodolfo Golluscio, personal communication). Restored rangeland would thus produce a maximum of 90 kg ha^{-1} of forage or maintain three sheep ha^{-1} month^{-1}.

After the grazing capacity is restored, sheep should be rotated to promote accumulated plant reserves, flowering and seeding, and seedling emergence. Rotational grazing might also benefit guanacos, since sheep are stronger competitors (Baldi et al. 2001). Guanacos and rheas can move over or through sheep fencing. It can be confidently concluded that this management approach can be sustained under a scenario of 20% individual plant mortality due to grazing.

Economic Analysis of Forage-Grass Restoration

Farming enterprises of three sizes were considered in the economic analysis. The relative influences of the different variables were evaluated from an economic perspective rather than predicting the likely outcome of the management. Based on the farming enterprise size distribution in western Chubut province (figure 13.2), farms carrying 1,000 (type A), 4,000 (type B), and 15,000 sheep (type C) were investigated. Total sheep numbers included different categories (ewes, lambs, hoggets), with an assumed 43% proportion of ewes in the three farm sizes. Type A farms are the most common size in the region. Type B is representative of the economic unit (economic units are farms that support one family, and their size depends on primary productivity and relative local and international prices) (H. Bottaro, personal communication). Finally, type C represents farms occupying most of the land area under analysis. In addition, the three farm types differ in supplies consumed and use of labor and capital, the data for which were obtained from earlier studies (Román et al. 1992; Bendini and Tsak-

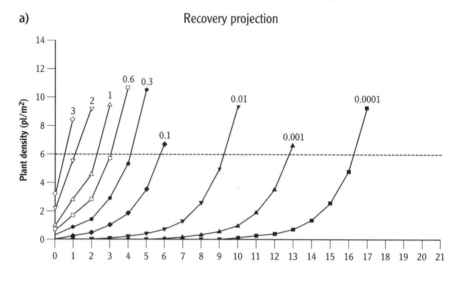

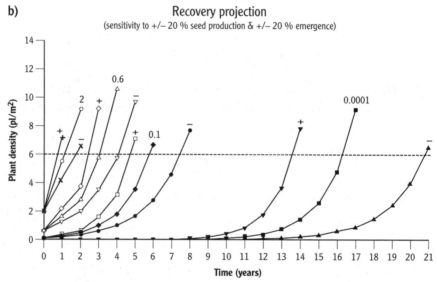

FIGURE 13.3. Simulated trajectories of plant density recovery of *Bromus pictus*. In panel (a) each solid line represents a trajectory starting with a different plant density. In panel (b) simulations consider a concurrent variation of +/-20% in both "number of seeds produced per plant" and "emergence rate."

oumagkos 1993; Román 1993; Golluscio et al. 2000). Prices were provided by regional and national sheep producer associations (*Corporación de Fomento del Río Chubut* [CORFO] and *Federación Lanera Argentina*) and represent the average between 1980 and 2003. These are expressed as the December 2003 values (in US dollars), which include all the federal taxes discounted from the producers' income. It was assumed that the three farm types had an initial 60% marking rate (an index of pregnancy, miscarriage, and lamb survival at two months), 10% lamb mortality and 7% ewe mortality, and that the personnel in type A were only the family; in type B, the family and one shepherd; and that type C employed three shepherds.

In terms of financial evaluation, it is considered that the farm, regardless of size, should be divided into four management units or sections. One section sequentially is closed to grazing from August to January. During three years, sheep graze the other three sections most of the year, except in January and February, when they return to the "in restoration" section and remain as long as forage is available. This four-section approach represents a way to implement rotational grazing and to analyze the viability of the project in an economically and ecologically heterogeneous region. A unit is considered restored when *Bromus* density reaches 6 plants m^{-2}. To implement this plan it is necessary to invest in fences and water points in the farm types A and B (US$20,637 and $11,646, respectively), while the considered costs of technical advice during the first four years of management are as follows: first year, ten days ($417); second and third years, three days ($174); fourth year, one day ($104). It was also assumed that during the first six years economic yields would decline because of reduced grazing area, while subsequent restoration would generate direct economic benefits because of increased forage availability. After twelve years, the *Bromus* density should have recovered and the restoration goal has been achieved. In addition, marking rate would increase gradually from 60% to 65%, with wool production expanding from 4.2 to 4.4 kg sheep^{-1} during a period of seven to thirteen years. These may appear to be moderate increments, yet in time they can promote an increase in total sheep numbers. It was decided to maintain constant sheep numbers on the farms and consider that the surplus was sold as wool and meat.

As restoration is inherently a slow process, an economic analysis spanning twenty years was applied. This period includes climatic extremes, though in the first attempt to evaluate this restoration project, climate was assumed to have had no influence (figure 13.3b). The economic indicators used were the *net benefit* (NB in $), *net present value* (NPV in $), and the *internal rate of return* (IRR) (%) (Gittinger 1983) with a discount rate of 4%. This is reasonable rent for money invested in other forms of extensive production farming comparable to wool production. The recovery time for capital invested was also calculated, while estimating the switching values of the NPV assessed the sensitivity of the project to changes in productivity indices. The switching value indicates the increment (%) in the index necessary for a negative result to turn into a positive one, or the reduction (%) required to get to zero if the NPV was positive. Two biological productivity indices (marking rate and wool production per animal), and two economic productivity indices (lamb price and percentage of the investment needing to be subsidized) were also tested.

The analysis indicates that net benefits in farm types A and B decrease during the first year because of the investment required for fences and water points. After year one, net benefit (NB) recovers slowly, with minimum increments occurring after year five. Type C maintained a positive NB for two reasons: (1) investments were not necessary; and (2) after year two, NB begins to increase slowly. In the twentieth year, the NPV and the IRR were as follows: type C > type B > type A. Farm types A and B do not recover their initial investment in fences and water points with an opportunity cost (i.e., discount rate) of 4% (table 13.1). Whereas the IRR in type C reached almost 9%, in types A and B, the IRR remained negative. The larger farms recover the investment in year seventeen, whereas the other two types did not during the twenty years considered in the analysis. Indeed, types A and B did not reach

TABLE 13.1

Economic results of the Patagonian restoration process during twenty years of analysis

Variables	Farms		
	Type A	Type B	Type C
Initial investment (US$)	11,645,5	20,636.8	0
Years with negative net benefits	1	1	0
Years with incremental net benefits	9	9	7
Net present value (NPV) (year 20) ($)	(−11,097)	(−15,519)	15,482
Internal rate of return (IRR) (%)	Negative	Negative	8.8
Result with no cost for assistance (%)	Negative	Negative	9.1
Result with no income tax (%)	Negative	NA (exempt)	9.0
Result with no cost for assistance and no income tax (%)	NA (exempt)	1.98	9.3
Recovery time of investment (years)	> 20	> 20	17
Switching values for NPV			
Marking rate (%): original value 60%	99	81	61
Wool production (kg): original value 4.2 kg/sheep^{-1}	6.7	5.6	4.1
Lamb price ($): original value $15/lamb^{-1}	181	40	8
Subsidy to initial investments (%)	92	77	NA

Note: Type A = farms with 1,000 sheep; Type B = farms with >4,000 sheep, which is the economic unit; and Type C = farms with 15,000 sheep.

the IRR of 4% even under a scenario of free technical advice and no income tax payments. The addition of fences to type C did not affect the NPV, and investment was recovered by year twenty. Switching value analysis indicates that biological parameters are not realizable for types A and B. Conversely, type C switches to nonviable with changes of 61% and 4.1 kg sheep^{-1} in the marking rate and per capita wool production, respectively. Farm types A and B results are not sensitive to changes in economic parameters (switch price is not reasonably achieved), whereas type C can tolerate a 47% lamb price decrease. In general, type A results need to be discussed in different terms since conventional evaluation criteria fail to capture nonmarketable benefits, such as subsistence (Ayalew et al. 2003). The analysis of the increase in net benefits (%) between years zero and nine appears more realistic, although subsistence consumption is still not accounted for. Nevertheless, this indicator suggests that financing the initial investments has a long-term impact on all farm types.

Contribution

Sheep husbandry is a key component of the Patagonian economy and society yet a major cause of natural capital degradation; without restoration efforts, negative sheep population growth rates will continue. Our restoration plan focuses on helping one forage species to recover, but other species (plants and animals) will also benefit. Guanacos compete with sheep, and the establishment of a rotational system will open grazing opportunities for them as well. This technology should now be tested with field trials before passing it on to farmers. It was assumed that managers of the three farm types would be willing to invest in restoration, though analysis of the financial and economic impacts indicates that only the larger farms are best suited to adopt this technology. Type B farms will probably need financial support to make the transition, although it is expected that this group could initiate a restoration process. Contrary to other subsidies that are paid continually (for unemployment or wool prices) a restoration subsidy is paid for a limited time, yet it represents a long-term benefit. Smaller farms (type A) are of the highest concern since they cannot economically support a family indefinitely, and the short-term pressure forces management to be opportunistic and generally environmentally destructive. The planning horizon should be widened so that sheep husbandry is not the sole economic activity, or farmers could implement changes cooperatively to reduce the negative effects of small farm sizes. Although restoration costs exceed the benefits in farm types A and B, the increase in natural capital will be enjoyed by future generations beyond the twenty-year analysis.

From the government perspective, the good news is that the analysis indicates that grazing restoration is viable in much of the area (i.e., type C farms). The bad news is that most of the farms are too small to undertake restoration without financial support. Moreover, these farms have more limiting inputs, such as land, sheep, and operational capital. This is not a new dilemma, and free market solutions such as land abandonment followed by land concentration under a single ownership will probably increase social problems. Whereas ethics guide individual behavior, policy guides societal behavior. It is necessary to develop an agreed-upon policy that tackles "investment in natural capital" (Ekins, Folke, et al. 2003). The focus of our analysis was marketable products, yet other ecosystem services will be provided by restored ecosystems. Hence, it appears necessary to start long-term negotiations encompassing both socioeconomical and biophysical perspectives to achieve a holistic view of agroecosystems. An analysis that includes marketable products will increase farmer support for a policy that promotes natural capital restoration and its maintenance.

Acknowledgements

We would like to thank R. Golluscio, J. Rotundo, P. Cipriotti, and P. Graff for their help throughout the project. This work was supported by grants from Agencia Nacional de Promoción Científica y Tecnológica (BID 1201 OC-AR PICT 01-06641 and PICT Redes 331) to MRA; and by University of Buenos Aires to MRA (UBACYT G-002); and to MER (UBA-CYT G-045). The International Liaison Programme of the National Research Foundation of South Africa funded the participation of MRA in the RNC Workshop in Prince Albert.

Chapter 14

A Community Approach to Restore Natural Capital: The Wildwood Project, Scotland

William McGhee

Whereas the previous three chapters dealt with technical approaches to restoration, this chapter and the next two focus on various approaches to community-based restoration projects. This chapter's focus is on the context, challenges, and some of the outcomes of such a community-based restoration project in Scotland. This restoration has been inspired, planned, and executed by a community group called the *Wildwood Group of Borders Forest Trust*. The Borders Forest Trust Wildwood is establishing a mosaic of seminatural woodland and associated habitats in a discrete catchment in the Southern Uplands of Scotland. *Seminatural woodland* is defined as native woodland established by planting and regeneration with little or no management after establishment.

The Context

Since the postglacial colonization Scotland's forest cover has declined from an estimated 75% of the land area, to less than 4% by the seventeenth century (Ray and Watts 2003) and 2% by the early twenty-first century (Stiven 2005), mainly as a result of increased agriculture, as will be discussed later. The loss of natural forest, its fragmentation and associated species loss has occurred in a number of other European countries, notably Ireland, the Netherlands, and Denmark, although probably not on the same scale as woodland fragmentation in the Scottish Borders (Badenoch 1994). Afforestation in the Borders, using exotic conifer plantations throughout the nineteenth and twentieth centuries increased the forest area to 17.1% (Forestry Commission 2005) of which about three quarters is composed of introduced species; 23% are native species including 10% native Scots pine.

Remaining *ancient seminatural woodlands* (ASNW) in the south of Scotland are fragmented, and because of their size (generally less than two hectares) and shape (long and thin), they do not function as woodland ecosystems. Walker and Kirby (1987) define ancient woodlands as having a proven continuity of woodland cover for at least the last ca. 250 years. Remnant ASNW in the Scottish Borders amount to less than 1,000 hectares and cover some 0.13% of the land area (Badenoch 1994). Native woodlands have been absent from the southeast of Scotland for many hundreds of years and paleoecological studies show that the original native woodland cover had been "utterly destroyed" by 1603 (Badenoch 1994) when James VI of Scotland and I of England established the Union of the Crowns.

Forest loss and subsequent lack of tree regeneration in Scotland throughout prehistory and in historic periods was mainly as a result of agricultural activity (Badenoch 1994), especially arable farming in the lowlands and intensive sheep grazing in the uplands. The Southern Uplands of Scotland are well suited to sheep farming and are almost completely denuded of their natural vegetation, arguably more so than any other part of the UK (Ashmole and Chalmers 2004). Ironically, however, there may currently be more tree cover in southern Scotland than at any time over the last two thousand years. However this tree cover is predominantly even-aged, exotic-conifer plantations of Sitka spruce (*Picea sitchensis*) or lodgepole pine (*Pinus contorta*).

The Political Context of Restoration

Due to the rapid expansion of plantation monoculture, commercial forestry faced hostility from environmentalists and the general public during the 1980s (Tompkins 1989). This phase of afforestation was fuelled by tax incentives from the UK treasury, and the widespread perception of forestry as a tax haven for the wealthy, at the expense of landscape and habitats, resulted in a moratorium on conifer plantations in England and conflict between environmental NGOs, the state, and the private forestry sector (Ramsay 1997). Concurrently there was a developing awareness of the threat to and decline of native woodland, specifically ancient woodland. The government's conservation agency, the *Nature Conservancy Council* (NCC), developed this awareness through research into the conservation and restoration of Scotland's native woodlands.

Individuals and groups unconnected with or disenfranchised from mainstream forestry began taking positive action to put native woodlands back into Scotland's landscapes and to reconnect communities and the public with forests and trees from a less industrial and exclusive perspective than was practiced by the state and private sectors. These groups included *Trees for Life* (a "wild land" group with strong spiritual links) and the *Loch Garry Tree Group* (a small group of individuals dedicated to establishing forests in remote Highland sites). Throughout this evolution of social and environmental forestry in Scotland, *Reforesting Scotland* (an environmental NGO) acted as a catalyst for alternative forestry thereby raising the profile of native woodland habitat restoration and championing the restoration of the "Great Wood of Caledon." The first community-owned woodland in Scotland (out with Common Land holdings) was bought in 1986 by *Borders Community Woodland*, and this provided a model for developing local community involvement in woodland ownership and management.

Restoration Through the Wildwood Group

Against this backdrop of changing attitudes and perceptions against commercial forestry, a local group called *Peeblesshire Environment Concern* (PEC) was established in 1986. It comprised people with varied background and skills who were concerned that large tracts of hill land in southern Scotland contained only a small portion of its former biodiversity. They also believed that campaigning in the developed world to halt tropical deforestation in developing countries was hypocritical in the absence of practical action to restore native forests in the United Kingdom.

The conceptual basis of creating a wildwood grew out of a conference held in the Scottish Borders in 1993 (Ashmole 1994) and crystallized in 1995 with the formation of a *Wildwood Group*. In 1995 the Wildwood Group initiated a search for a suitable area of land. There were four criteria for site selection: (1) the site should constitute a discrete visual and ecological entity; (2) it should be on the order of eight hundred hectares (two thousand acres) in extent; (3) it should rise to at least six hundred meters; and (4) it should not contain large conifer plantations or intrusive human-made structures. As a response to large-scale funding from the National Lottery, the *Millennium Forest for Scotland Initiative* (MFSI) was created in 1995 to disburse monies to organizations and local groups. *Borders Forest Trust* (BFT) was created in 1996 to act as a conduit for the MFSI funding and to bring together groups and individuals whose objectives were the re-creation of a woodland culture through community and habitat restoration projects (Jeanrenaud 2001). The Wildwood Group was a founder member of BFT and has retained a distinctive and devolved position within the organization. The Wildwood Group's mission is set out it the Carrifran Wildwood Policy Statement, 30 March 2005: "The Wildwood project aims to re-create, in the Southern Uplands of Scotland, an extensive tract of mainly forested wilderness with most of the rich diversity of native species present in the area before human activities became dominant. The woodland will not be exploited commercially and human impact will be carefully managed. Access will be open to all, and it is hoped that the Wildwood will be used throughout the next millennium as an inspirational and educational resource."

The Wildwood Site

The Wildwood Group investigated approximately ten sites within the Ettrick Forest area in detail, and Carrifran Valley was selected on the basis of its availability and suitability. Carrifran valley lies in the west of the Moffat-Tweedsmuir Hills in the central Southern Uplands of Scotland. These hills form the western part of the area traditionally known as Ettrick Forest.

Carrifran is a spectacular U-shaped valley, carved by glaciers out of the Ordovician-Silurian rocks of the Tweedsmuir and Moffat Hills. It straddles the marches of Dumfriesshire and Peeblesshire (Scottish county boundaries). The valley extends to 670 hectares (~1650 acres), rising from 180 meters near the Moffat Water to the 821-meter summit of White Coomb, one of the highest points in the area. It lies entirely within the *Moffat Hills Special Area of Conservation* (SAC) site, which is scheduled under a European Union Designation for its geomorphological interest and for the arctic-alpine plants. These rare plants survive mainly on the highest crags and ledges, while the lower part of the valley was almost entirely denuded of its natural vegetation. Instead of a wide range of native woodland, montane scrub, and heathland habitats, the valley contained impoverished grassland, overgrazed heather, and eroded bare ground, with only a few relict trees to indicate that woodland existed in the past.

Negotiations with the landowner, a businessman and farmer, concluded with an option to buy the valley. This would represent approximately one-third of the farm at a price equivalent of £388,000 (US$720,083). This money was raised by the Wildwood through private subscription. No public (government) money was used in the land purchase, and six hundred individuals, as Wildwood "Founders," contributed sums of £250 ($474) to £35,000 ($66,323) per person. A further eight hundred individuals donated smaller sums (< £250 [$474] per person) with the result that the Borders Forest Trust Wildwood Group took possession of Carrifran on the first day of January 2000.

The discovery in 1990 of a bow, the Rotten Bottom Bow, found on peat hags (Rotten Bottom) at a height of 660 meters, is significant. The bow is the oldest found in the UK, dated to the Early Neolithic (4040–3460 BC) and was broken and presumably discarded by a hunter. The bow is on display in the new Museum of Scotland in Edinburgh. As a result of this discovery, an analysis of the fossil pollen at Rotten Bottom was conducted by Tipping (1998), providing the longest unbroken sequence known from any upland British site, recording plant assemblage data back to the end of the last ice age, some 10,000 years ago. The discovery of the yew bow and the subsequent palynological work contributed to understanding woodland composition in the Neolithic era and provided the Wildwood project with guidance for woodland restoration.

The Restoration Approach

The objectives in creating a wildwood can be interpreted as restoring the land within Carrifran Valley to a state of natural woodland. Defining which type and state of natural woodland that could or should be developed in Carrifran valley was the subject of much academic and practitioner debate. In the Scottish Highlands, southern England, or much of Continental Europe there may be no need to plant trees to establish new native woodland. Trees will regenerate naturally as has been the case on abandoned agricultural land in northern Italy (Conti and Fagarazzi 2005). In a regeneration scenario, local seed sources and pioneer tree species such as birch and alder could be supplemented with colonizers of gaps and open ground (e.g., oak). Restoring the natural capital of a site, which for millennia has been devoid of its natural cover, presents a number of practical challenges; asking ecologists, botanists, foresters, and geneticists to agree on a strategy for such a restoration added considerably to the challenge. A conference held in the Royal Botanic Gardens of Edinburgh (Newton and Ashmole 1998) considered the merits of different ecological restoration paradigms to assist with decisions. Various issues were discussed, such as the most appropriate tree species to be planted, genetic considerations in tree provenance selection, the native woodland types to mimic, the patterning and stocking of trees in the valley, the potential for reintroductions of plants and animals, and the management of associated habitats. The restoration of native woodland can take various forms, and the approaches are arranged approximately in descending order of naturalness (after Hunt 2003):

1. Nonintervention (apart from control of grazing animals)
2. Expansion of existing seminatural woodland by encouraging natural regeneration (through fencing, burning, scarification, and bracken control)
3. Improving conditions of seminatural woodlands by active intervention (exotic conifer removal, restructuring, and wetland restoration)
4. Expansion or species diversification of existing seminatural woodland by planting

5. Conversion of *Plantations on Ancient Woodland Sites* (PAWS) to native species
6. Conversion of other plantation woodland to native species
7. Creation of new native woodland by planting on bare land

Of the above approaches, only the last, "planting on bare land," was a realistic option for the Wildwood. The typology of natural woodland that could potentially be restored at Carrifran was outlined by Tipping (1998) and Peterken (1998), using UK-specific terminology and knowledge of past woodlands. It consisted of (1) "past natural," the woodland that would have been on the site if people had never influenced the valley; (2) "present natural," the woodland occurring in the region at present; or (3) "future natural," the woodland that would develop if people left the site to natural processes.

The establishment of new native woodlands is difficult due to a lack of knowledge and understanding of past woodland composition, structure, and function (Tipping 1998). The use of present natural woodland is equally fraught, with many arguing that there are no surviving woods in southern Scotland that can be usefully used as models (or reference sites) to mimic ancient woodlands. It was decided that elements of all three woodland "states" would inform the choice of woodland to be established at Carrifran. The original woodland (>6,000 BP year before present) proposed by Tipping (1998) was dominated by hazel accompanied by oak and elm on the drier glacial drift mounds and on the well-drained vegetated scree slopes. Alder, birch, and willow would occupy the damper soils and stable alluvium, and Scots pine may have occupied the colder east-facing slopes at higher elevations (tree line on poor soils). Mid-elevations would have been occupied by hazel, birch, ash, rowan, holly, aspen, blackthorn, and wild cherry. On the plateaux, juniper would have persisted with subalpine scrub.

The Wildwood project opted to take a pragmatic approach to selecting woodland type and tree-establishment techniques. The accessibility of recent palynological work from Carrifran and from sites nearby, combined with description and analysis of different woodland communities provided by Rodwell (1991) as part of the *National Vegetation Classification System* (NVC), allowed the project to develop a framework with which to plan woodland species composition (Rodwell and Patterson 1994), spatial distribution, and structure.

The Restoration

To date (September 2006) the Wildwood project has planted over 450,000 trees on approximately 280 hectares of land below 500 meters. Woodland types include birch woodland, upland oak-birch woodland, upland broadleaved woodland, and juniper woodland. The tree-planting techniques, the stocking densities, and the random tree spacing have been done in such a manner as to approximate natural woodland species composition and distribution and was executed with minimal disturbance to soil and vegetation.

The planting has been staged over a number of years with the first trees planted on 1 January 2000 and the major phase of valley bottom tree planting is due for completion in 2007–2008.

Voluntary Effort

Much of the day-to-day management of the Wildwood site has been and is carried out by paid BFT staff, and local contractors have carried out most of the physical work. It is notable that the Wildwood project comprises a large voluntary effort with respect to the planning, fund-raising, management, and physical work. This voluntary effort has been coordinated through the Wildwood Steering Group on behalf of the Wildwood Group, and it is this group that has driven the restoration effort.

Voluntary contributions to the project can be measured in person hours of time contributed by unpaid staff. Since 2000 there has been a volunteer day on the third Sunday of each month, which can attract between six and thirty volunteers who contribute to site-based tasks for up to six hours. Volunteers also assist in site management on a weekly basis, with an average of five individuals attending for up to eight hours. There is a volunteer deer-stalking rota of a dozen individuals who are prepared to carry out deer control at least four times a year. There are also volunteer "boundary wardens" who are organized on a rota to ensure that the high-elevation perimeter fence and internal fences are checked on a monthly basis.

Taking the site-based input of volunteers and combining it with the thousands of hours it takes to plan tasks, attend meetings, collect seed, organize conferences, represent Wildwood at seminars, and contribute articles to journals, the voluntary input to Wildwood from a relatively small community group is impressively large. The success of the project is the result of this large-scale voluntary effort, a characteristic shared by various other conservation and restoration projects across the UK but not often on the scale of the Wildwood project.

Funding

As of June 2006, Wildwood had raised a total sum of £700,000 (US$1,328,660) from private donors and trusts with contributions from approximately 1,800 individuals, of whom 615 are Wildwood Founders and 233 are Wildwood Stewards. This level of fund-raising and the large number of donors toward a relatively modest project in a small valley in the south of Scotland is due to a number of factors: (1) the dedication and effort of the Wildwood Group and many associated individuals and groups; (2) the visionary nature of the project; and (3) the assistance from others in the sector, such as the John Muir Trust, who allowed BFT Wildwood to fund-raise through their membership newsletter.

Challenges

The practicalities of creating a "wild" forested area in a highly artificial and intensively managed landscape have posed a number of challenges. Like many forestry projects, getting trees to grow in a hostile environment is not easy without a highly interventionist and mechanistic approach. Conditions are aggravated by the fact that the removal of grazing stock facilitates rapid grass and bracken growth, which compete with tree saplings. Wildwood has had to contend with feral goats, sheep, roe deer, hares, and voles. The low-key "ecological" approach to restoration adopted by the Wildwood group differs from mainstream forest establishment; the cost of such an approach has been high tree mortality due to deer browsing and competition from vegetation and, subsequently, slower establishment of trees.

The physical challenges of the Carrifran site have been exacerbated by anthropogenic influence. The decision to remove all feral goats from the valley by the Wildwood Group prompted an intense and vociferous local campaign against the Wildwood in 2000. Campaigns against the removal or the management of wild, feral, and domestic animals are not uncommon in relation to restoration efforts in the UK. The arrival in the south of Scotland of foot-and-mouth disease in 2001 meant delayed planting, negatively affecting some 30,000 trees and resulting in the exclusion of all management from the site for the better part of a year. This exclusion of management resulted in high tree mortality from browsing roe deer and unchecked weed competition. Muir burning, to remove rank grass and heather, is a feature of upland management in the UK, and in spring 2003 a fire set by the neighboring farmer ran out of control and killed 10,000 trees.

In spite of these challenges, trees planted in the valley are growing and the woodland is progressing well. The establishment of trees in the valley will take a number of years and the goal of minimal human intervention is not envisaged until post-2020. It is hoped that by that time the valley will contain enough woodland to give the appearance and feel of a wildwood.

Contribution

The Wildwood project is an expression of the wider realization that in areas where natural ecosystems have almost completely disappeared, conservation of surviving relict fragments needs to be complemented by restoring natural capital through the establishment of habitats that function without the intervention of humans. The Wildwood name and the Wildwood ideal have provided an inspiration to people who wish to be associated with "wildness." Many of the people who donate to the Wildwood may never see or set foot in Carrifran valley but, like those who contribute to saving rain forests and whales, they believe that an area of wildwood should exist. Donations to Wildwood are a response from many individuals in the UK who wish to see an area of wild land restored in the British Isles.

Dogged determination and self-belief in the "ecological cause" has characterized much of the Wildwood Group's efforts, and the project has shown that a community group can undertake the large-scale restoration of natural capital and provide inspiration to many.

Chapter 15

An Adaptive Comanagement Approach to Restoring Natural Capital in Communal Areas of South Africa

CHRISTO FABRICIUS AND GEORGINA CUNDILL

South Africa's communal areas are home to the majority of its rural population (Shackleton et al. 2000), many of whom rely heavily on natural ecosystem services (Cavendish 2000). For some inhabitants the income from natural resources comprises up to 35% of their household income (Rangan 2001). This "invisible economy" (Campbell and Luckert 2001) often forms the backbone of people's livelihoods in remote rural areas where jobs are scarce. Indeed, natural capital is an important buffer against economic, ecological, and political change (Shackleton and Shackleton 2004a). Nowhere is the interdependent relationship between people and natural capital more evident than in communal areas where traditional knowledge, local institutions, management practices, and ecosystem services and functioning are intertwined (Fabricius et al. 2004). To rural communities this connection is obvious, as ecosystems form part of their identity by providing ancestral links, "shopping centers," and sources of spiritual inspiration. In many traditional societies, local knowledge systems have coevolved with ecosystems, and complex sets of local rules have developed around the management and maintenance of ecosystem services (Berkes 1999; chapter 16).

Many of southern Africa's communal areas have, however, suffered from social, institutional, and ecological degradation (Ainslie 1999; Hoffman and Ashwell 2001). The reasons are complex and relate to the following factors:

1. The high human population density in communal areas resulting from forced removals and other forms of social engineering during the apartheid era (Hoffman and Ashwell 2001)
2. Weak tenure security in communal areas, where land belongs to the state and local communities have use rights but not ownership (De Wet 1995)
3. The complexities of common pool resource management when infrastructural and local management capacity are weakly developed (Ostrôm et al. 1999)
4. The virtual collapse of local institutions, such as traditional leadership structures and their accompanying codes of conduct, in South Africa's communal areas (Manona 1998)
5. The legacies of political and economic policies and interventions, which precipitated the collapse of local institutions and undermined customary land management practices (Ainslie 1999)

6. Weak technical and financial government support, for example, a collapse of agricultural extension services following decades of heavy-handed intervention in rural agriculture (Cundill 2005)

In addition, land degradation makes people vulnerable to change, consumes time, and decreases material well-being (iKhwezi 2003). Clearly, the restoration of natural capital could mitigate land degradation and associated social problems. However, high population densities, lack of security of tenure, and low income levels frustrate individual initiatives. It is therefore important for communities together with government and nongovernment institutions to develop policies and strategies to restore natural capital in communal areas, as failure to achieve this will negatively affect the well-being of millions of South Africans.

The aims of this chapter are to provide a rationale for natural capital restoration in communal areas by documenting its importance for human well-being and to provide a framework for natural capital restoration. We focus on the indirect and intangible values of natural resources, as these aspects have been neglected in the academic literature (MA 2005e); examples are mainly from the Eastern Cape. Our proposed framework for intervention is qualitative, based on experience and lessons learned during scores of case studies conducted over more than a decade in southern Africa's communal areas (e.g., Fabricius et al. 2001, 2004).

Is Natural Capital Important to People in Communal Areas?

In most communal areas, "off-ecosystem" sources of income—for example, money sent by family members working in cities, and from social grants and jobs—contribute the lion's share of formally recorded household incomes (Palmer et al. 2002). It might therefore appear relatively easy to discount the role of ecosystems in rural livelihoods on the basis of their comparatively small contribution. Records of household incomes, however, do not take into account the value of "free" services, and they disregard the important nonfinancial services provided by ecosystems at critical times (Campbell and Luckert 2001). Economists, policymakers, and planners therefore undervalue ecosystem services in communal areas (Cousins 1999).

The Direct Economic Value of Ecosystem Services

The direct value of ecosystems in the form of fuelwood, building materials, medicinal plants, implements, and food is well documented for communal areas (Rangan 2001; Cocks et al. 2004). These "everyday" or "informal" resource uses (Shackleton and Shackleton 2004a; chapter 17) are often central to the livelihoods of the very poorest rural people. The value of everyday resources is conservatively estimated at US$550–$600 per household per annum, equal to or exceeding that of an old-age pension or disability grant (Shackleton and Shackleton 2004b).

Livestock is another important form of capital in many southern African communal areas (Cousins 1999). The value of 7,670 cattle units in a rural settlement in the Eastern Cape was calculated at more than $3 million (iKhwezi 2003). Cattle and other livestock are integral to communal-area management, institutions, and traditions. They function as a "bank" while also providing food, traditional medicines, fuel (from dung), and traction to plough fields.

Buffer Resources During Crises

Although certain resources may have a low direct-use value, as they are infrequently used or not part of daily livelihood strategies, they may nevertheless be critical during crises. For example, natural stream usage is declining due to substitutes such as piped tap water (Shackleton et al. 2000), and it may be easy to discount their current value. However, when municipal water supplies periodically fail, people return to natural water resources. Similarly, when households cannot afford electricity or paraffin, they use fuelwood for heating and cooking. Livestock are another key buffering resource in rural livelihoods in South Africa; their slaughter or sale is reserved for food crises, financial crises, and ancestral rituals (Rhodes University et al. 2001). For this reason the condition of the rangeland ecosystem in communal areas is important to local people.

At Mount Coke State Forest, communities participating in a forest monitoring program listed twenty items supplied by the forest's natural capital (including building materials, bushmeat, fungi, and fruit, as well as religious and cultural values) that were vital to their livelihoods. Many of these products were infrequently used and would therefore have low direct-use values in standard economic analysis. In fact, the local people considered that certain products, even infrequently utilized, were as important as those in daily usage (Rhodes University et al. 2001). People use a wide variety of food sources, and in times of scarcity even eat insects, seeds, and leaves not normally included in human diets (Madzwamuse and Fabricius 2004). In addition to obtaining food from the wild or buying it, surveys at Mount Coke indicated that 94% of households maintained home gardens, cultivating thirty-three types of crops (Cundill 2005).

Strengthening Social Capital Through Customs and Traditions

Local traditions, such as ancestral veneration, initiation ceremonies, traditional healing, and rituals, are difficult to value but are essential for maintaining social cohesion and a sense of identity (Berkes 1999). Natural ecosystems are vital for providing the basis of these traditions, with most rituals directly linked to ecosystem services.

Natural ecosystems also provide a "sense of place" to local communities. In the Kat River valley, for example, wild olive trees and the complex landscapes associated with them were so highly prized that the local people were unwilling to be relocated or sacrifice these trees (H. Fox, personal communication).

Diversifying People's Livelihoods

Rural people in communal areas engage in diverse livelihood strategies (Scoones 1998; Campbell et al. 2002), enabling them to be flexible and to cope with the risks associated with living in a communal African area (Ellis 2000). Natural resources such as fuelwood, building materials, food, and artifacts obtained from ecosystems make a major contribution to livelihood diversification (Shackleton et al. 2000), as do livestock and crops (Timmermans 2004). At Nqabarha, for example, 79% of the fifty-eight tree species were identified as having some use, and 28% had multiple uses (McGarry 2004). For this reason, a landscape mosaic of old fields, pastures in different successional stages, and dense forest patches provide a rich array of goods and services.

How People Are Affected When Natural Capital Becomes Degraded

Despite households relying on many different income sources and livelihood strategies, ecosystem degradation can have severe consequences, as people lose access to essential renewable natural resources of substantial financial value. For example, in the Eastern Cape's Emalahleni District, land degradation has lead to major livestock losses (iKhwezi 2003). Furthermore, land degradation can affect social capital as local institutions weaken or strengthen in response to the condition of the natural resource base.

When natural capital is lost, people are forced to invest more time collecting fuelwood, medicinal plants, and construction materials. At Pikoli in the Peddie District, the density of preferred fuelwood species (sneezewood and sweet thorn, *Acacia karroo*) decreases linearly with distance from the edge of the village (Biggs et al. 2004). Hence, women, who have the role of collecting fuelwood, have to walk long distances and complain that it is more difficult to find than in the past (Ainslie 2003). Another effect of forest degradation is reduced groundwater recharge, with participatory workshops revealing that natural spring water availability in the communal area has decreased over the past three decades, relative to that in the state forest (Rhodes University et al. 2001). Food security in rural areas is also compromised by crop failures, wildlife scarcity, and stock losses, resulting in greater reliance on home gardens than on pastures and large fields. At Macubeni, people commented that medicinal plants had become so scarce that it was not possible to treat their livestock, let alone people (Fabricius, Matsiliza, et al. 2003).

Other more indirect impacts of land degradation are emigration of economically active people from the community; loss of pride, identity, and traditional ecological knowledge; institutional collapse; and, ultimately, rural stagnation and poverty. The combination of these negative impacts can allow a degraded rural area to enter into a self-perpetuating cycle of institutional neglect and reduced government support.

Interventions Can Restore Natural Capital in Communal Areas

Land use in the communal areas of South Africa is embedded within a history of state intervention, subsequent political isolation, and more recently in improvements in basic service delivery (Cundill 2005). An understanding of the history of dispossession and alienation of local people from their daily utilized resources should guide the search for appropriate interventions and strategies for resource management.

Despite South Africa's complicated history of communal areas, it is possible to halt or even reverse natural capital degradation. No single organization such as an NGO, however, has the capacity to achieve this, nor is it conceivable without the cooperation of the people who use and interact every day with the natural resources. The answer lies in adaptive comanagement, that is, a partnership between local communities, government, municipalities, technical advisors, and funders.

An Adaptive Comanagement Approach to Restoring Natural Capital

Adaptive comanagement, defined as "an inclusive and collaborative process in which stakeholders share management power and responsibility" (Carlsson and Berkes 2005, 73), is based on the principles of cooperative governance, with direct community involvement in decision making (Olsson et al. 2004). The incentives for adaptive comanagement are (1) the realization by local people that natural capital is becoming scarce, as a result of overuse of resources by locals and outsiders, reduced government capacity to provide technical assistance, and climatic change; (2) local concern about insecure land tenure arrangements in communal areas.

At this time there is the potential to capitalize on the lessons learnt from more than a decade of participatory natural resource management, including community conservation (Hulme and Murphree 2001), social ecology (Fabricius 2003), integrated conservation and development (IIED 1994), community wildlife management, and community-based natural resource management (Fabricius et al. 2004). These initiatives often appear to be discredited for lack of delivery to communities and poor biodiversity conservation track records (Magome and Fabricius 2004). However, adaptive comanagement holds promise to improve on conventional community conservation interventions, mainly because it is community driven, focuses on the capacity of ecosystems to produce goods and services rather than on biodiversity conservation per se, draws on the knowledge and capacity of support networks at different levels, and relies on adaptive learning and monitoring rather than a blueprint approach. The two examples that follow show how rural communities and other institutions combined forces to restore the natural capital, resulting in improved benefit flows to communities from clean water and tourism-based livelihoods.

At Macubeni in the Eastern Cape, local communities have mobilized themselves around a vision of sustainable natural resource use and ecosystem restoration, articulated as "a better life for all, by managing our natural and manmade resources sustainably, in order to improve our livelihoods, health, education, and economy, while still maintaining our traditional culture and values, so that there will be a brighter future for the people of Macubeni" (Fabricius and McGarry 2004). A land use planning steering committee has been formed, overseeing the development of a long-term land use plan. Local people are organizing themselves to restore the severely degraded watershed around the Macubeni dam with the assistance of a project steering committee, comprising twenty-eight community representatives, two local municipalities, and local government departments concerned with agricultural, environmental, health, and economic issues. In conjunction, a university research program is helping with fund-raising and facilitation, technical support, capacity development, and lobbying for resources.

On the Wild Coast of the Eastern Cape, community action has led to the formation of a community tourism organization, a community-based conservation plan, and emerging partnerships with the private sector. A community trust and three institutions created by the local community, that is, a forest management committee, a craft production committee, and a medicinal plant user's group, fall under the umbrella of the Nqabarha Development Trust. They have zoned their land to register a conservancy; formulated rules and penalties for community-based law enforcement; and developed strategies for income generation, fund-raising, and training. The community has also established a vegetable and medicinal plant nursery and a craft workshop, with plans to put certain tourism opportunities out to tender to private investors. The local municipality, Rhodes University, the national government, and the German agency for technical cooperation, Deutsche Gesellschaft für Technische Zusammenarbeit (GTZ), support their work.

These initiatives are, of course, not without their problems, including high turnover of officials, weak capacities of all stakeholders to participate meaningfully, and communication problems (Fabricius and McGarry 2004). There are also many unknowns, such as how people will deal with conflicts centered on benefit flows once real revenue is generated, whether the business side will be managed appropriately, and if efforts can be sustained when the current intensive facilitation levels are reduced. There is, however, hope that the local municipalities will play a central role in sustaining these social, economic, and restoration initiatives once their capacity has been strengthened and their roles clarified.

Guidelines for Adaptive Comanagement

Through trial and error over more than a decade of applied research and learning (Fabricius et al. 2001; Fabricius, Matsiliza, et al. 2003), a number of guidelines for adaptive comanagement of restoration initiatives in communal areas have been developed. This checklist of issues to be addressed for reducing the incidence of failure is based on ten design principles:

1. *Conceptualize the system as integrated, complex, and adaptive.* Social-ecological systems respond to feedbacks at different spatial and temporal scales, for example, contemporary and historical political trends, policy changes, climatic fluctuations, ecosystem shifts, and macroeconomic trends, as well as localized availability of infrastructure, labor, and alterations in land tenure. To understand these feedbacks and interactions it is useful to start by constructing a provisional model as an abstraction of the complex and adaptive nature of the system.

2. *Combine formal and informal sources of knowledge and information.* Local knowledge can make a valuable contribution to validate initial conceptual models. During the process of collecting information, it is essential to broaden local awareness about natural capital trends, as well as to reassure the role of local people and the value of their traditional knowledge for restoring ecosystems. Local knowledge must, however, be validated and combined with formal knowledge. Combining low technology methods, such as participatory mapping, with high technology computer techniques, such as *geographic information systems* (GIS), is very useful for delineating land boundaries, recording trends in land use, and affirming land and resource ownership and access.

3. *Identify stakeholders in advance and work closely with them.* Adaptive comanagement requires the development of strong horizontal and vertical social networks between participating individuals and organizations. Strong and functional networks reduce vulnerability through redundancy, while strengthening the resilience and stability of comanagement initiatives. Those stakeholders who are directly affected by the outcome of the initiative, or have claims to access or ownership, are the preferred participants in these social networks.

4. *Agree on a clear and shared vision.* The participation of visionaries at all levels of the adaptive comanagement process is essential. All stakeholders should agree on a future vision of the social-ecological system's condition, which should be jointly developed and reaffirmed in every document and at every occasion. A shared vision will usually combine ecological and socioeconomic benefits, while referring to the link between livelihoods and ecosystem management.

5. *Ensure that the benefits are understood and shared by all stakeholders.* The benefits from restoration are mostly indirect, intangible, and long term. It is, however, vital for primary stakeholders to experience tangible short-term benefits, such as temporary employment, poverty relief grants, infrastructural repair, or markets for ecosystem products. Other rewards may include conflict reduction and newly acquired skills provided through stakeholder participation. From the outset, however, all stakeholders should understand and accept that the real benefits from restoration are longer term. There is, however, a risk of creating dependence on interim benefits such as poverty relief grants. Project managers should acknowledge this and develop strategies to gradually reduce such dependency.

6. *Identify key individuals to take responsibility for each major function.* The seven key functions in adaptive comanagement are (a) maintaining and affirming the vision; (b) developing capacity; (c) harnessing knowledge and human resources; (d) monitoring, learning, and feedback; (e) maintaining the ecosystem; (f) maintaining and administering comanagement institutions and organizational structures; and (g) financial management and fund generation. Each of these functions should have at least one highly motivated individual and an assistant, with the latter being able and ready to take on the responsibility easily, if necessary.

7. *Foster flexibility and diversity to reduce vulnerability and increase resilience.* Most adaptive comanagement initiatives are plagued by a lack of resilience and have often failed in their early stages due to the departure of key individuals, unexpected changes in local and national politics, unmet expectations in relation to benefits, or conflicts between role players. An effective mechanism to promote resilience is to achieve human diversity and minimize the redundancy in stakeholders, developing highly motivated individuals and institutional structures while promoting a culture of learning and adaptive renewal. However, flexibility made possible through human diversity may have disadvantages when attempting to attract large investors who could be deterred, perceiving this necessary complexity as disorganization in communal areas. Potential investors therefore need to be briefed in advance about flexible strategies and procedures, while stakeholders should invest in a range of restoration strategies and promote a variety of small enterprises. Similarly a number of different sustainable land use options should be initiated to enhance landscape and species richness.

8. *Provide ongoing and professional facilitation.* Experienced facilitators are needed to manage conflicts before they escalate, break deadlocks, help role players with technical information, and assist with the formulation of plans and proposals, while sharing their wider experiences. Adaptive comanagement is a complex and long-term process, and hence facilitation should be ongoing. Although local governments are ideally positioned to provide permanent facilitation services, they often lack the capacity and human resources to fulfill these functions. The Emalahleni and Mbhashe local municipalities, based in Lady Frere and Dutywa, respectively, employ only two local economic development officers each, who have to assist hundreds of villages with advice. A system of community development workers, intended to extend ca-

pacity of the municipalities at the village level, has recently been initiated and might alleviate this problem.

9. *Develop the capacity of all stakeholders to contribute meaningfully*. Few natural resource professionals are trained to integrate economic, political, social, and ecological processes. Their capacity to participate in adaptive comanagement must therefore be developed. Even more so, rural people are in greater need of engaging with other stakeholders and capacity development, including maintaining and administering organizational structures and funds. Local governments have only recently been given natural resource management responsibilities and hence require much capacity development in this field.

10. *Monitor, learn, and respond to feedback*. Adaptive comanagement is thus a form of "learning by doing," representing a perpetual cycle of action, monitoring, adaptation, and response, which implies that the managers are integral to the system. Participatory monitoring systems (Rhodes University et al. 2001) are an essential part of any initiative, and consequently adaptive comanagement means that stakeholders should frequently revise and adapt their plans, strategies, and approaches in response to new information.

Contribution

This chapter has shown that it is possible to restore natural capital in communal areas through community-led adaptive comanagement. It is essential that participants recognize the value of the natural capital; that is, the focus should be on ecosystem productivity and livelihoods rather than on biodiversity conservation per se. The process should include broad stakeholder involvement, constant monitoring and adaptation, and building and enhancing the resilience of the social-ecological system through promoting adaptability and flexibility in cooperative governance. Adaptive comanagement initiatives are, nevertheless, in their early stages, and there are many unknowns, such as how communities will deal with inevitable conflicts caused when the material benefits start flowing; whether the social networks will function during economic and political setbacks, and if local government is committed to providing long-term facilitation services.

Chapter 16

Participatory Use of Traditional Ecological Knowledge for Restoring Natural Capital in Agroecosystems of Rural India

P. S. Ramakrishnan

Large areas of forest cover in the developing tropics have been converted to other uses or severely degraded. These forests are the natural capital that provides subsistence livelihoods for many people, the so-called traditional societies living close to natural resources. Of the estimated 14.6 million ha yr^{-1} of global deforestation loss, 14.2 million ha occurs in the tropics (FAO 2001). During 1980–1990, an estimated 15.4 million ha yr^{-1} of tropical forests and woodlands were destroyed or seriously degraded, largely through agricultural expansion, excessive livestock grazing, logging, and fuelwood collection (FAO 1993). More recent evaluations indicate that in tropical Asia secondary forests in various degradation levels (Chokkalingam et al. 2000) and a rapidly expanding plantation forestry sector (FAO 2001) are steadily replacing the remaining primary forests.

Globally, plantation forestry has become the most common replacement for natural forests; the developing tropics are no exception. This is particularly significant since tropical reforestation has not kept pace with logging (WRI 1985). Plantation forestry does have a role to play in making productive use of degraded landscapes and meeting socioeconomic goals (Lamb 1998; Parrotta 2001). Unfortunately, this conversion often means replacing species-rich natural ecosystems with species-poor monocultures, usually of nonnative (or alien) tree species. The long-term ecological sustainability and economic viability of such conversions, particularly in the case of short-rotation, fast-growing species, remains doubtful. With alien species, sustainability is also questionable due to possible invasiveness (Ramakrishnan 1991; Ramakrishnan and Vitousek 1989). However, if carefully chosen, aliens can play a positive role in catalyzing natural vegetation regeneration by improving soil conditions during the initial phases of plantation or rehabilitation (Ewell and Putz 2004). Moving toward mixed plantation programs that address biodiversity concerns and include multipurpose native species traditionally used by local people can increase the value of restoration.

These traditional societies are directly dependent upon forests for nontimber forest products (Patnaik 2003), and indirectly for many services to ensure food security (Ramakrishnan 2001). In many developing countries, rural forestry programs have played a vital role in supplying societal needs, such as fuel, fodder, timber, and even organic residues for sustainable agriculture. In addition, these forests provide numerous nontimber products, such as medicines, spices, and lesser-known food items (Depommier 2002). It is unfortunate that at present only a few well-known, mostly alien sylvicultural species are being planted in the name of agroforestry and rural forestry programs. These have little interest for rural communities be-cause they lack options for multipurpose use, equitable distribution of benefits between rich and marginalized farmers, or employment opportunities for the landless poor (see, for exam-ple, chapters 15, 19, and 20).

Thus, in a developing country context, community participation is essential during the restoration process, particularly when involving societies with a rich traditional knowledge in-tegrally linked to biodiversity and natural resource management. This is the context in which *traditional ecological knowledge* (TEK) assumes significance for community participatory res-toration of natural capital. TEK connects ecological and social processes when appropriately linked with *formal knowledge* and can be incorporated into management systems, such as cheap and decentralized water control. In this chapter, using case studies linked to soil fertil-ity and water management, I will illustrate the vital role of participation in ensuring sustain-able, landscape-level restoration of natural capital.

Restoring Natural Capital by Integrating Various Knowledge Systems

Many attempts to restore natural capital in India and other developing countries have been based upon textbook ecological knowledge of ecosystem structure and function, tree biology, and silviculture (Wali 1992; Lamb and Tomlinson 1994). Formal knowledge has been the basis for designing restoration strategies, which implies that extrinsic actors are forcing the changes. However, integration of formal and traditional ecological knowledge systems offers a broader foundation for building natural capital through community participation (Rama-krishnan 1992a, 2001).

Types of Traditional Ecological Knowledge (TEK)

Traditional ecological knowledge is largely derived through societal experiences and percep-tions accumulated by traditional societies during their interaction with nature and natural re-sources. While not excluding the universality of knowledge, TEK combines location speci-ficity and a strong human element, with emphasis on social emancipation (Elzinga 1996). The challenge is to achieve valid generalizations across socioecological systems.

There are three basic kinds of TEK: (1) *economic*—traditional crop varieties and lesser-known plants and animals of food and medicinal value, as well as other wild resources; (2) *ecological/social*—biodiversity manipulation for coping with environmental uncertainties, hydrological control, and soil fertility management; and (3) *ethical*—unquantifiable values centered on cultural, spiritual, and religious systems, operating on three scales: sacred spe-cies, sacred groves (habitats), and sacred landscapes (Ramakrishnan et al. 1998). As TEK links ecological and social processes, this discussion focuses on its relevance in soil fertility–linked and water management–linked issues for restoring natural capital with community participation.

Case One: Restoration of Shifting Agricultural Landscapes

For centuries, forest farmers in many tropical areas have managed a range of traditional shift-ing (slash-and-burn) agricultural systems, including those known in India as *jhum*. In the past, these small-scale perturbations helped enhance forest biological diversity, while crops and associated organisms benefited from the extra nutrients released. With increasing external pressure on forest resources, larger populations, and declining soil fertility, forest cover is rapidly being lost (Lanly 1982; FAO 1995) and agricultural cycles shortened, causing marked productivity reduction and other negative impacts. Reduced system stability and resilience can result in social disruption, biodiversity decline, biological invasions, increased CO_2 emissions, and long-term desertification. However, all current attempts for finding alternatives to shifting cultivation have had little or no influence on farmers, and more holistic approaches are required for tackling the complex issues involved (Ramakrishnan 1992a, 2003).

A participatory approach was developed for finding sustainable solutions to the jhum problem, which would contribute toward restoring natural capital. This approach was originally developed in the hill areas of northeast India, where large-scale timber extraction, in response to external demand and increasing population pressure, has lead to livelihoods below subsistence levels in highly degraded landscapes.

Role of TEK in Landscape Restoration

Natural capital restoration with community participation involves a more holistic approach, which implies that sustainability needs to be viewed not only from a biophysical perspective, but also from the livelihood angle in the social system. Jhum has a rich TEK embedded within these cropping and forest-fallow practices in the northeastern hill region of India (Ramakrishnan 1992a). Besides the forest successional stages (climax and successional, broad-leaved forest types, arrested bamboo and weed communities), there is a range of agroecosystem types linked to jhum. These include wet-valley rice cultivation, traditional plantation systems, and home gardens, which appear forestlike yet have been created by local people and comprise economically important plants. Indeed, jhum is linked to forest ecosystems because slash and burn favors early successional plants. Therefore, an integrated view of natural and human-managed systems is needed if natural capital reconstruction is to be meaningful to local people. Where one is seeking community participation, TEK is a powerful tool because it links ecological and social processes and helps to design strategies for sustainable land use.

The practices described here give some idea of the integration of TEK in this agroecosystem:

1. Jhum is a complex multispecies system with over forty crop species under longer cycles and at least six to eight species under shorter ones. Based on experience, and to optimize productivity, traditional societies plant nutrient-efficient crops higher up the slope and less-efficient species along the bottom to match the soil fertility gradient.
2. During shorter cycles, nutrient-efficient tuber and vegetable crops are planted, while less nutrient-efficient cereals require longer ones, which is an adaptation to soil nutrient status under different cycle lengths.
3. Although all species are sown shortly after the first monsoon rain, harvesting occurs sequentially during several months of maturation, which reduces interspecific competition. After harvesting, the remaining biomass, including weeds, is recycled into the agricultural plot to maintain fertility.

4. Rather than "weed control," jhum farmers practice "weed management." Irrespective of farmers' sociocultural background or ecological conditions, about 20% of the weed biomass is left in situ. Field trials have shown that it conserves nutrients, which otherwise could be lost through erosion or hill-slope leaching, but it has no negative impact on crop yield, 20% being below the threshold where weeds out-compete crops for nutrients.

5. Earthworms form an important component of many traditional agricultural systems, and farmers generally view them as keystone species and soil fertility indicators.

6. Nepalese alder (*Alnus nepalensis*) is conserved within the jhum system, and being nitrogen-fixing, it forms part of the traditional fallow-management practices. Other socially valued species, including bamboos (*Dendrocalamus hamiltonii, Bambusa tulda* and *B. khasiana*), have been shown to conserve nitrogen (N), phosphorous (P), and/or potassium (K), and locals consider such species attributes important during forest-fallow selection.

Key Role of Socially Selected Species

The rich biodiversity in these agricultural systems contributes to system resilience and economic productivity, yet the significance of key species has not been adequately explored. Certain species may be essential for managing primary and secondary forests at different degradation levels (Ramakrishnan 1992a, 1992b) as well as increasing biodiversity in human-managed ecosystems (Ramakrishnan, Purohit, et al. 1994). Socially selected species are often key species, such as nitrogen-fixers, within the jhum ecosystem (Ramakrishnan et al. 1998). For example, in many highly degraded areas, a lesser-known tuberous crop, *Flemingia vestita* (Fabaceae), locally known as *soh-phlong*, is planted under a two- to five-year rotational fallow system, with yields up to 3,000 kg ha^{-1} (Gangwar and Ramakrishnan 1989). Use of *F. vestita* in the rotational bush-fallow system allows reduced slash-and-burn intensity, and the shortening of the jhum fallow cycle (Ramakrishnan 2001). By fixing 250 kg N ha^{-1} yr^{-1}, this legume ensures sustainability of these low-input agroecosystems under conditions of extreme land pressure and low soil fertility. In addition, fallow management is made possible by varying organic residues and promoting earthworms (Senapati et al. 2002).

A second example of the importance of socially valued species is the Nepalese alder used by traditional societies in shifting agricultural plots (Ramakrishnan 1992a). Alder trees may also be planted during the cropping and fallow phases, with an ability to fix up to 125 kg N ha^{-1} yr^{-1}, and a potential to recover 600 kg of N lost over a five-year cropping cycle; under natural fallow regrowth this would otherwise take a minimum of ten years. Besides soil fertility improvement, during a five-year cycle alder can provide a cash income of US$100 ha^{-1} yr^{-1} (using a conversion rate of 45 rupees to 1 US$) for its timber, and up to $444 for the associated crops, compared to about $133 in its absence. There are no financial costs because the only input needed for improving soil fertility and harvesting timber is the labor supplied by the villagers themselves. Using this system, farmers can dispense with shifting agriculture associated with slash and burn, so that forest farming becomes sedentary. In addition, alder is easily pollarded and agriculture can continue for up to three years before shading becomes a problem for crop growth.

Participatory Jhum-Redevelopment Pilot Study

Jhum redevelopment, using participatory fallow management (Ramakrishnan 1992a), is being tested in the state of Nagaland as part of a decentralized village development plan (the NEPED project) (NEPED and IRR 1999). Management of forest (Ramakrishnan 1992a) or grass fallows (Lal et al. 1979), depending on the ecology, is a cost-effective alternative to shifting agriculture. The major element for long-term success is that TEK and farmers' perceptions of different tree species form the basis for tackling problems associated with the declining jhum cycle and natural resource degradation. Participation of local people is facilitated by identification with such a value system.

Over a thousand villages have been organized into Village Development Boards taking account of traditional organization of the given cultural group. Using this mechanism, the highly mobile, shifting agricultural systems currently operating at or below subsistence level are being redeveloped by strengthening the tree component that has been weakened because of extreme deforestation. The basis for this natural capital reconstruction is the rich TEK of these hill societies, with the Nagaland government aiming at augmenting traditional agriculture rather than radically changing it. Indeed, alder-based agroforestry systems have been maintained for centuries by some local tribes, such as the *Angamis*, which formed the impetus for this initiative.

During preliminary trials about a dozen tree species were investigated in more than two hundred test plots covering 5,500 ha. Subsequently, local-based trials have been carried out by farmers in 870 villages, in a total area of about 33,000 ha (38 ha/village). In these plots, local adaptations and innovations for activities, such as soil and water management, were emphasized, which lead to a three- to fourfold increase in agricultural productivity, in contrast to a typical five-year forest jhum cycle ($133 ha^{-1} yr^{-1}). It is too early to evaluate the economic returns from timber sales, since governmental regulations forbid tree felling. A policy revision that provides incentives for local people combining agroforestry with agriculture is clearly needed.

In summary, the objective of the NEPED program is to redevelop the agricultural component (jhum system) with greater productivity and diversified cropping patterns for a better quality of life, while accelerating forest successional processes through human-mediated, fallow-management procedures and improving forest biodiversity.

Case Two: Water as an Incentive for Natural Capital Restoration

During a research agenda initiative for the Indian Himalayan region, scientists and developmental agencies produced a long list of activities whereas participatory interaction with local communities identified one critical issue, namely, dry-season water availability. Water was identified as a key element for a community-participatory restoration program, as the annual average rainfall of 1,200 mm could vary between 300 and 2,400 mm (Ramakrishnan 1992a). Indeed, many of the Himalayan rural societies have traditional water-harvesting and distribution technologies (Agarwal and Narain 1997), which are unfortunately disappearing under modern influences.

By reviving available water-harvesting systems and capturing rainy-season surface runoff, supplemented by subsurface seepage into rainwater tanks lined with high-density polythene

(Kothyari et al. 1991), a variety of ecosystem restoration/redevelopment efforts could be combined, which elicited enthusiastic community participation. Traditional water-harvesting restoration was inexpensive, as the communities had the infrastructural facilities and supplied the labor freely, while polythene-lined tank construction was US$200.

Participatory Bamboo-based Plantation Forestry

At higher elevations in the Kapkot region of Kumaon Himalaya, in the central Himalayas, a number of early successional bamboo species are important for local communities (Ramakrishnan, Purohit, et al. 1994), including *Thamnocalamus apathiflorus, T. falconeri, T. jaunsarensis,* and *Chimonobambusa falcata* (*ringal*). Due to overexploitation of wild bamboos and lack of knowledge about cultivating them, these species were in short supply, causing conflict among the villagers. In a short period of time, a cooperative project focused on bamboo restoration successfully involved over a hundred local villagers. Both pure and mixed stands of bamboos with other broad-leaved species, occurring on private land and village commons, were considered for social and plantation forestry. Subsequently, this effort was extended to planting bamboos along field edges.

The four, locally available, broad-leaved tree species identified for mixed plantation systems were oak, *Quercus leucotrichophora,* a fodder and fuelwood tree; walnut, *Juglans regia,* with edible dry fruits and of medicinal and natural dye value; horse chestnut, *Aesculus indica,* a fast-growing fuelwood tree that enhances soil fertility; and ash, *Fraxinus micrantha,* used for making agricultural implements and household items. An additional important component for landscape restoration was the introduction of medicinal plants (*Picrorhiza kurrooa, Orchis latifolia, Angelica glauca, Thalictrum foliolosum, Mentha arvensis, Rheum emodi, Aconitum heterophyllum, Swertia chirata,* and *Nardostachys jatamansi*).

Based on intensive interactions with local communities, water was identified as a limiting factor for landscape restoration. As these communities lacked traditional water-harvesting systems, dugout water-harvesting tanks were constructed. As an experiment, about ten hectares of land were "restored" in one year with voluntary local labor. The local communities gained economic benefits (table 16.1), though the longer-term implications of this project require a detailed cost-benefit analysis.

Agriculture-linked Forestry Plantation in Village Common Lands

By participatory appraisal, water was identified as a key resource in short supply at Banswara village in the Chamoli District of the central Himalayas. Most of the villagers (87% of those questioned in 256 households) wished to be involved in the rehabilitation planning, as they had lost confidence in the government-sponsored village council, and, indeed, mass participation is more effective for conflict resolution. By contrast to government-sponsored tree plantations, using barbed-wire fences and costly stone-wall trenches for protection, the villagers opted for cheaper *social fencing* (a community decision to recognize grazing or any other encroachment as an offence), complemented by *biofencing* (use of the alien, fiber-yielding, nonpalatable *Agave americana*). On the whole, community participation ensured successful rehabilitation in a cost-effective way, in spite of fears of government annexation to achieve its 66% target of land set aside for forest as part of the national forest policy.

TABLE 16.1

Ten-year, cost-benefit analysis of plantation in the Kapkot region of Kumaon Himalayas

Item	Total in US$
Costs	
Plantation	1,666.67
Maintenance[1]	2,308.89
Total cost	3,975.56
Benefits	
Fodder	4,066.67
Bamboos[2]	122.22
Medicinal plants[3]	800.00
Total benefit	4,988.89
Cost:benefit ratio	1.25

Source: Ramakrishnan, Purohit, et al. 1994.
Note: Rupee (Rs) values have been converted to US$ at the rate of Rs45 per US$.
[1]Includes expenditure for medicinal plants ($0.03/ha).
[2]The actual cost of the bamboo is not included; values are only the fees that would have been charged by the forest department for collection of bamboos from the wild; with bamboo cultivation, this resource is perceived as a free commodity accessible to the farmers.
[3]Returns are from only 0.03 ha area where medicinal plants were planted.

Ten tree species, chosen by the community as a whole, formed the basis for mixed plantations, and included *Dalbergia sissoo, Ficus glomerata, Grewia oppositifolia, Albizia lebbek,* and *Alnus nepalensis.* These multipurpose tree species contrast with the *Pinus roxburghii* plantations established earlier by the government (through formally elected village councils) and which proved of little direct value to local communities. Availability of dry-season water has improved soil fertility status under the mixed plantations (table 16.2) and increased local economic benefits (table 16.3).

Community Participation in Restoration Through Adaptive Socioecological Management

Traditional societies in the developing tropics view themselves as an integral component of a *cultural landscape* that they create around them, based on the socioecological conditions in which they operate. If the social dimension complexities in a country such as India are superimposed on the physical template, the issues involved become more complex. All land-use systems within a given landscape, both natural and human managed, should be considered in any integrated management plan. Such a multifaceted and participatory management plan obviously requires detailed knowledge of local people, their needs, and how they interact with their environment. Developmental organizations need to be sensitized to these approaches.

Understanding the processes operating within and between social and ecological systems from an agricultural plot and family level up to the landscape and community level is highly complex. This can be achieved only by an interactive, and hence participatory, process of exchanging ecological and social information (figure 16.1). This information provides the

TABLE 16.2

Soil characteristics before and after five years of rehabilitation, Banswara village,
Chamoli District, central Himalayas

Characteristic (%)	Before rehabilitation	After rehabilitation	
		Irrigated system	Unirrigated system
pH	6.40±0.05	6.27±0.05	6.20±0.03
Water-holding capacity	21.49±0.87	37.04±0.65	27.20±0.60
Organic carbon	0.83±0.04	1.57±0.05	1.16±0.05
Total nitrogen	0.03±0.01	0.05±0.01	0.05±0.01

Source: Maikhuri et al. 1997.
Note: Mean±standard deviation.

TABLE 16.3

Monetary inputs and outputs of land rehabilitation in the Banswara village,
Chamoli District, central Himalayas

	Costs or benefits (US$)	
	With water	Without water
Inputs		
Tools and implements	37.78	39.78
Water-harvesting tank (material costs)	175.07	0.00
Labor	1,214.09	1,153.78
Other costs, including transport, insecticides	80.00	33.33
Total	1,506.93	1,226.89
Outputs		
Wood and fodder	353.33	201.11
Agronomic yield	2,471.89	1,152.53
Total	2,825.22	1,353.64
Net return over five years	1,318.29	126.76
Cost:benefit ratio	1.87	1.10

Source: Ramakrisnan, Purohit, et al. 1994.
Note: Rupee (Rs) values have been converted to US$ at the rate of Rs45 per US$.

framework for identifying the critical catalysts and incentives to mobilize community restoration.

Contribution

Traditional ecological knowledge (TEK) links ecological and social processes, catalyzes sustainable management, and provides a value system for natural capital that traditional societies understand and appreciate, including participatory species selection. TEK provides strong incentives for community participation in restoration of natural capital, as demonstrated by two case studies. Through these South Asian case studies, we have shown that socially valued plant species can be equivalent to ecological keystone species within a given ecosystem (Ramakrishnan et al. 1998). Integration with "formal" knowledge-based approaches to forest ecosystem management, such as sylvicultural practices, is still possible (Ramakrishnan et al. 1982; Ramakrishnan 1992b; Shukla and Ramakrishnan 1986). Community identification of water as a critical resource for their livelihoods is reasonable since long

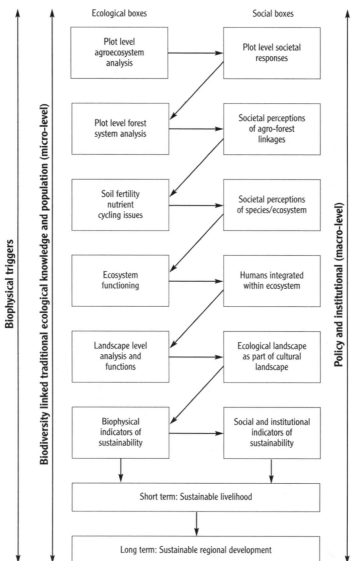

FIGURE 16.1. Integrative socioecological system approach toward sustainable natural resource management of traditional societies.

dry periods typify the tropical Asian monsoon climate (Ramakrishnan, Purohit, et al. 1994; Ramakrishnan, Campbell, et al. 1994).

Institutions play a key role in ensuring effective community participation (Ramakrishnan 2001). For this reason, it is essential for the success of participatory natural capital restoration and other development and conservation initiatives that local institutions be based on cultural traditions from the very beginning.

Chapter 17

Overcoming Obstacles to Restoring Natural Capital: Large-Scale Restoration on the Sacramento River

Suzanne M. Langridge, Mark Buckley, and Karen D. Holl

The floodplain forests and wetlands that are the natural capital of the Sacramento River, the largest river in California, have been lost and damaged by deforestation, river canalization, dam building, and water diversion. Restoration of these riparian ecosystems is a crucial goal for improving important ecosystem services, such as water quality, fisheries, and terrestrial wildlife habitats (Postel and Richter 2003). The Sacramento River, a critical breeding and migratory habitat for wildlife, is essential since only 4% of the original riparian forest remains due to hydrological changes and deforestation for agricultural development. However, the amount and pattern of riparian restoration is shaped by competing uses of the floodplain and river, such as water supply, flood control, and agricultural land, and restoration success is further influenced by stakeholder perceptions and socioeconomic dynamics.

In this chapter, we discuss the Sacramento River restoration project as an example of the biological, physical, and social barriers and bridges that can exist when attempting large-scale riparian restoration (Gore and Shields 1995; Wohl et al. 2005). Large-scale river restoration must take into account the complex social and biophysical interactions that occur between different patches across landscapes. For example, stakeholders bring multiple frames of reference to restoration projects, including reasonable objections to restoration and valuing natural capital (Pahl-Wostl 2006). Although restorationists generally view restoration as having only positive effects, many stakeholders view it as having local negative effects. Whether these negative effects are real or perceived, large-scale restoration projects must incorporate these concerns as part of their research and management programs. We discuss potential methods for restoring the Sacramento River, the conflicts that have arisen due to perceived negative impacts, and methods for resolving these conflicts. We also discuss the lessons learned and how they may be applied to other restoration projects facing social conflicts.

Historical Perspective on the Natural Capital of the Sacramento River

The Sacramento River originates in the Klamath Mountains of northern California (figure 17.1) and is bordered by the Coast Ranges on the west and the Sierra Nevada to the east, both of which supply many tributaries. The river descends from the mountains into the upper Sacramento Valley, where it flows across a broad flat floodplain. Historically, during winter storms, the river regularly spilled over into flood basins, causing inundation up to eight kilo-

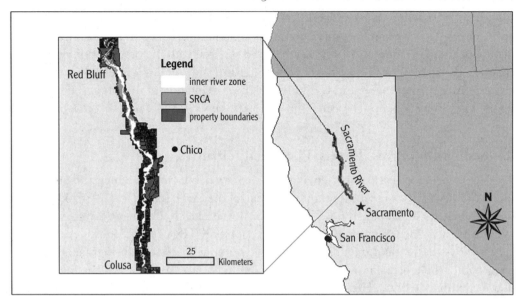

FIGURE 17.1. Orientation Map for the Sacramento River restoration project. Map shows the location of the Sacramento River restoration project within California and the location of the Sacramento River Conservation Area (SCRA) and adjacent properties surrounding the inner river zone of the Sacramento River.

meters on either side. During major flood events, the upper Sacramento River valley filled with sediment-rich waters that formed natural levees where riparian forest flourished. These natural levees also prevented many of the tributary streams from entering the Sacramento River, and instead distributed the water into extensive wetlands that filtered water and stabilized river flows (Thompson 1961).

Before European settlement, the expanse of riparian forest and the heterogeneity of habitats along the Sacramento River led to a diverse biota. The valley was rich in wildlife, including predators such as grizzly bears, game animals such as antelope and tule elk, large populations of migratory birds, and extensive migrations of anadromous fish. Plant communities included expanses of marshes, upland oak forests, and tangles of cottonwood, willow, oak, and other species in widespread riparian forest habitat (Thompson 1961).

Current Pressures and State of the River

In 1848, gold was discovered in the region, resulting in a rapid population increase and forest destruction (Thompson 1961). By 1866, two-thirds of the land around the Sacramento River was under cultivation for orchards, row crops, and perennial pasture. Riparian forest was further removed to power steamboats and to supply domestic and industrial uses in Sacramento and San Francisco. In 1850, a large flood event caused widespread damage to the many valley towns and agricultural developments located on the floodplain. This and other damaging floods led local communities to construct the first human-made levees, altering the natural hydrologic systems (Kelley 1989). In addition, groundwater pumping, river channelization, dam building, and water diversion were developed for flood control and to meet agricultural and urban demand for water (Kelley 1989). By the late 1970s, over 96% of riparian forests had been removed, with fragments surviving in an agricultural matrix under pressure from human populations (Thompson 1961; Katibah 1984). Even with major alteration of the riparian forest and natural hydrology, the Sacramento River still contains some of the most diverse and extensive riparian habitat in California. Hence, restoration of this important river is essential to conserve remaining species and ecosystem services.

Restoration Objectives, Targets, Plans, and Initiatives

In 1986, recognition of the importance of this riparian habitat to threatened and endangered species and ecosystem services led to the passage of the State Senate Bill 1086. This legislation designated the Sacramento River Conservation Area (SRCA), a 160-km floodplain between Red Bluff and Colusa (figure 17.1), for restoration (CRA 2000). Holistic restoration of large river systems should include the hydrogeomorphic processes causing spatial and temporal habitat heterogeneity through erosion and deposition as the river channel migrates (Gore and Shields 1995). Human alterations to the Sacramento River have changed the frequency, magnitude, timing, and duration of these hydrogeomorphic processes, such as the construction of Shasta Dam in the 1940s, which regulates water flow through the restoration project area. Hydrogeomorphic processes have also been altered by bank revetment and levees. These changes have reduced natural river processes, including meandering, channel and bank erosion and sedimentation, river branching, channel cutoff and oxbow formation, which affect the associated vegetation succession, structure, and wildlife use (Buer et al. 1989). However, completely restoring these processes to their natural states is incompatible with current human settlement patterns and appropriation of water for agricultural and household use (Golet et al. 2006).

Given the limitations for restoring large-scale hydrogeomorphic processes, the Sacramento River project is pursuing the following reach-scale strategies: (1) acquiring land from voluntary sellers, particularly flood-prone areas bordering remnant riparian habitats; (2) revegetating those properties with native trees, shrubs, understory plants, and grasses; and (3) restoring some natural river processes (Golet et al. 2006). To date, using federal, state, and private sources, approximately 2,000 hectares have been planted with riparian species by two main nongovernmental organizations, The Nature Conservancy (TNC) and River Partners. The majority of the planting has been with riparian forest species, although more recently some native grasslands have been restored to reestablish natural heterogeneity and minimize flooding (Efseaff et al. 2003).

Given the large scale of the restoration, the costs are substantial. These include land acquisition (approximately $2,500–$10,000/ha) (Hunter et al. 1999) and site preparation, planting, and maintenance (approximately $10,000/ha) (R. Luster, TNC, California, personal communication). Major funding has come from a state-federal cooperative program (Calfed) aimed at restoring the San Francisco Bay–Delta and Tributaries, as well as other state, federal, and private organizations and individuals, such as the Wildlife Conservation Board, U.S. Fish and Wildlife Service, Ducks Unlimited, and private landowners (Golet et al. 2006). Restoration has also been paid for by bond measures, in which California residents vote on whether to borrow money from the state to pay for restoration. However, most of these measures have passed in urban counties, where voters often believe that restoration will benefit them through clean drinking water and watershed protection, yet they have received well below the majority vote in the counties where the restoration is being done.

Resistance to the Restoration of Natural Capital

Landscapes are connected not only ecologically but also socioeconomically. While conservationists highlight the economic and ecological benefits of restoration, many landowners in the Sacramento River region perceive net negative impacts of restoration (Golet et al. 2006). These transboundary influences between adjacent landscape elements can lead to tension and conflict. In several surveys conducted in the SCRA, many members of the farming and larger regional community perceived the restoration of natural habitat as locally providing mostly negative effects (Wolf 2002; Singh 2004; Buckley 2004; Jones 2005). These perceptions of negative externalities have led to efforts by the regional community to stop or reduce restoration. For example, in March 2002, the SCRA Board voted to reduce the SRCA from 86,000 to 32,000 hectares at a meeting where more than one hundred landowners spoke out against the SRCA (Martin 2002). Furthermore, the county and city of Colusa, located within the SCRA, voted in 2006 to enforce more stringent protection for private landowners when approving restoration projects (Hacking 2006).

Farmers' concerns associated with natural habitat restoration include the possibilities of increased numbers of vertebrate pests such as squirrels and deer; endangered species use of their land and consequent critical habitat designation; agricultural weeds; and the general loss of farmland and farm culture (Wolf 2002; Singh 2004; Buckley 2004; Jones 2005). Moreover, many of the farmers have lived for generations along the river and feel strong bonds with the land and their neighbors, while respecting the management strategies of their ancestors. There is also general concern that cheap agricultural imports, suburban sprawl, and large-scale corporate farming imperil the farmers' lifestyle. Such threats are more difficult to influence locally than restoration activities, leaving the latter as a more accessible target for local farming communities. The larger regional community has also voiced concerns regarding security, loss of local tax revenue, and flooding.

Concerns about increased flooding are valid in some situations, as reforestation can slow the movement of water during high-flow events due to increased surface roughness, leading to higher flood levels (Sellin and Beesten 2004). However, the type of habitat and spatial positioning of restored areas can affect the possible flooding effects. In contrast, little research exists to support or refute some of the negative perceptions farmers hold about restoration. For example, weed species of concern to farmers are often found in restored or remnant riparian habitat, as well as in road verges and small patches at the edges of farms, where herbicides are not applied. Although weeds are not planted in the restored riparian habitats, they often colonize these areas (Efseaff et al. 2003; Holl and Crone 2004). In fact, there is extensive research on invasive species movement from agriculture to restored habitat (Fox et al. 1997), yet little research has focused on the opposite direction. Additional fears within the wider community are the potential for increased trespassing and security problems associated with greater recreational use of natural habitat (Jones 2005), which may require more police activity, although these issues are largely unquantified.

Concerns about restoration costs to the larger regional community through revenue loss are controversial. When land is removed from private ownership and agricultural production, corresponding agriculture revenues are lost from the local economy and tax base, reducing available funding for community services such as education and fire protection. However, farmers are often employed to plant trees during restoration activities and thereafter there are potential recreational uses, which may have some local multiplier effects (Adams and Gallo 2001). Nonetheless, compensation for lost local tax revenue by state and federal governments has not been perceived by local landowners to adequately compensate for losses from restoration in the SRCA (Bharvirkar et al. 2003).

Resolving the Conflict Concerning the Restoration of Natural Capital

The distributional differences in expected costs and benefits among stakeholders have made implementation challenging for local restoration organizations. The Sacramento River project, which includes several nongovernmental and governmental agencies, has used several approaches to resolve some of the conflicts, including better communication and coordination, early and consistent local involvement in planning, and research and management to quantify and reduce transboundary impacts.

Communication and Coordination

The Sacramento River Conservation Area Forum (SRCAF) was established in 2000 to bring together communities, individuals, agencies, and organizations within the SRCA to make the restoration efforts more sensitive to the community (SRCAF 2003). The SRCAF is a locally based, nonprofit organization overseen by a board of directors with diverse representation. It considers itself a "voice for all interests," and has incorporated the concerns of local landowners into its guidelines (SRCAF 2003). The SRCAF includes several programs and committees that focus on specific issues to decrease conflict and increase communication. For example, the SRCAF started a Landowners Assurances program in 2001, and through this developed a *Good Neighbor Policy* that was adopted in 2003. Guidelines for the policy state that differences exist between riparian habitat and farming, and that the "challenge is to understand the various land uses to the extent that each can be managed to remove or minimize the negative impact on the others" (SRCAF 2006). The policy also recognizes that in situations where "conflicts and harm are unavoidable, there should be a mechanism established to determine the extent of impacts," as well as resources available to compensate or to find acceptable solutions to the impacts (SRCAF 2006).

Organizations that are conducting restoration along the Sacramento River have incorporated the Good Neighbor Policy into their approaches. For example, TNC proposed to eradicate weeds on their restored properties and to minimize flooding on neighboring lands. In addition, River Partners found success with neighbors through integrating buffers and flood-neutral revegetation (grasslands) planning into their projects. The SRCAF and restoration organizations have developed several specific policy actions including increased dialogue regarding changes in land use, consideration of buffer zones or fences between farms and restored habitat, "safe harbor" and incidental take permits for endangered species, a grievance procedure, and compensation for economic losses due to habitat restoration.

Targeted Research

Data on both negative and positive transboundary effects are critical to resolving stakeholder conflicts. General concerns raised about restoration have led to several surveys of landowners in the Sacramento River watershed between 2001 and 2005 by academic, government, and restoration groups to clarify these issues. These surveys highlighted concerns by farmers about pest species, including mammals, weeds, and arthropods (Wolf 2002; Singh 2004; Buckley 2004; Jones 2005), as well as general stakeholder concerns about flooding.

To address farmer concerns, several research projects have been initiated to quantify transboundary patterns and effects of pest species (S. M. Langridge, University of California, unpublished data; G. H. Golet, TNC, personal communication). However, little is known about potentially important positive transboundary effects such as pollination and pest control (Kremen 2005), which may help to offset negative restoration effects. One study has initiated research on the possible beneficial effects of restored riparian forest providing habitat for birds and arthropods that forage within agricultural areas and reduce pests (S. M. Langridge, University of California, unpublished data). Researchers and restoration project managers are also testing means to lessen negative impacts on farms, such as adding bird or bat boxes to reduce both mammal and insect pest damage.

Research targeting stakeholder concerns can also be incorporated by restoration organizations in their planning procedures. For example, TNC addressed flooding concerns by using hydraulic models to determine the effects of different restoration scenarios on flooding along a proposed restoration project on 13 km of the river. TNC integrated scientific and stakeholder information to determine the appropriate input parameters for the models (Golet et al. 2006). These studies engaged with stakeholders to include their observations and knowledge of the river, leading to cooperation by stakeholders in restoration of this reach of the river (Golet et al. 2006).

Modeling to Determine Strategies for Focusing Investment and Management

Restoration ecologists typically recognize that their projects confront various social constraints, such as government priorities, limited funding, and long-held beliefs (e.g., McIver and Starr 2001). Acknowledging these limitations allows restorationists to achieve a more stable and socially acceptable outcome for specific projects and improves overall project success. A greater degree of regional restoration might be achieved by considering social constraints in a systematic, strategic fashion rather than by pursuing full restoration at every local site. While positive correlations between compromise and stable stakeholder agreements have long been recognized, modeling methods can more precisely identify the specific issues and criteria for voluntary and maintained cooperation. While all parties may not achieve everything they seek, outcomes can be more desirable for all involved than would occur without such agreements and can produce a more optimal overall strategy (Buckley and Haddad 2006).

One approach has been applying modeling techniques from game theory to allow consideration of how decisions by individuals with potentially opposing goals and beliefs (farmers and restorationists) interact to determine outcomes (Buckley and Haddad 2006). Respondents to a mail survey of farmers in the SRCA reported a high likelihood of carrying out defensive investments if a restoration project were to occur adjacent to their property

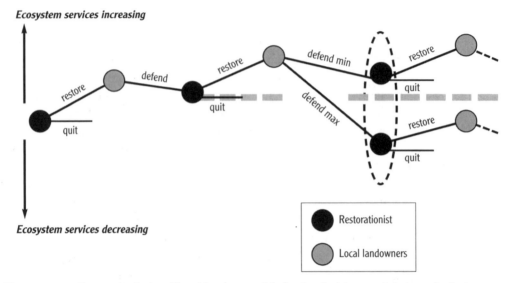

FIGURE 17.2. Restorationists' and local landowners' defensive decisions and their ecological consequences. With decisions by restorationists (black circles) from their indicated decision nodes, local landowners (gray circles) make corresponding decisions to defend against negative impacts. At the second decision node for local landowners, the net effect on ecosystem services can have two different responses (indicated by black dotted oval). Overall ecosystem services could increase if the local landowners have a minimal defensive response. If the local landowners respond with a maximum defensive action, the net effect on ecosystem services could be negative, indicated by the potential for ecosystem services to decrease below the level prior to the last restoration project (indicated by gray dotted line).

(Buckley 2004). Such defensive investments included revetment of riverbanks, removing natural vegetation, fencing, and increasing chemical usage on their property and sometimes on adjacent restoration sites without permission. Defensive investments might also take the form of political activity, such as lobbying, voting, or opposition publicity campaigns, if it is expected to be more effective than on-site activities. Generally, if restoration is perceived to potentially elicit negative social feedbacks, then there might be certain points where restoration elicits responses that offset some or all of the ecological gains (figure 17.2). Identifying these points before they occur is critical to achieving socially desirable outcomes for farmers and restorationists, and spatial positioning, on-site project design, and total landscape-project concentration can all influence the magnitude of response.

Farmers will take defensive actions if they expect that the value of prevented damage to their livelihood and lifestyle is greater than the cost of defense (figure 17.2). The best response for the restorationist is to restore the piece of land if the net ecological gains outweigh the financial and opportunity costs of restoration (figure 17.2). Including social planning and maintenance efforts to remove or reduce negative transboundary effects might increase total restoration costs and reduce on-site ecological gains. However, by preventing negative effects that elicit defensive behavior, total regional ecological-function gains may increase the possibility of reaching the overarching goal (Buckley and Haddad 2006). Directly applying these models would require extensive parameterization and further characterization of the social dynamics and marginal costs and benefits for landowners and restorationists. However, this approach has the potential as a framework for incorporating social impacts and feedbacks when planning large-scale restoration.

Contribution

To restore natural capital in socioecological systems, such as most large river systems, it is necessary to build bridges among stakeholders early in the process. Restoration is generally viewed and approached by practitioners as having only positive benefits, yet there may be negative local effects that are not addressed when social impacts are considered. This can lead to conflicts among stakeholders in the region that are difficult to resolve after restoration and any potentially damaging effects have occurred. As has been demonstrated on the Sacramento River, negative perceptions by landowners can ultimately reduce and hinder restoration efforts.

Some conflicts over the local effects of restoration projects are inevitable. While the ecological flows (such as nutrient transport, water quality maintenance, and species migration) from restoration of natural capital along the Sacramento River are distributed on large scales (with possibly higher local benefits), the costs of negative externalities are locally concentrated. Local stakeholders also have multiple frames of reference in which they perceive restoration, leading to valid, unknown, or invalid objections. As part of a successful restoration program, managers and policymakers must address these objections through mitigation, research, and communication, ideally prior to implementing restoration. Hence, a socially strategic approach, which focuses on determining and preventing negative effects that could elicit defensive behavior, can lead to local acceptance of activities that improve regional ecosystem services and processes.

Chapter 18

An Approach to Quantify the Economic Value of Restoring Natural Capital: A Case from South Africa

James N. Blignaut and Christina E. Loxton

The biodiversity component of natural capital (including all living plants, animals, and microbes) supplies people with an array of environmental goods and services, including food, clean water, atmospheric regulation, and the development and protection of soils, as well as nutrient cycling (Nunes et al. 2003; MA 2005a). Environmental degradation has a negative impact on biodiversity and is therefore likely to reduce the quality, quantity, and variety of goods and services from natural areas.

In South Africa, much of the environmental degradation used to be in former "homelands," that is, reserves for Black African people under the former apartheid regime (DEAT 1997; Hoffman and Ashwell 2001). Degradation occurred because people were forced to live on marginal land with little or no infrastructure and/or means for economic survival, causing overgrazing and unsustainable biomass harvesting for energy and construction purposes (Hassan 2002). Although a stable democracy has replaced the apartheid regime, the majority of people remain poor (earning less than US$1 a day) in these heavily impacted areas (SARPN 2003).

In this chapter we consider the question of whether community conservation, coupled with community involvement in restoring natural capital, could be a feasible alternative to subsistence agriculture, as carried out elsewhere (Barnes et al. 2003; Luckert and Campbell 2003). Here, we present alternative economic scenarios (with and without restoration of natural capital) for an impoverished rural community living outside a South African national park.

Background and Discussion of the Natural Capital

The Bushbuck Ridge (BBR) District in the Limpopo Province, South Africa (31°00′ to 31°35′E; 24°30′ to 25°00′S), comprises 234,761 hectares, including 184,301 hectares of communal land not used for cultivation or habitation but openly available to some 500,000 community members for resource harvesting. Of the communal areas, at least 43% is heavily degraded (CSIR 1996). In 2000 the gross geographic product (GGP) per capita, or alternatively, the average income earned per person in the district, was estimated at R (Rand) 3,400 (US$485) per annum within the context of a 65% unemployment rate and formal employment declining 1.2% annually between 1995 and 2000 (Limpopo Government 2002). Hence, alternatives to alleviate poverty need to be considered.

The BBR communal area and the bordering Rooibos Bushveld zone of the Kruger National Park have the same climate and would have shared the same vegetation and animal life before human impact. However, the natural capital of the park area remains intact, delivering a wide range of ecosystem goods and services, while the communal area is becoming increasingly degraded. To assess whether community conservation in the BBR area is a viable alternative to subsistence requires a comparison between the total economic value of ecosystem goods and services provided by the Rooibos Bushveld area and the value of products extracted from the communal area. Using these data, a potential value for the communal area, but under low-impact conservation, can be calculated. This potential value is based on the premise that subsistence agriculture could be replaced by community conservation, while allowing sustainable resource harvesting. Indeed, community resource harvesting in a protected area is not uncommon and is permitted in World Conservation Union (IUCN) Category VI protected areas that by definition are managed mainly for conservation of species or habitats through management intervention that allows for restricted resource harvesting (e.g., Mulongoy and Chape 2004). In practice, this probably requires fence realignment between the park and the communal area to incorporate part of the latter into a larger conservation zone. This extended conservation area, operated by the community as a private nature reserve, while sharing wildlife with the park, would generate local income.

Results and Discussion in Measuring the Value of the Natural Capital

The results from this comparative analysis between the park and the communal area will be done by focusing on the stock of natural capital valued if all tradable species are liquidated and sold off, and natural capital's ecosystem goods and service (or function) values using a range of valuation methods.

Stock of Natural Capital

Because little game exists on the communal land and no livestock survey has been undertaken, the value of animals could not be calculated.

For the adjacent area of the park, densities of the main tradable mammal species were obtained from Zambatis and Zambatis (1997), with numbers adjusted to 2002/03 levels (SANParks 2003) and weighted to reflect the relatively high animal density in the Rooibos Bushveld area. Based on 2003 auction prices (differentiating between trophy animals and breeding herds), the market value of the tradable mammal stock was estimated at $25.37 million or $155.74/ha.

A list of tradable plant species was assembled from various sources (Shackleton and Shackleton 1997, 2000; Botha et al. 2001; Hassan 2002; Van Zyl 2003). Based on Shackleton and Scholes (2000), Netshiluvhi and Scholes (2001), and Scholes et al. (2001), the biomass per plant species and per hectare and for the whole Rooibos Bushveld area was calculated, with the biomass percentage of each useful plant species specified. The estimated value of tradable plant species, should they be harvested completely, was $481.3 million or $2,954.70/ha (based on 2003 market prices). Though this hypothetical amount is considerable, it only accounts for the value of the standing biomass traded and does not incorporate nontraded species. Since 43% of the communal area was considered degraded, its tradable plant stock value was taken as 57% of the park value per unit area.

Function Values of Natural Capital

Function values distinguished here are direct-use values, nonconsumptive values, and indirect consumptive-use values. Direct-use values often refer to ecosystem goods, whereas nonconsumptive and indirect consumptive-use values refer to ecosystem services.

COMPARING DIRECT-USE VALUES

The direct use of plants for timber, fuelwood, medicines, and livestock is very important to the local people in the BBR communal area. If operated as a community conservation area, with livestock removed, controlled game hunting could be exploited very profitably, while many of these other practices could be sustainably managed (table 18.1).

However, the park, according to the IUCN's classification, is a Category 2 national park, which excludes natural resource utilization, and hence the direct-use value for Rooibos Bushveld area is zero. Despite this, the potential value of harvestable goods, if sustainable resource use was allowed, can easily be ascertained by examining the actual and potential direct-use values from the slightly larger BBR communal area (table 18.1).

BUSHBUCK RIDGE COMMUNAL AREA: ACTUAL AND POTENTIAL DIRECT-USE VALUES

Various studies, based on primary household survey data, have been carried out to calculate the actual value of resource harvesting in the Bushbuck Ridge communal area (Shackleton 1998; Shackleton and Shackleton 1997, 2000, 2002; Shackleton and Scholes 2000; Netshiluvhi and Scholes 2001; Scholes et al. 2001; Botha et al. 2001; Hassan 2002; Van Zyl 2003). Household heads were asked about which products they were harvesting, their harvest rates, and the market prices, should these products be bought rather than harvested. Subsequently, the combined data from these studies were adjusted to 2002/03 levels using the consumer price index (table 18.1).

The direct consumptive-use value is estimated to be $220/ha ($40.63 million for the whole study area), which implies $81.26 per person based on a beneficiary population of 500,000 (Hassan 2002). The major contributors to value from resource harvesting are the sales of livestock, edible fruit, herbs, and vegetables, as well as thatch and fuelwood.

If the communal area was incorporated into the park and managed as an IUCN Category 6 protected area, sustainable natural resource use, mainly to support local livelihoods, would be allowed under strict guidelines. Shackleton and Shackleton (1997, 2000) argue that the biomass production of the area under consideration is 3% per annum, though it is not all suitable for economic use. The sustainable harvest was conservatively assumed to be 1% of the biomass for fuelwood, construction timber, and branches, and, predicting more limited market options, only 0.5% for crafts and medicinal products. Interestingly, edible fruit harvesting comprises 50% of the full annual production. To calculate the tradable biomass volume that

TABLE 18.1

Comparison of direct-use values for the Rooibos Bushveld area of Kruger National Park and communally owned land (BBR)

	Rooibos Bushveld			BBR (Actual)			BBR (Potential)		Difference (potential less actual)
	ha	$ millions	$/ha	ha	$ millions	$/ha	$ millions	$/ha	$/ha
Fuelwood	162 904	0	0	184 301	5.76	31.24	3.50	18.96	−12.28
Timber	162 904	0	0	184 301	2.70	14.65	4.41	24.01	9.36
Crafts	162 904	0	0	184 301	0.25	1.34	51.22	278.22	276.89
Medicinal	162 904	0	0	184 301	4.78	25.92	47.11	255.38	229.46
Edible fruit, herbs, and vegetables	162 904	0	0	184 301	9.28	50.36	1.51	8.19	−42.17
Thatch	162 904	0	0	184 301	7.01	38.02	0.61	3.19	−34.82
Livestock	162 904	0	0	184 301	9.38	50.88	0.00	0.00	−50.88

	Rooibos Bushveld			BBR (Actual)			BBR (Potential)		Difference (potential less actual)
	ha	$ millions	$/ha	ha	$ millions	$/ha	$ millions	$/ha	$/ha
Wild animals	162 904	0	0	184 301	0.00	0.00	4.3	23.4	23.4
Other: reeds, sticks, grass, brushes, birds, etc.	162 904	0	0	184 301	1.49	8.08	0.00	0.00	−8.08
Total direct consumptive use	162 904	0	0	184 301	40.63	220.48	112.6	611.35	390.88

Sources: Adapted from Shackleton and Shackleton 1997, 2000; Scholes et al. 2001; Netshiluvhi and Scholes 2001; Hassan 2002; and Van Zyl 2003.
Note: $ = US$.
BBR: *actual values* under subsistence management; *potential values* following restoration of natural capital, as valued in 2002/03.

can be harvested, the biomass per species and by product was multiplied by either 1% or 0.5% (or the production volume) and the market price.

Based on these assumptions, the potential direct use value is $611.35/ha, much of which is derived from crafts and medicinal products having a value-added component. The total size of the market is unclear, and though it would be possible to generate the returns indicated (table 18.1), achieving these values over the whole study area is questionable because of market saturation. No livestock value has been estimated since domesticated herbivores would be excluded from the area and replaced by game. Trade in game (including hunting) would be restricted to 50% of the total new births per species per year, due to reduction through predation and other natural causes and to allow for replacement.

Nonconsumptive Values

These values comprise those direct-use values that are nonextractive in physical terms, with tourism providing a useful example. Currently, tourism within the communal area is zero, and to calculate its potential value, this was extrapolated from the adjacent area of the park, as it is assumed that tourism in the restored communal area is likely to be equivalent to that of the protected area.

Although the Rooibos Bushveld area comprises only 8% of the Kruger National Park, it contains 24% of the park's tourist accommodation facilities, and calculation of the total tourism value for the area (table 18.2) is based on this proportion (SANParks 2003). The number of visitors to the Rooibos Bushveld area was calculated as 254,189 per year, and the total number of bed nights is estimated to be 213,207 per year. Indeed, the total turnover value of visitors to this part of the park, inclusive of gate fees, overnight accommodations, and expenditures at park stores is $8.54 million, or approximately $70 per visit. Furthermore, the total cost of travel, which is an acceptable method to determine visitors' willingness-to-pay for the unaccounted amenities for a recreation site (Dixon et al. 1994), is $7.46 million; this implies a total tourism value of $16 million, or $98/ha.

Indirect Consumption Values

These values comprise (1) environmentally produced goods and services useful to people (including livestock grazing, soil nutrient recycling, honey production, and carbon sequestration); and (2) option, existence, and bequest values, which capture the possible future use of environmental goods and services from ecosystems. It was considered inappropriate to include livestock grazing, since the value of livestock sales is included under direct consumptive-use values in the current communal areas, whereas these activities would be excluded after restoration. No values for soil nutrient recycling were found. There are currently no formal honey production activities in either the park or communal area, but based on an average of 20 kg/hive (Turpie et al. 2003) and one hive/5 km^2 (R. Crewe, University of Pretoria, 2003, personal communication), with an average price of $4.56/kg, the potential retail value of honey production is estimated to be $0.33 million or $0.18/ha.

No formal market for carbon currently exists in South Africa. Carbon trading in the park would also not be feasible given the additionality principle, which implies that the existing biomass does not count since it does not contribute to additional carbon storage. The communal area, however, has a good carbon-trading potential. Based on a carbon absorption capacity of 4 t/ha (Scholes and Van der Merwe 1996; Scholes and Bailey 1996), and an average price for carbon of $15.7/ton or $4.2/ton CO_2, the potential carbon sequestration value amounts to $12.31 million or $66.87/ha.

Option, existence, and bequest values are estimated simultaneously since distinguishing between them is seldom possible. There has been no comprehensive study estimating the willingness-to-pay for conservation, either by contingent valuation or conjoint analysis, in South Africa. Results of two regional studies by Turpie (2003) and Turpie and Joubert (2001) indicate an average value of $60.83 ha^{-1}, which is the value used in this study (table 18.3).

TABLE 18.2

Value of tourism for the Rooibos Bushveld area of the Kruger National Park, 2002/03.

Overnight visitors	Day visitors	Foreign visitors	SA overnight visitors	Total number of visitors	Total number of bed nights	Turnover (gate fees, shops, and accomm.) ($ million)	Average expenditures / person ($)	Average expenditures / night ($)	Average expenditures / visit ($)
121 377	132 812	67 345	54 032	254 189	213 208	8.54	33.59	40.12	70.36

Total vehicle km traveled (millions)	Cost of travel / unit, ($ / km)	Travel cost ($ million)	Average travel cost / visitor ($)	Total willingness-to-pay ($ million)	Tourism modes			Total average value / person ($)	$/ha
					Passive ($ million)	Adventure ($ million)	Ecotourism ($ million)		
28	0.27	7.46	29.33	16	13.54	0.38	2.08	62.92	98.18

Source: Calculations based on SANParks (2003).
Note: $ = US$.

TABLE 18.3

Comparison of the economic value of Rooibos Bushveld area of Kruger National Park with communally owned land (BBR)

Value of the standing stock at prevailing market prices

	Rooibos Bushveld			BBR Subsistence (Actual)			BBR Restored (Potential)		BBR diff ($/ha) Potential less actual
	Size of area	Total value ($ million)	Value ($/ha)	Size of area	Total value ($ million)	Value ($/ha)	Total value ($ million)	Value ($/ha)	
Mammals	162 904	25.38	155.74	184 301	n/a	n/a	28.72	155.74	n/a
Vegetation	162 904	483.43	2967.98	184 301	311.70	1691.49	546.96	2967.98	1365.50
Total value	162 904	508.81	3123.72	184 301	311.70	1691.49	575.68	3123.72	n/a
			Biodiversity function or flow values						
Direct consumptive	162 904	0	0	184 301	40.58	220.48	112.6	611.35	390.88
Direct nonconsumptive: Tourism	162 904	15.96	98.25	184 301	0	0	18.09	98.25	98.25
Total indirect consumptive use	162 904	20.85	127.7	184 301	0	0	23.89	127.88	127.88
Indirect consumptive									
Honey production	162 904	0	0	184 301	0	0	0.33	0.18	0.18
Carbon sequestration	152 904	10.94	66.87	184 301	0	0	12.31	66.87	66.87
Option and existence values	162 904	9.91	60.83	184 301	0	0	11.25	60.83	60.83
Function: grand total	162 904	36.81	225.95	184 301	40.58	220.48	154.58	837.48	617.01
Function: total of alternative scenario	162 904	36.81	225.95	184 301	40.58	220.48	90.78	491.32	270.72

Source: Authors' analysis, drawn from tables 18.1 and 18.2.
Note: $ = US$.
BBR: *actual values* under subsistence management; *potential values* following restoration of natural capital in 2002/03.

Summary

Though it was not possible to establish an actual value for the animal stock in the Bushbuck Ridge communal area, the value of tradable vegetation is considerably below its potential (table 18.3). The actual extraction value of biodiversity function-related activities is $220.48/ha, whereas the potential value is $837.48/ha.

An alternative conservative scenario in which the estimated worth of crafts, medicines, and tourism, and the less tangible option and existence values, have been reduced by 50% yields an economic return of $491.32/ha. This is $270.72/ha more than the actual current return.

Contribution

The potential total economic value of the BBR communal area is considerably higher than that of the actual land use value. This is based on the premise that the area could be incorporated within the Kruger National Park, though with unchanged land tenure and allowing selective resource-use access. The actual returns from land use practices are estimated at $220/ha, a portion of which would be benefits in kind. However, the potential total economic value of community conservation has been estimated conservatively at $837.5/ha and $491.32/ha. Therefore, the increased value to be gained from restoring degraded land appears to be considerable.

Unfortunately, there are five problems, any one of which has the potential to spoil the viability of the proposed scheme. First, the total economic value does not imply "money in the pocket." It would be necessary to introduce a national system that rewards rural communities for providing ecosystem goods and services by creating a market for them. The second potential pitfall is that market penetration for either direct or indirect consumptive-use products might be low. The third problem relates to management structure (see also Olukoye et al. 2003). Though it is foreseen that the protected area will be managed by a professional service provider and the proceeds (after cost) centralized into a community-conservation fund before local distribution, this arrangement would have to be negotiated and allow for community buy-in, which can be a complicated process. A fourth hurdle that would have to be overcome is that of insurance risk, uncertainty, and the resultant costs. Finally, restoration costs need to be calculated. This was not possible, except by carrying out the management and restoration plan under close monitoring. Nevertheless, it is expected that the long-term, wide-ranging benefits obtained after restoration will easily justify economic expenditures.

Despite these challenges, the opportunities for community-based nature conservation are ample and plausible given an appropriate institutional structure and the will to implement such a strategy. The economic scenarios presented in this chapter can be used as a basis for collaborating with local communities, government institutions, and nongovernmental organizations to develop better futures for impoverished rural communities living in proximity to protected areas.

Acknowledgements

The authors gratefully acknowledge the assistance of Nolan Loxton and the information and perspectives provided by Professor R. Crewe of the University of Pretoria.

Chapter 19

Capturing the Economic Benefits from Restoring Natural Capital in Transformed Tropical Forests

Kirsten Schuyt, Stephanie Mansourian, Gabriella Roscher, and Gérard Rambeloarisoa

According to the World Wide Fund for Nature (WWF) Living Planet Index, the tropical forest species index declined by 25% in the last thirty years, primarily due to land transformation. Typically, agriculture initially replaces forest and later remains the preferred alternative to forest restoration on already degraded land. It is estimated that between 1975 and 2000 approximately 370 million hectares of natural forests were deforested in the tropics, mostly for commercial and small-scale agricultural expansion (Kessler and Wakker 2000). Indeed, most of the world's oil palm, soya, and cotton plantations occur within these converted areas.

Cash crops may provide a quick economic return, yet their long-term environmental impact is rarely taken into account in balancing the costs and benefits of different land uses. Forest cover loss can lead to a reduction in soil quality, lowering of water tables, and increases in erosion and land-surface temperatures. In addition, often within a few years of intensive land use, cash crop productivity may markedly decline and investment in fertilizers is necessary, thus raising the overall production costs (Bickel and Dros 2003). According to the International Food Policy Research Institute (IFPRI) nearly 40% of the world's agricultural land is seriously degraded, which could undermine the long-term productive capacity of soils (IFPRI 2000).

The goods and services that flow from natural forest have a significant social, cultural, and economic value, sustaining livelihoods and contributing to biological and genetic diversity. For example, the World Bank (2004) has estimated that 350 million people living in or close to forests are reliant on them for subsistence or income. In addition, people also depend on forests for less tangible benefits, such as carbon sequestration, flood protection, and erosion control. Therefore, sustainable management and restoration of the natural capital of tropical forests is essential to maintain the livelihoods and quality of life for millions of people, in both the developing and the developed world.

This chapter argues that restoration of forest natural capital is possible within a broad landscape context. We use four case studies from around the world to show that the restoration of natural capital can be funded and promoted through marketing and holistic accounting methods, and through engagement of the forest-product sector in better practices and standards that include restoration.

Long-term Benefits of Natural Capital in Forests

In the short term, forest transformation often appears to be profitable. Decisions to plant cash crops are influenced by available markets, cheap (forest) land, and subsidies or other financial aid, while restoring or conserving forest cover is frequently not perceived to have economic benefits.

With increasing human pressure, restoration and protection of forests are vital to reduce increasing fragmentation and to maintain forest functions (see also chapter 8). The value of a protected area (or well-managed forest patch) can be greatly reduced if there is no remaining forest around it. For example, in China over 50% of the panda population does not remain within the protected-area boundaries but roams outside into forests that are severely fragmented (WWF/China 2005). Consequently, designating and managing protected areas alone will not be sufficient to conserve pandas. Moreover, if the only remaining trees occur in protected areas, people with no alternative fuel sources will seek to exploit these areas as well.

The reason that forests and many of the goods and services they provide are not recognized as having economic value is that their consumption and production often fall outside the marketplace. It is only with the decline in the quality and quantity of forests that there is growing awareness of the economic consequences of forest loss. For example, downstream stakeholders such as hydroelectric and water purification companies are directly impacted by severe flooding or erosion caused by upstream deforestation. Such costs are, however, often recognized only in the long term, when restoring the natural capital of damaged forests is either impossible or very costly.

The key lies in making conservation and restoration of forests pay so that decisionmakers recognize this as a serious alternative to other land uses. Tools that include economic valuation of forest goods and services are increasingly applied to better understand and highlight the economic value of forests, while the development of payment mechanisms for environmental services continue to grow in importance as a way of making sustainable forest management financially attractive. For example, in 1997, $14 million was invested in the payment for environmental services in Costa Rica, which resulted in the reforestation of 6,500 hectares, the sustainable management of 10,000 hectares of natural forests, and the preservation of 79,000 hectares of natural forests, all of which are privately owned (Pagiola et al. 2002).

Forest restoration is usually on a small scale and is rarely perceived as economically appealing to a land user or society. Therefore, a broader natural capital restoration approach is required, which includes multipurpose forests in a landscape mosaic.

> The concept of multi-functionality is more than just a fine-tuning of existing approaches to land use planning. If forests are distributed optimally in the landscape, and if the different elements of the landscape mosaic complement one another, then the total area of forest needed to provide a given yield of forest benefits is less. (Sayer et al. 2003)

For the restoration of forest goods and services to be applied and widely adopted, an approach both recognizing multiple values and allowing for economic incentives is needed that targets multiple land users.

Making the Restoration of Natural Capital in Forests Economically Attractive

Forest landscape restoration (FLR) seeks to restore the goods and services that forested land-scapes provide to both people and biodiversity and is formally defined as "a planned process that aims to regain ecological integrity and enhance human well-being in deforested or de-graded landscapes" (WWF/IUCN 2003). A landscape in this context is an area that is physi-cally and socially heterogeneous, with an overall quality more complex than the sum of its parts. WWF and the IUCN (The World Conservation Union) are promoting FLR under their joint forest strategy since FLR considers the landscape scale that offers an optimal bal-ance of land uses and helps to negotiate tradeoffs.

Implementing forest restoration should take account of the social, economic, and biolog-ical context within a landscape. It does not mean planting trees across an entire landscape but implementing strategic restoration necessary to achieve an agreed-upon set of functions, such as suitable wildlife habitat, soil stabilization, or the provision of building materials for local communities. In this way, FLR has both a socioeconomic and an ecological dimension, with local people as the stakeholders engaged in improving the state of their landscape. Res-toration can sometimes be achieved simply by removing whatever caused forest loss, such as perverse incentives and overgrazing. However, unless the causes are clearly identified and re-moved, any restoration effort will be in vain.

FLR opts for a package of solutions, as there is no single restoration technique; each ap-proach must be determined by the local conditions. The FLR package may not only include practical techniques, such as agroforestry, enrichment planting, and natural regeneration at a landscape scale, but also embraces policy analysis, training, and research. It will involve a range of stakeholders in planning and decision making to achieve a solution that is accept-able and more likely to be sustainable. Setting the long-term restoration target should in-clude representatives of different interest groups in the landscape. If a consensus cannot be reached (as can often be the case), interest groups will need to negotiate and agree on what may seem like a less-than-optimal solution if taken from a single stakeholder perspective; that is, it may be necessary to make tradeoffs.

FLR places the emphasis on both quantity and quality of forest. All too often forest quan-tity is what decisionmakers think about when considering restoration, yet improving forest quality can yield greater conservation benefits for a lower cost. Because FLR aims to restore a range of forest goods, services, and processes, it is not just the trees themselves that are im-portant, but all attributes of healthy forests, such as nutrient recycling; soil stabilization; plant products, including medicines; and species habitat. The focus on these functions helps to di-rect the restoration response (techniques, location, species, etc.), as well as allowing for more flexibility in tradeoff discussions with stakeholders by providing a diversity of values.

FLR therefore targets multiple land users and recognizes the economic value of forests in addition to their sociocultural and ecological values. Hence, it provides the optimal approach to make forest restoration economically attractive by (1) recognizing the economic value of forests (beyond timber), and (2) allowing for the use of economic incentives in restoration ac-tivities. Both are necessary if restoration is to be perceived as a viable option for landowners, land managers, and society as a whole. Recognition of the economic values of forests lets land users incorporate these into their decision-making processes.

Economic values of forests are derived from the numerous functions delivered in terms of provisioning, regulating, and providing cultural services to people (UNEP and IISD 2004).

In India, for example, the value of forests in regulating river flow is estimated at US$72 billion/yr, while nontimber forest products generate approximately $4.3 billion/yr. Once the economic benefits of forests are recognized and taken into account in decision-making processes, sustainable forest management may become an economically viable option. In West Africa, on Mount Cameroon, a study (Yaron 2001) has shown that the total economic value of sustainable forest management amounts to $2,570/ha/yr, as opposed to $1,084/ha/yr for conversion to oil palm or $2,114/ha/yr from small-scale agriculture.

Recognizing the economic value of forests facilitates the application of economic incentives for forest conservation, including restoration. One such mechanism is *payments for environmental services* (PES), which has the basic principle of rewarding those who provide environmental services, such as carbon sequestration, watershed protection, landscape beauty, and biodiversity conservation. Four case studies show how different economic incentives have been used to promote forest restoration.

Case Study 1: Economic Incentives for Forest Restoration in Malaysia

WWF/Malaysia and the Sabah Wildlife Department have been collaborating since 1998 to establish the Kinabatangan Wildlife Sanctuary in the floodplain of the Kinabatangan River Basin. This area contains forest fragments rich in wildlife, including orangutans, elephants, and proboscis monkeys, but much of it has disappeared due to agricultural impact, particularly oil palm cultivation. As a result, animal movements, such as those by elephants, are problematic since the elephants must pass through oil palm plantations to reach the forest remnants.

The philosophy of this program is built on the identification of options that create linkages between conservation and development. Commercial partners, who are aware of the potential benefits yet are prepared to collaborate and make the necessary investments to satisfy the different stakeholders, have subsequently been sought. Based on an action plan for establishing a connected forest landscape along the lower Kinabatangan, several parcels of land have been identified as critical corridors. In collaboration with oil palm companies, through memoranda of understanding, various tree-planting schemes are being implemented on privately owned land. Indeed, some companies have taken the initiative to restore forests in flood-prone areas to reduce the negative impact of flooding on their estates. This action is driven not only by the need to protect their plantations but also by the potential and real marketing motive at a time when buyers are becoming more selective. The government-owned company Sawit Kinabalu Berhad has established a tree nursery and set aside more than one thousand hectares of land for forest restoration, having failed to cultivate it owing to floods. This area is an important link for wildlife and is large enough to demonstrate better than small fragments could the economic benefits of forest restoration.

Another example of collaborative restoration on a smaller scale is a reforestation program by the company Borneo Eco Tours that enables tourists to participate in tree-planting activities along the Lower Kinabatangan River. As of 2005, sixty-four hectares of riverine forest reserve area have been adopted for this purpose. Unfortunately, few of the planted trees at that time survived due to compacted soil conditions and the destruction caused by elephants. While these efforts are praiseworthy, larger areas of restored natural forests are needed so that the total wildlife habitat available in Kinabatangan increases. Asian elephants for instance are known to range over 200–400 km². Though small-scale restoration efforts may help other species, the long-term sustainability of the Kinabatangan elephants is not guaranteed. WWF, in cooperation with the Sabah Foundation, is considering how the critical elephant sites can remain connected by natural forest corridors through land-use planning interventions within the mosaic of oil palm plantations and forest patches. Information on elephant movements, based on radio tracking, should make it possible to establish oil palm plantations in such a way that conflict is reduced and vital corridors are not blocked.

Simultaneously, economic incentives for catalyzing these types of environmental initiatives are also being provided on the demand side. Consumer awareness in Europe is growing with regard to the adverse impact of oil palm expansion at the expense of natural forests. As a result, retailers are responding by requesting more sustainably produced palm oil from their suppliers. An example is the Swiss supermarket chain Migros, which has developed its own criteria for sustainable palm oil. The company pays a higher price for this product but does not pass on this extra cost to its customers. Instead, the company appreciates the public relations benefits and uses sustainable palm oil as a "green" marketing argument. Economic incentives on both the demand and the supply sides provide a solid basis for increasing sustainable forest management, including substantial forest restoration activities.

Case Study 2: Economic Incentives for Forest Restoration in Mexico

Starbucks Coffee Company and Conservation International (CI) collaborate on this program to promote the sustainable production of coffee in the endangered cloud forest of Chiapas, Mexico, which includes the El Triunfo Biosphere Reserve, considered extremely important to the conservation of global biodiversity. The program helps conserve traditional coffee farms and provides ecological benefits to the reserve by supporting coffee production under the protection of a shade-tree canopy, which creates and maintains a forested buffer zone.

In this region, CI works with cooperatives and producer organizations representing hundreds of families for whom coffee contributes most of the annual household income. CI provides farmers with technical assistance in the growing, processing, and marketing of high-quality coffee. An issue arising, while defining best practices with the stakeholders, was the possible expansion of plantations as a consequence of better prices. *Conservation coffee best practices* are socially and environmentally sustainable practices that reward farmers economically and benefit the biodiversity that surrounds their farms. As a best practice, the coffee farmers are encouraged to increase the plant density per hectare in conjunction with restoring lands degraded due to other crop production, such as maize. Participating farmers also have to maintain any forest they own when entering the conservation coffee program.

In 2005, conservation coffee practices were well established in the communities of Puerto Rico and Colombia participating in the program. In 2004, participating cooperatives received and repaid more than $330,000. This was made possible through the establishment of a financing mechanism in Chiapas called *Eterno Verde*, which provides credit to farmers to finance their crop. CI's own investment fund, Verde Ventures, and Ecologic Finance, a credit fund based in Cambridge, Massachusetts, provided the loan, which was 100% repaid each year over a period of three years under Eterno Verde. Farmers producing shade-grown coffee received a 44% price premium over local prices for their product. In 2005, there were 694 farmers and almost 2,200 hectares involved in the program. The combination of credit access to farmers adopting coffee conservation practices and a premium-paying market provides powerful incentives.

Starbucks is promoting shade-grown coffee in its stores. As a marketing tool, the packaging bears such comments as "Starbucks and CI have made a difference in farmers' lives with the sale of this exceptional coffee" and "By paying a premium price for this shade grown coffee, Starbucks improves the well-being of coffee farmers and encourages them to preserve the forest environment." Furthermore, the company has developed an interactive online experience entitled "On Good Grounds" that brings the region of Chiapas to life. Internet users everywhere are able to watch, listen, and learn about the people and animals living in this unique protected area. In October 2003, Starbucks committed an additional $1.5 million over three years to support the replication of the project in Costa Rica, Panama, Peru, and Colombia. This partnership proves that a leader in a commodity industry such as coffee can integrate conservation and restoration of natural capital into its business, creating a net benefit for the environment.

Case Study 3: Economic Incentives for Forest Restoration in China

In 1998 and 1999, China experienced serious river flooding resulting in thousands of deaths, probably caused by a combination of climate change and insufficient forest cover in the upper watersheds. Forest-cover loss meant that the functions of water and soil retention performed by these forests were significantly reduced, which in conjunction with excessive rainfall produced major flooding and landslides. Subsequently the government decreed a logging ban and drew up a legal framework to support reforestation. To encourage farmers to engage in restoration activities, the government, under the *Grain for Green* program, donates tree seeds and seedlings, as well as between 1,500 and 2,250 kg of grain for every hectare of forest planted, and pays $37.50 per year for each hectare returned to the forest. Indeed, it was estimated that, in 2000, over two thousand hectares were reforested.

However, the Grain for Green program has shown mixed results (Perrin 2003). Positive effects were accelerated afforestation and reforestation and natural forest protection. Furthermore, households were provided with the opportunity to diversify out of agriculture into other income-generating activities, and in some cases soil erosion decreased. Negative side effects include a focus on a few marketable species (particularly those providing fruits) for afforestation and reforestation, which has resulted in saturated markets for these products, as well as the creation of relatively homogeneous tree cover rather than a diverse forest. In addition, the program may have created a culture of dependence on government handouts.

The important lesson from this case study is that perhaps the will of a government or decisionmaker to achieve restoration, coupled with an incentive program, can bring back forests on degraded land in the short term. Hence, some basic ecosystem functions can be reinstated, such as increasing soil and water retention, but this is not the same as ensuring long-term restoration of natural vegetation or providing sustainable livelihoods for local people.

Case Study 4: Economic Incentives for Forest Restoration in Madagascar

Between 2000 and 2005, Madagascar lost four million hectares of forest due to slash-and-burn crop cultivation, one of the highest deforestation rates in the world (FAO 2005). This ancient practice is an important cause of poverty in the country and creates a vicious cycle

(Programme Dette Nature 2003). Forest loss to plant rice (the staple food) causes soil degradation and sedimentation on the rice fields, necessitating further deforestation.

In northeastern Madagascar, the expansion of vanilla plantations was one of the driving forces of forest loss. Vanilla is the most important export crop of Madagascar, and its plantations attract an increasing number of farmers. Since 1996, the Malagasy government has fought against the extension of vanilla plantations into the forests.

In traditional vanilla plantations, forests are generally cleared except for a small number of large trees that are used to shade vanilla plants. Using this method, 4,000 to 5,000 vanilla vines are planted per hectare, with each vine producing three hundred grams of vanilla per year. In fact, this destructive approach is also temporally shortsighted, with each plantation lasting only six to eight years before the producers need to move to another forest parcel (Andriatahina 2003). With the support of the European Union and a Malagasy development research center, the Malagasy government has stimulated the application of a new technique of vanilla production outside forests. This "semi-intensive" technique, applied in nonforested areas, uses the planted tree species *Gliricidia maculata* as shade for vanilla plants. The result is a rise in vanilla production with up to 1,500 grams per vine, and although only eight hundred vanilla vines are planted per hectare, these plantations can be sustained for more than twenty years if well managed, thereby reducing the pressure to clear natural forests (Union Européenne 1998).

The economic benefits of this new system of vanilla production are substantial for protecting primary forests and promoting the livelihoods of local communities. Furthermore, with farmers having productive plantations closer to their homes, they suffer fewer logistical problems and attain greater security from vanilla thieves.

Contribution

In the tropics, large forest areas have been converted to cash crops, such as soya, sugar, cotton, or oil palm, driven by anticipated short-term financial returns. However, after conversion, a range of problems appear related to the loss of forest-related goods and services, such as lack of timber and nontimber forest products, increased landslide vulnerability, negative soil-structure changes, habitat disappearance, and reduced water quality. In the long run, these constitute major costs to society. However, the short-term financial arguments for cash crops versus the long-term, less economically tangible benefits of forest-cover restoration make it difficult to reverse the status quo of increasing land degradation without clear economic incentives to counter this trend.

The key lies in recognizing and capturing the economic value of forests, thereby making landscape restoration more financially competitive with alternative land uses. Restoration of forests can be possible within a broader landscape context using strategies such as marketing, quantification of real benefits from plantations and small-scale farming, and engagement of an entire sector in better practices and standards that include restoration. The common denominator in all of these strategies is the utilization and better representation of economic incentives for restoration in decision-making processes on land use.

The integration of economic incentives and restoration requires an approach that recognizes the economic value of forests and links conservation and development, such as in forest landscape restoration. Economic arguments and incentives should make restoration attractive. The case studies presented here, from Malaysia, Mexico, China, and Madagascar, illustrate the various levels of success achieved to date but also represent pilot studies for making restoration economically competitive with cash crops. The next step is large-scale recognition of the economic value of forest goods and services, by both developed and developing countries, and the implementation of economic incentives that will allow forest restoration activities to compete with other land uses. More generally, the issue of better representation of the economic arguments for restoration options in decision-making processes needs to be addressed.

Chapter 20

Restoring Natural Forests to Make Medicinal Bark Harvesting Sustainable in South Africa

Coert J. Geldenhuys

Africa's forest resources provide many useful products and services, including timber, construction and fencing poles, fuelwood, traditional medicines, foods, craft wood and fibers, household goods, and implements. In South Africa, for example, the forests include 568 woody plant species, of which 365 species (64%) are economically used in one or more ways (Geldenhuys 1999a). In addition, South African forests have many indirect values (McKenzie 1988). However, little information is available on the relative use-value of the various species or the impact of harvesting natural resources. Recognizing the values of forest products and the effects of their extraction on forests could play a major role in reducing conflict in land use options.

This chapter focuses on just one form of natural capital in South Africa's forests, namely, tree bark used for traditional medicine, and the implications of its current extraction rates and methods for future yields. The shift from subsistence use to commercial trade of medicinal plants has led to increased harvesting from wild habitats (Mander 1998). Although the bark of many species is used, only a few are in high demand commercially.

This is a first attempt to quantify the capital and flows of bark harvesting in South Africa's limited, natural, evergreen forests. Overexploitation of this resource is identifiable by its impact on the natural environment, society, and economy. However, restoration is achievable by simulating the natural disturbance-recovery processes and applying adaptive research and management in collaboration with resource users.

Rationale and Concepts for Utilizing and Restoring Natural Capital

The practices of utilizing and restoring natural capital require understanding the flows of the capital during natural disturbance and recovery processes, and during resource use, for successful implementation in sustainable rural development.

Natural Capital and Its Flows in Natural Forest

The natural capital of a forest comprises the sum of the physical (substrate and atmospheric environment) and biotic components (plants and animals). However, it is flows from this capital that provide the values to the environment, society, and individuals. The forest environment is nested within the larger natural environment, which is itself inside or part of the larger human-made environment. Here, the forest capital is considered separately from the capital of the larger natural and anthropogenic environments.

The interactions between the forest components (such as disturbance, recovery, nutrient cycling, and reproductive processes) represent internal flows within the forest capital. The forest also interacts with the natural and anthropogenic environments, and these represent external flows, such as seed dispersal to nearby nonforest habitats or the sale and utilization of harvested forest products. Invasive alien plants can also arrive through seed dispersal, while fires generated externally can have a major impact. Furthermore, forests play a major role in absorbing carbon dioxide and releasing oxygen.

Disturbance and Recovery Processes

Natural disturbances (e.g., fires, tree falls, lightning, landslides, browsing by wild animals) or stress events (droughts, frosts, flooding, chemical extremes) can disrupt ecosystem, community, and population structure and change resources and substrate availability, as well as other aspects of the physical environment (Hansen and Walker 1985; White and Pickett 1985). Areas receiving more than 800 mm of rain per annum (that is, ca. 7% of South Africa) are potentially suitable for forests, but fires appear to have been a major factor in fragmenting and confining them to refuge sites (Geldenhuys 1994). Indeed, indigenous high-canopy forests are limited to less than 0.1% of the country—much as they have been for the past four hundred years. When areas are protected from fire, invasive alien trees are usually the first to become established (Geldenhuys et al. 1986). Nevertheless, like pioneer tree stands, they nurse the establishment of more shade-tolerant and diverse indigenous forest stands. Inside the remaining afrotemperate and mistbelt forests, natural disturbances generally cause relatively small gaps, although heavy storms can create larger openings. However, browsing and seed consumption by insects, birds, and antelope limit species recruitment and recovery.

The disturbance regime (frequency, intensity, seasonality) of forest resource-use practices should preferably simulate natural processes determining the floristic and structural composition of a particular forest development stage. Appropriate disturbance-recovery regimes for various forest plant species may be determined through forest grain analysis (relative abundance of canopy tree species regenerating) and stem diameter distributions of key canopy tree species (Midgley et al. 1990; Geldenhuys 1996). In a fine-grained forest canopy, species regenerate in the shade of established trees, whereas in a coarse-grained forest, canopy dominants are shade-intolerant and regenerate episodically in disturbed sites.

Sustainable Rural Development

The sustainable utilization of natural capital from forests requires integrating four components (Geldenhuys 2004): (1) ecological, to maintain the composition and processes of the natural forests; (2) social, to satisfy the sociocultural and livelihood needs of all stakeholders; (3) economic, to provide direct and indirect potential benefits; and (4) policy, to provide a legal framework and empowerment to the relevant institutional structures established.

Four other concepts are important for successful and sustainable rural development (Geldenhuys 2004).

- Diversified and integrated development is needed to satisfy the interdependent and diverse needs and interests of the rural community and to buffer failures in any one type of development. For example, a focus on tourism will satisfy only one component of community interest, and interests in traditional medicines, fibers, wood for crafts, and fruits for juices and jams need to be considered.
- A business concept has to be developed, step-by-step, to assist people to improve their business skills and master associated technology for producing higher investment returns. Introducing technology for development at a level above the skills and affordability of the rural entrepreneur will lead to failure, despite good intentions.
- Resource-use needs should be matched to resource availability, and developers should be aware that urbanization of rural communities increases pressure on forests through unsustainable commercialized activities such as bark harvesting.
- Alternative resources or products should be developed if the natural resource is in short supply or cannot recover from the harvesting rates and practices.

Use of Bark for Traditional Medicine

Despite legislation, uncontrolled resource-harvesting practices continue. Lack of research coordination and effective resource management prompted the formation of the Commercial Products from the Wild Consortium (CPWild 2003), funded through the South African Innovation Fund. Following a country overview, development projects were implemented involving fibers for crafts; fruits for juices and jams; and roots, bulbs, herbs, and bark for traditional medicine.

Bark use for traditional medicine affects species, ecosystems, and the future business of the bark traders. An integrated action plan for sustainable business was developed through adaptive management research (Geldenhuys 2004). Four evergreen forest-tree species (*Ocotea bullata, Curtisia dentata, Rapanea melanophloeos,* and *Prunus africana*) were selected on the basis of their diverse use values, observed resource-use impacts, available information, and success likelihood.

The case used here focuses on the Umzimkulu forest patches in South Africa and addresses the process and key issues in developing sustainable practices for harvesting medicinal bark. In 2000, the forests and populations of certain nationally and internationally protected species were severely impacted. The resources could not supply the large market demand for bark, and bark stripping had wasted valuable timber.

Definition of the Product

The user (medicinal plant trader or herbalist) determines the species and the type of product used (fresh or dried bark, from old or young trees, leaves or roots), the volumes required, and the timing of harvesting. The product also influences the harvesting technique and the impact on the resource. In this study the people selling bark at the Durban Herbal Market indicated that they preferred fresh bark from live trees, as bark from dead trees is difficult to remove and process.

Size, Condition, and Value of the Resource: The Capital

Forest surveys were conducted in May and June of 2000 to determine the location and extent of forests harvested and the condition and size range of target tree species. The Umzimkulu forest patches vary between one hundred and eight hundred hectares and occur on steep slopes in fire refugia outside the fire zone of grasslands. Surveys in thirteen forests using 388 circular plots (0.04 ha each) indicated that bark was harvested from thirty-six of the ninety-five species, and 6.2% of the 7,280 stems recorded. Although this impact was relatively small, some species were severely stripped: *Curtisia dentata* (50 stems recorded, 60% harvested); *Ocotea bullata* (359 stems, 57.4%); *Prunus africana* (10 stems, 70%); *Pterocelastrus rostratus* (29 stems, 86.2%); *Rapanea melanophloeos* (124 stems, 38.7%). On average, 20% to 40% of the bark was removed from the main stem of the most intensively harvested species. Of these, stem diameters of *O. bullata*, *P. africana*, and *R. melanophloeos* showed strong bell-shaped distributions, suggesting episodic seedling recruitment, particularly noticeable in *O. bullata*, with no trees smaller than 15 cm diameter. *P. africana* seedlings germinated below the canopy of parent trees but did not become established due to insufficient light. In *R. melanophloeos*, typically a forest margin species, there were no saplings inside the forest. Within these species, bark had been harvested from most of the trees >30 cm DBH (diameter at breast height). When such trees die as a result, over 60% of the bark on each main stem and all the bark on the branches is wasted.

The potential value (in U.S. dollars) of the bark and timber capital was calculated for the Malowe and Nzimankulu forests (total area of 1,114 ha), for four selected bark-harvested tree species (table 20.1). In 2003, timber (auction prices from the Southern Cape, or general estimates) and bark value (retail prices per bag at the Durban Herbal Market), respectively, were as follows: *Cryptocarya myrtifolia* ($91/m^3, $106/bag); *O. bullata* ($304/m^3, $121/bag); *P. africana* ($152/m^3, $121/bag); *R. melanophloeos* ($76/m^3, $106/bag).

The estimated timber volume for the four key bark species ≥30 cm DBH from the two forests totaled 57,957 m^3 (76% *O. bullata*) with a potential timber value of $15.17 million (88% *O. bullata*). Estimated bark volume for trees ≥10 cm totaled 254,604 bags with 35 kg/bag (79% *O. bullata*) and a potential bark value of $30.53 million (80% *O. bullata*), that is, a much higher value than the timber. However, only an estimated 63,519 bags, worth $7.61 million (85% *O. bullata*), were harvested. Unfortunately, the illegal nature of the trade and loss of bags in the forest results in not all the harvested bark reaching the market, while the destructive collecting methods cause much of the capital to be lost. In fact, many of the trees were already dying at the time of the survey in May and June of 2000. A resurvey of the two forests in January 2004 showed that one of the nine harvested trees of *P. africana* died with no coppice regrowth. Of the 126 *O. bullata* trees recorded, 80 were harvested standing and a few were cut; 66 of the standing trees were dead, of which only 34 (51%) showed signs of basal sprouting (root system still alive). Hence, by allowing dying trees to die without cutting, all the remaining bark on a tree was lost, together with the timber, as well as halving the population of this legally protected species within the forest.

Impacts of Uncontrolled Bark Harvesting on Tree Populations

Collectors generally peeled small pieces of bark but removed large sections from species in high demand, often ring-barking to heights of 2–10 m. Sometimes trees were felled to harvest bark from the upper parts, but the timber was not used. Occasionally neighboring trees were even felled onto standing trees to enable harvesters to climb higher. The degree of ring-barking had a greater negative effect on crown condition than the total percentage of bark removed from the stem; trees with <40% living crowns were considered to be dying.

O. *bullata* was observed to quickly develop coppice shoots at the base of the stem (basal sprouts) or around the debarked wound (stem sprouts), though the coppice shoots were heavily browsed. If severely debarked, the tree and its coppice shoots, particularly stem sprouts, died. In 2000, most standing debarked trees in Nzimankulu and Malowe forests were dying, and by January 2004, 82.5% of these were dead, with more than 50% developing no vegetative regrowth. However, if such trees were cut down, they developed vigorous coppice regrowth from the stump (i.e., from an established root system). Debarked *P. africana* trees rapidly regrew bark through cambium development on the wound and also developed coppice shoots. However, *R. melanophloeos* and *C. myrtifolia* did not easily develop basal or stem sprouts or callus tissue around debarked wounds.

Regulating the Use of Natural Capital and Associated Practical Measures

The production rate of a resource determines how much can be used sustainably. Like the interest rate on invested capital, if the amount used exceeds this, the invested capital is eroded and future benefits decline. For sustainable bark harvesting, both tree growth and bark recovery rates are important considerations. Information on mean growth rates (stem diameter or plant height) within a specified time period, the range in rates and causes of rate fluctuations are not yet known for the Umzimkulu forests (Geldenhuys 1999b). Bark-harvesting experiments in South Africa, Malawi, and Zambia will soon provide baseline data on rates of bark regrowth on wounds caused by removal of vertical strips of bark 5–20 cm wide and 1 m long (Vermeulen and Geldenhuys 2004). It was also noted that marked O. *bullata* stumps produced multiple coppice shoots of 3–4.5 m height in eighteen months, that is, within the range reported by Lübbe (1990). Such regrowth offers the potential for future rotational harvesting of individual stems.

Sustainable harvesting practices require an understanding of the ecological processes operating in a natural habitat. Observations during the forest surveys, combined with other information and knowledge, indicated how management could be improved. For most tree species, the best practice is to remove bark in long, vertical strips about 10 cm wide, with a thin, flexible blade such as a bush knife, without lifting the bark edge. Subsequent application of "tree seal" prevents the wounded wood from drying out, but it does not appear to prevent insect boreholes or fungal development on the wound, or to facilitate bark recovery, though responses varied among species (Vermeulen and Geldenhuys 2004). Species for which bark does not readily recover should be selectively felled as part of a regulated system, so that all the bark (and the timber) can be used. The forest margin tree *R. melanophloeos* could easily be managed in this way, while also facilitating stand development toward mixed forest.

TABLE 20.1

Estimated volume and potential value of timber and bark in Malowe and Nzimankulu forests, Umzimkulu District (June 2000)

Species	Timber		Bark			
	Volume m³	Value, $	Estimated volume (bags)	Harvested volume (bags)	Value, total ($)	Value, used ($)
Cryptocarya myrtifolia	2.59	235.79	9.73	0.85	1,034.96	90.41
Ocotea bullata	39.85	12,112.96	182.09	48.58	22,139.16	5,906.30
Prunus africana	7.12	1,082.56	21.08	1.96	2,563.40	238.30
Rapanea melanophloeos	2.47	187.34	15.64	5.63	1,664.04	598.86
Total	52.03	13,618.65	228.55	57.02	27,401.56	6,833.89

Sources: Calculations based on bark-yield data in Geldenhuys et al. (2002) and Geldenhuys and Rau (2004).
Note: US$1 = SA Rand 6.58 [31 December 2003]; one bag of bark weighs 35 kg.

It has been proposed that species with the ability to coppice, which are dying due to excessive bark harvesting, should be felled where to do so does not endanger other trees. This would enable the remaining bark to be harvested, processed, and stored for gradual release into the market. The main-stem timber (for high-quality furniture) and branches (for tourist industry wood crafts) would provide an additional income source to local entrepreneurs. Management of the coppice regrowth could ensure future productivity. Such an interim, holistic-harvesting approach would have enabled the gradual establishment of a regulated system for extraction, based on tree and bark production rates. Sadly, the resource managers did not implement this suggestion, and the capital value of the timber and bark was lost (table 20.1).

A simple and practical approach to protect coppice shoots against cattle and browsing wildlife was to stack branches from the cut tree or other fallen dead trees over the stump, or around developing coppice shoots at the base of a debarked tree.

Development of Alternative Natural Capital Resources

Alternative resources have to be developed if natural resources cannot satisfy the demand. When planning supplementary planting of indigenous species, it is necessary to consider finding planting stock that will not cause genetic pollution in the natural environment.

PLANTING STOCK

Resource management to increase yields can reduce harvesting pressure on limited natural resources that usually have low productivity (Geldenhuys and Delvaux 2002). Besides coppice management, techniques include the establishment and protection of seedlings and the planting of mixed stands of high-demand species. These economically important species can be planted close to the villages or home gardens as live fences, without impacting on existing grazing and crop land. Otherwise, they could be planted in small open areas (mainly in riparian zones) within timber plantations or other productive land use systems.

Alternatives to bark harvesting include the less destructive collection of leaves from planted trees, and their use in pharmaceutical production of medicines, which would be a long-term solution provided that they are affordable for the rural poor (Mander et al. 2006).

Seedlings of several bark-harvested species in the Umzimkulu forests that lacked regeneration were found naturally occurring in the adjacent timber plantations. Birds dispersed seeds over considerable distances away from the forest margin into the plantation. A survey in the twenty-one-year-old *Pinus patula* stand adjacent to Nzimankulu forest recorded between 6,250 and 6,380 seedlings per hectare of twelve forest species (Geldenhuys and Delvaux 2002). These seedlings were strong and grew fast because of the better light conditions. This plantation was used as a nursery. Traditional healers from the nearby Cancele village collected seedlings for planting inside the fence around their primary health care center. Forest managers transplanted seedlings to rehabilitate forest gaps and margins.

GENETIC VARIATION AND ACTIVE INGREDIENTS OF HARVESTED SPECIES

Many of the forest species investigated here have widely distributed but disconnected populations, resulting in distinct provenances (Von Breitenbach and Von Breitenbach 1995). The genetic differences in *O. bullata* (Van der Bank 2000) are reflected in its external morphology (leaf shape and size, growth habit) and growth variability (vitality of seedlings and vegetative regrowth) across its geographical range (Geldenhuys 2004), therefore, as a precautionary general rule to maintain genetic integrity, cultivation of a species from the wild, in the wild, should use local material and not import this from elsewhere.

The active ingredients in the bark associated with healing are also present in the leaves, but in lower concentrations (Drewes and Horn 2000). The leaves can therefore be used for the same healing purposes as the bark. However, this requires an attitude change in the traditional healers, some of whom were willing to try this alternative.

Monitoring and the Adaptive Management of Natural Capital

In the development of sustainable bark-harvesting practices there are many unknown variables. The least favorable option was to implement research and wait for results, because in the meantime the resource may be lost. The better option was to apply an adaptive management approach, using conservative harvesting practices based on available information in conjunction with research and monitoring. Harvesting methods can then be adapted as new information becomes available.

In the bark studies, much information was obtained during the resource survey. Some dying *O. bullata* trees were cut, with some of the stumps experimentally covered with branches. This showed that the assumptions were correct from the survey observations regarding vigorous sprouting of cut trees and the need for protection against browsing. The experimental bark studies also provided refinement of the initial observations of bark recovery responses, and thus this monitoring and reevaluation will be continued as an integral part of adaptive management.

Institutional Structures for Participatory Forest Management

A participatory, rather than a top-down, approach was used to resolve the issue of uncontrolled bark harvesting. Following the bark-resource survey, discussions were held with a group of twelve bark harvesters from the Umzimkulu District selling their products at the Durban Herbal Market. The harvesters were largely willing to participate in discussions for seeking a solution, if it allowed them to continue earning a living. They were mostly women from the Umzimkulu District, and they depended almost entirely on bark harvesting and trade for their livelihoods.

In all meetings, resource managers of the Department of Water Affairs and Forestry (DWAF), the state department responsible for forest management in the area, were present to facilitate open discussions, shorten administrative procedures, and assist both groups in reaching a common understanding. Meetings were held at the Durban Herbal Market to clarify the intentions and objectives, and in Nzimankulu forest to identify and discuss the harvesting issues. The harvesters were assisted to form an association through which an agreement could be negotiated with the DWAF, and follow-up meetings were held to maintain regular communication.

The Sizamimpilo Association for medicinal plant harvesters was formed as a legal entity to interact with the DWAF. The members, with the assistance of an external facilitator, participated directly in drafting their constitution. The key components, in terms of sustainable resource management, were (1) to develop the business skills of members; (2) to train them in bark-harvesting techniques that would ensure sustainable tree use; and (3) that all members sign an agreement between themselves and the association that bound them to a set of agreed objectives, standards, and rules (Geldenhuys 2004).

DWAF granted permission to the Sizamimpilo Association for harvesting bark under guidance of the management plan for natural forests in the Umzimkulu District. Practical training sessions were organized with the forestry and nature conservation authorities and the bark harvesters. Each member carried a visible identification tag with her or his name and the association logo. Membership grew from the initial twelve in 2000 to over two hundred by the end of 2003.

The Umzimkulu Forest management plan provides guidelines for sustainable resource use and stipulates the arrangements between the DWAF and the Sizamimpilo Association. Guidelines distinguish between immediate action in degraded areas and long-term management. In terms of immediate action, the purpose of tree harvesting was to salvage damaged and dying trees. By contrast, the long-term management is for sustainable timber and bark harvesting, based on natural turnover, the recovery rate of the debarked wounds and cut stumps, and the development of alternative resources. The monitoring program that aims to provide longer-term information feeding into sustaining the bark-harvesting practices complements this.

Contribution

The partial capital value for timber and bark of selected tree species in Malowe and Nzimankulu forests is large: $40,000/ha. Assessment of the size and condition of the capital has shown that much loss occurred through wasteful, illegal-harvesting practices. The external flow of bark, from the forest to the market, was much higher than the production and recovery rate of bark inside the forest. This damaged the remaining trees as the source of bark and also destroyed the remaining bark through the death of harvested trees and their inability to recover by vegetative regrowth. The root causes were a lack of control by state authorities; commercial bark-harvesting practices that ignored traditional conservative practices; and a lack of understanding of the species-specific ecological processes that maintain the natural capital, namely tree growth, bark regeneration, and tree recruitment.

However, the study showed that the forest is not necessarily a "museum object" but a dynamic system. Analyses of the forest survey data showed that the target species had population structures typical of tree species requiring some disturbance and more light for their seedlings to become established. Disturbance seems to be necessary but has to be controlled and managed. Recommended restoration practices include the felling of dying tree species that coppice and the translocation of seedlings to the available canopy gaps. Fire exclusion and protection from browsing are necessary for securing the investment in species restoration. Even for species with poor bark recovery there are options for harvesting them sustainably, if their requirements for recruitment and bark recovery are understood.

Human needs can be harmonized with the potential of natural capital production in forests through sustainable resource-harvesting practices. The women of the Sizamimpilo Association showed that they are willing and keen to adapt to the new practices and to become part of the forest management action, including the control of illegal bark harvesting from these and other forests. An adaptive research and management approach, combined with the participation of local traditional resources users, should be considered by resource management agencies. This study has shown that forest natural capital can potentially be restored through an integrated approach focusing on sustainable business development, with short-, medium-, and long-term strategies to address critical issues, and through involvement of all relevant stakeholders from the start. An appropriate survey of the bark resources and harvesting impacts provided insight that improved resource management practices. Working with the bark harvesters, who had vested interests in finding a solution, rather than with the community at large, facilitated good collaboration. However, this collaboration between the resource managers and bark harvesters needs further development and frequent discussions to maintain the best working relationship. Training of bark harvesters and resource managers was part of the exercise from the start. A legal framework within which the bark harvesters could operate facilitated the establishment of a management and monitoring program. This integrated approach should have a more positive long-term effect on the flows of medicine bark and the forest natural capital than forest protection by exclusion and prosecution of harvesters.

Acknowledgements

This research was partially funded by the South African Innovation Fund, and project R8305 of the Forest Research Program of the United Kingdom Department for International Development (DFID). The views expressed are not necessarily those of DFID.

Chapter 21

Assessing Costs, Benefits, and Feasibility of Restoring Natural Capital in Subtropical Thicket in South Africa

Anthony J. Mills, Jane K. Turpie, Richard M. Cowling, Christo Marais, Graham I. H. Kerley, Richard G. Lechmere-Oertel, Ayanda M. Sigwela, and Mike Powell

South African subtropical thicket, a tall thorny shrubland, is the principal form of natural capital in the southwestern part of the Eastern Cape (ca. 33°S, 25°E). Numerous benefits accrue from this vegetation. It supports an exceptionally high natural diversity and abundance of large browsing mammals, such as black rhinoceros (*Diceros bicornis*), elephant (*Loxodonta africana*), and antelope (Skead 1987; Kerley et al. 1999); is often intensively harvested by local people for wood, fruit, and medicines (Cocks and Wiersum 2003); can sustain appropriately managed goat pastoralism (Aucamp 1976; Stuart-Hill and Aucamp 1993); is the center of a growing tourism industry (Kerley et al. 2002); and, for a semiarid region, stores an unusually large quantity of ecosystem carbon (Mills et al. 2005).

The subtropical thicket is composed of succulent (e.g., *Portulacaria afra*) and spinescent shrubs (e.g., *Azima tetracantha, Gymnosporia polyacantha, Putterlickia pyracantha, Rhus longispina*), as well as small trees (<5 m) (e.g., *Pappea capensis, Euclea undulata*, and *Schotia afra*). Despite a long association with large, indigenous herbivores (Midgley 1991; Kerley et al. 1995), the thicket is surprisingly sensitive to injudicious goat pastoralism (Stuart-Hill 1992). Heavy browsing by goats can transform the dense, closed-canopy shrubland into an open savanna-like system (figure 21.1), a process that can occur possibly within a decade (Hoffman and Cowling 1990b; Lechmere-Oertel et al. 2005a). Of the 16,942 km^{-2} of solid (unbroken canopy) thicket (with a *P. afra* component), 46% has been heavily impacted and 36% moderately impacted by domestic herbivores, while only 1.8% and 0.5% have been transformed by cropping and urbanization, respectively (Lloyd et al. 2002).

Excessive goat browsing in this ecosystem depletes natural capital by reducing species diversity (Moolman and Cowling 1994; Johnson et al. 1999; Lechmere-Oertel et al. 2005a), above- and belowground carbon stocks (Mills et al. 2005), soil quality (Mills and Fey 2004), and plant productivity (and hence livestock stocking capacity) (Stuart-Hill and Aucamp 1993). Differences in plant productivity between transformed and intact thicket are especially apparent during drought years (Stuart-Hill and Aucamp 1993). Transformation also reduces the availability of wood, fruit, and medicines for local communities, with a potential financial loss of approximately US$150 per annum per household (Cocks and Wiersum 2003). In this chapter, we discuss proposed methods for restoring the natural capital of subtropical thicket, the ecological thinking underlying these methods, and the economic viability of restoration at a landscape scale.

FIGURE 21.1. A fenceline contrasts semiarid thicket transformed by goat keeping (left) with intact thicket (right).

Restoring Degraded Natural Capital in the Eastern Cape

Unlike indigenous browsers, such as black rhinoceros, elephant, and kudu (*Tragelapus strepsiceros*), which prune the shrubs from above, goats browse from within and below the shrub canopy (Stuart-Hill 1992). Goats are able to select leaves among the thorns and appear to be less affected by the thorny defenses of the thicket plants than are larger indigenous mammal species (Wilson and Kerley 2003). Unfortunately, rapid restoration (achieved within a human lifetime) is not as simple as just removing the goats. Regeneration in formerly heavily impacted thicket is slow or nonexistent (Stuart-Hill and Danckwerts 1988), being primarily hampered by a lack of shrub recruitment (Sigwela 2004).

Restoration, therefore, requires active intervention to establish shrubs; yet there is no consensus on the most practical and effective methods. Sowing seeds is unlikely to be effective (Todkill 2001), as the harsh microclimate of the exposed soil in transformed thicket appears to limit seed germination and also prevents seedling recruitment of thicket plant species that normally establish in protected microsites beneath the shrub canopy (Holmes and Cowling 1993; Sigwela 2004).

A potentially cost-effective, practical restoration method is planting cuttings of the succulent shrub *P. afra* (spekboom) (Swart and Hobson 1994). This shrub is dominant across large areas of the thicket biome (Vlok et al. 2003), especially in arid and valley thickets. It propagates vegetatively from branches that reach the ground at the canopy edge or those broken off by large browsing mammals (Stuart-Hill 1992) and is able to switch between C3 and CAM photosynthetic pathways (Guralnick et al. 1984a, 1984b), an unusual and useful adaptation to arid conditions. The use of C3 photosynthesis when soils are wet probably enables *P. afra* to be more productive than succulents that use only CAM. Several land managers have used *P. afra* cuttings to restore the shrub cover of transformed thicket. At Krompoort (between Uitenhage and Steytlerville), for example, cuttings were planted in 1976 at 1–2-m intervals. By 2005, a thicket composed of shrubs over 2 m high covered 90% of the experimental site (Mills and Cowling 2006).

Subtropical Thicket Restoration Project

Initiated by the Working for Water program of the South African government, the Subtropical Thicket Restoration Project aims to demonstrate the logistical and practical feasibility of restoring thicket at a farm scale (i.e., hundreds of hectares); to provide quantitative information on biodiversity gain and carbon sequestration rates in restored sites; to quantify the financial costs of restoration; and to establish the protocols and methods for sourcing funding that will initiate restoration on a landscape scale (i.e., thousands of hectares) (figure 21.2).

Currently, cuttings of *P. afra* and other easily propagated succulent taxa typical of subtropical thicket (e.g., *Crassula, Aloe, Euphorbia,* and *Cotyledon* spp.) are being planted in various densities (1–3-m intervals) and patterns (e.g., clumped or scattered) at a farm scale. Exclosures have been constructed to determine the effect of browsing by indigenous herbivores on plant establishment; and pioneer shrubs are being propagated in a nursery. Pioneer shrubs are relatively resistant, being the last component to disappear under heavy goat browsing (Hoffman and Cowling 1990b; Stuart-Hill 1992), and are likely to create suitable microsites for establishment of larger shrubs or trees (e.g., *E. undulata, P. capensis,* and *Sideroxylon inerme*) (Cowling et al. 1997). It is anticipated that pioneer shrubs bearing bird-dispersed fleshy fruits (e.g., *Lycium* spp., *R. longispina, A. tetracantha, Carissa bispinosa*) and *P. afra* cuttings with mistletoe infestations (e.g., *Viscum crassulae* and *V. rotundifolium*) will attract birds that, in turn, will accelerate plant diversity reestablishment into restored sites via dispersal of seeds from the surrounding intact subtropical thicket.

Can Ecological Integrity Be Restored?

Restoration implies the return of ecological integrity and the full pattern of biological complexity and diversity, together with the ecosystem processes that maintain this pattern (Hobbs and Norton 1996). Planting cuttings of *P. afra* and other succulent plant species does not qualify on its own as restoration. We hypothesize, however, that *P. afra* in particular will im-

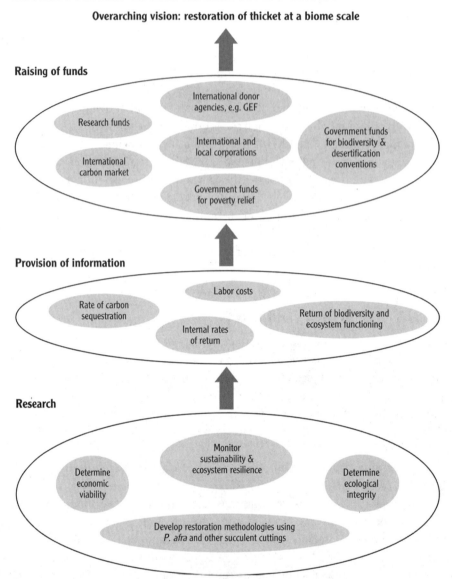

The Subtropical Thicket Restoration Project

FIGURE 21.2. Objectives of the natural capital restoration project. Biological and economic information will be provided to access various sources of funding to initiate a landscape-scale restoration project across the thicket biome. (GEF: the Global Environment Facility)

prove the microclimate of the planting sites for plant growth and will provide cover for seed-dispersing animals and birds, thereby facilitating natural ecosystem recovery.

At present, many transformed thicket landscapes appear to have abiotic barriers that restrict seedling establishment. These barriers include extreme soil-surface temperatures (up to 50°C), reduced soil water-holding capacity (Lechmere-Oertel et al. 2005b), and soil crusts (Mills and Fey 2004). Planting *P. afra* and other succulents could remove such barriers by shading soils and returning organic matter. Remnant shrubs are likely to benefit from the effects of *P. afra* establishment, though seedlings do not establish readily under *P. afra* canopies (Sigwela 2004), possibly due to root competition, excessive shading, rainfall interception, allelopathy, or a combination of these mechanisms. It is not known to what extent seedlings will establish via natural dispersal adjacent to *P. afra* canopies or below planted pioneer shrubs. An adaptive management approach is advocated whereby restoration methods will be fine tuned as research during the implementation phase yields knowledge on shrub establishment.

In some restored sites, *P. afra* may show a greater dominance than it does in pristine thicket (Mills and Cowling 2006). We suggest, however, that although *P. afra*–dominated thicket may produce a new stable state (i.e., different from the pristine state), this plant community is preferable to the present transformed landscape because it provides a food source for livestock or indigenous herbivores during drought, and its value is likely to increase through time as soil carbon reserves accumulate and additional plant and animal species colonize the site. Whereas the natural capital of a site restored with *P. afra* is likely to appreciate, that of a transformed and denuded landscape depreciates with time due to ongoing death of remnant shrubs and trees (Lechmere-Oertel 2005a), loss in soil quality, and reduced ecosystem productivity. Milton (2003) discusses the concept of "emerging ecosystems" (ecosystems that emerge from land that has been cleared of natural vegetation for agricultural, industrial, or commercial use) in a South Africa context and notes that if society decides to manage these emergent states it is possible that their social, economic, and ecological value may be enhanced. Transformed thicket restoration using *P. afra* cuttings may be an example of such management.

Can We Create a Hyperbeneficial Thicket?

Transformed landscapes where *P. afra* used to be dominant should be the initial target areas for restoration programs using cuttings of this shrub. However, landscapes where *P. afra* occurs, though not as the dominant plant, should not necessarily be precluded. Decisions by landowners could be taken to create new *P. afra* ecosystems that are both sustainable and productive. Such a designer ecosystem (Palmer et al. 2004) could potentially provide more benefits than the original thicket, as is achieved in some tropical forest systems (Gadgil et al. 1993; McNeely 1994), where useful planted species occur in greater abundance than naturally. In the case of thicket, these could include fruiting species (e.g., *C. bispinosa* and *P. capensis*); valuable browse species (e.g., *E. undulata*); species utilized for cultural practices (e.g., *Olea europaea* subsp. *africana*) (Cocks and Wiersum 2003); medicinal plants (e.g., *Gasteria bicolor*, *Bulbine* spp., *Haworthia* spp., and *Aloe* spp.); and threatened species, such as cycads (*Encephalartos latifrons* and *E. arenarius*) and succulents (e.g., *Haworthia* and *Euphorbia* spp.) of conservation significance and horticultural importance (Victor and Dold 2003). This type of restoration could be viewed as *ecosystem farming*, rather than just returning biological integrity to a landscape.

Potential Benefits of the Restoration of Natural Capital

Thicket vegetation provides a variety of ecosystem goods and services that contribute to the economy; these include livestock keeping, nature-based tourism, and goods such as plant products used for domestic consumption and sale. Other benefits such as pollination and water flow have yet to be quantified. Restoration will in addition promote the sequestration of carbon, which in time could be sold or traded in international markets.

Livestock Keeping

The stocking capacity for mammalian herbivores, measured in large stock units (LSU), is greater for intact *P. afra* thicket (0.14 LSU ha^{-1} in wet years; 0.08 LSU ha^{-1} in dry years) than for transformed landscapes (0.07 LSU ha^{-1} in wet years; <0.01 LSU ha^{-1} in drought years) (Stuart-Hill and Aucamp 1993). The shrubs in intact thicket buffer the stocking capacity of the rangeland (Kerley et al. 1995) and enable effective stock management and planning strategies (Stuart-Hill 1989; Stuart-Hill and Aucamp 1993). By comparison, the stocking capacity of transformed landscapes is less buffered and is likely to decrease with time as further ecosystem decline occurs (Lechmere-Oertel et al. 2005a). Restoration with *P. afra* cuttings does not preclude goat keeping as long as stocking rates and browsing periods are carefully managed. Indeed, the ecosystem recovery rate is likely to increase if *P. afra* is browsed rather than protected from herbivory (Aucamp et al. 1980). Further research is, however, required to ascertain the appropriate stocking rates during restoration.

Nature-based Tourism

Conservation efforts should aim to capitalize on the Eastern Cape's inherent attractions for tourists, such as high densities of large mammals (Kerley et al. 1995; Boshoff et al. 2002), spectacular scenery, and malarial absence. The tourism potential of restored thicket is likely to be greater than transformed landscapes. Restoration, for example, will increase the wildlife stocking capacity. Animals such as elephants and black rhinoceros (at appropriate and carefully managed densities) may even be vital for restoration success in terms of achieving biological integrity (Sigwela 2004).

Plant Products

Many thicket species are harvested for household or commercial use. For example, in a rural community near Peddie, 103 plant species are harvested on a regular basis by local people for uses such as kraal construction (enclosures for livestock), fuelwood, rituals, fencing, wild fruit, traditional medicines, timber, wild vegetables, sticks, tools, and fodder (Cocks and Wiersum 2003). The mean gross direct-use value of thicket has been estimated to be US$150 per annum per household. Transformed landscapes are likely to yield considerably fewer medicinal and nutritional benefits for local people as the plant communities, at their extreme impoverishment, consist of herbaceous vegetation dominated by an Australian *Atriplex* species (Fabricius, Burger, et al. 2003).

Carbon Sequestration

The sequestration of carbon in biomass and soils during thicket restoration assists in the mitigation of global climate changes arising from an elevated atmospheric CO_2 concentration.

The sequestration rate will vary according to climate, planting density, herbivory intensity, and soil type. Mills and Cowling (2006) quantified sequestration rates at Krompoort, a farm near Kirkwood (250–350 mm annual rainfall), and in the Great Fish River Reserve near Grahamstown (400–450 mm), where restoration using *P. afra* cuttings began in 1976 and 1982, respectively. The difference in the estimated sequestration rates at Krompoort (4.2 t C ha^{-1} yr^{-1}) and the reserve (1.2 t C ha^{-1} yr^{-1}) has been ascribed to greater herbivory in the latter. Predicted costs associated with vegetation destruction and subsequent carbon losses to the atmosphere (causing climate change) amount to about $12 per ton of carbon (Turpie et al. 2004).

Poverty Relief

The Eastern Cape is afflicted with 49.4% unemployment of people between the ages of 18 and 65, compared with the national rate of 41.8% (SAIRR 2004). Restoration of 1 ha of transformed thicket utilizing *P. afra* cuttings at a spacing of approximately 1.5 m requires approximately fifty labor days. A project that restores 10,000 ha per year would consequently employ approximately 2,000 laborers. There are about 4 million ha of transformed thicket, so there is unlikely to be a shortage of land for implementing projects of this size. In addition to the provision of jobs, there will be numerous long-term benefits to local communities, such as improved pasture and sustained access to fuelwood, timber, fruit, and medicines.

Is the Restoration of Natural Capital Economically Viable?

Restoration costs depend on the initial planting density, which in turn is dependent on the remaining shrub cover. Costs also depend on whether restoration involves planting cuttings of *P. afra* alone or establishing a variety of species. A preliminary estimate of the present value of all costs for restoration of thicket, with less than 25% of the original biomass remaining, is approximately $722 ha^{-1} with *P. afra* only, and $862 ha^{-1} using a variety of species. Included in these costs are sourcing of reproductive material; seedling propagation; initial establishment of vegetative material; replacement of dead cuttings during the first two years; custodianship, including invasive alien plant and domestic livestock control; and project management, administration, monitoring, and evaluation.

We performed a cost-benefit analysis over a fifty-year period using a discount rate of 8% to determine whether these restoration costs were justifiable in terms of the benefits yielded. Benefits were a function of (1) the aboveground dry biomass of thicket, and (2) the extent to which recovering thicket could be utilized. Biomass at any time period was in turn dependent on the growth function used for recovering thicket.

We estimated the growth function of recovering thicket based on the findings at Krompoort by Mills and Cowling (2006). The growth function

$$B_t = B_{t-1} + B_{t-1}.r \left(1 - \frac{B_{t-1}}{K}\right)$$

was dependent on (1) the initial aboveground dry biomass (B_{t-1}) of the area to be restored (including the biomass of planted cuttings), (2) the maximum biomass (K), and (3) the intrinsic growth rate (r), and was calibrated such that biomass reached 67 t ha^{-1} in the twenty-seventh year. The biomass of planted cuttings was assumed to be about 0.15 t ha^{-1}. With a very low initial vegetation biomass of 0.125 t ha^{-1}, to which the cuttings are added, and a K of 100 t ha^{-1}, the growth coefficient r equals 0.28. This suggests that a restored patch could reach a biomass approaching 100 t ha^{-1} within about thirty-five years, though further research on growth rates is required to verify this model. The limited data available suggest that growth rates in the model are realistic. Aucamp and Howe (1979) measured a net primary production of approximately 4.5 t ha^{-1} yr^{-1} (aboveground dry biomass) in *P. afra* thicket (assuming a total dry:wet ratio of 0.4), and data presented by Mills et al. (2005) indicated that 100% *P. afra* cover could generate 4.5 t biomass ha^{-1} yr^{-1} in leaf litter alone. Benefits considered within the model were livestock and game production, harvesting plant products (assuming natural recovery of biodiversity), and carbon sequestration. Herbivore-stocking capacity is a function of aboveground biomass, and it was assumed that 85% of this is utilized for small stock and the remainder for game. Stocking capacity was estimated as a function of aboveground dry biomass (B), based on Stuart-Hill and Aucamp (1993) and Turpie et al. (2003), as follows:

$$CC(LSU/ha) = 0.0005 * B$$

The value of animal production was estimated as \$220 LSU^{-1} yr^{-1} for livestock and \$724 LSU^{-1} yr^{-1} for game (Turpie et al. 2003). In addition, the harvesting value (for medicinal plants and firewood) was estimated as \$0.02 t^{-1} biomass (Turpie et al. 2003). Both animal production and plant harvesting potentially affect the vegetation recovery rate, with differences probably linked to browsing pressure. The high recovery rates at Krompoort were attributed to herbivore exclusion for about seventeen years, followed by low stocking densities. The model thus incorporated an allowable-use factor for browsing and harvesting that was initially assumed to be 0% of the browsing capacity for biomass <10 t ha^{-1} equated to roughly the first seventeen years in the initial model), followed by 25% for biomass <60 t ha^{-1}. These assumptions were varied in the sensitivity analysis.

Carbon sequestration (t ha^{-1} yr^{-1}) (C) was estimated from aboveground dry biomass, (B) using the ratio of total ecosystem carbon to aboveground biomass of 1.8:1.0 (Mills and Cowling 2006) as follows:

$$C_{seq(t)} = 1.8(B_t - B_{t-1})$$

The value of carbon accumulated was estimated as \$12 t^{-1}, based on estimated damage costs to vegetation in South Africa (Turpie et al. 2004). This also falls within the range of potential income (\$5–\$25 t^{-1}) that could accrue from carbon credit sales. Transaction costs (including verification of carbon stocks) were included in the model.

The model results suggest that the financial benefits are potentially positive, although the internal rate of return (IRR) on investment could be fairly low. The default scenario, which most closely resembles the situation at Krompoort, had an estimated IRR of 9.2%. Hence, the project could be considered financially viable only with a discount rate of less than 9.2%. Projects implemented by multilateral funding bodies normally require an IRR of 12% to be considered viable. IRR was sensitive to the growth rate of thicket, the degree to which the area under restoration can be used for animal production, and the price of carbon. As the growth coefficient r changed from 0.23 to 0.32, IRR rose from 7.9% to 10.3%. Changing animal production to 10% of stocking capacity for aboveground dry biomass <10 t ha^{-1} and to

50% for aboveground dry biomass >10 t ha^{-1} resulted in an IRR of 11.2%. Results from the Great Fish River Reserve, however, suggest that browsing reduced the recovery rate by two-thirds (Mills and Cowling 2006), in which case returns would decrease rather than increase. In addition, the carbon market value in the future is highly uncertain, being dependent on international policy agreements. IRR changed from 8.1% to 10.8% across a range of $5–$25 t^{-1} carbon. If labor costs are excluded, then the economic rate of return for the baseline scenario increases from 9.2% to 12.3%. Indeed, the more positive economic outlook lies in the employment opportunities created by the project.

Given the implications for use, the quality and quantity of the initial biomass in the restoration area is of critical importance to the outcome. With a starting biomass of 12.5 t ha^{-1} (within the most degraded thicket category), and assuming this biomass could be used by domestic livestock or game, financial IRR's could be potentially as high as 24%. Another possible source of error in the economic viability estimate is in the carbon sequestration rates, which have been measured only on a very small scale at a few sites. It is quite possible that these rates are much higher than realistically achievable on a large scale. This error would largely be associated with the amount of carbon stored in soil. When below ground carbon storage was excluded, the IRR decreased from 9.2% to 7.9%. Our results thus suggest that while restoration on a large scale could be a viable option, further research is urgently required to improve the model.

Contribution

Subtropical thicket in the Eastern Cape, South Africa, has been transformed by unsustainable farming practices over the last century. This has reduced the flows of benefits for pastoralists, local communities, game farmers, tourism operators, and conservationists. The large-scale restoration of transformed thicket using labor-intensive methods to plant cuttings of *P. afra* and pioneer shrubs would provide temporary employment, thereby assisting in the alleviation of rural poverty and would simultaneously increase the landscape value for all land users. Preliminary analysis suggests that, although the costs of restoration are high, the returns could make it viable at a large scale. The viability is dependent on the rate of carbon sequestration, the price of carbon, and the potential for restored land to support livestock, all of which are highly uncertain variables at present. The Subtropical Thicket Restoration Project aims to develop effective methods for restoration and to reduce the uncertainty within the present economic model.

Chapter 22

Costs and Benefits of Restoring Natural Capital Following Alien Plant Invasions in Fynbos Ecosystems in South Africa

PATRICIA M. HOLMES, DAVID M. RICHARDSON, AND CHRISTO MARAIS

Fynbos ecosystems are located in the Cape Floristic Region (CFR) in the Western Cape Province of South Africa, one of the floristic "hotspots" of the world (Myers 1990). The soils of this area are too poor for crop production, and the vegetation has no value for livestock husbandry; however, the area has a rich natural capital of fine-leaved evergreen shrubs and reeds unique to the area. These include striking flowers (Proteas, Leucadendrons, Ericas, and others) that have a high value for tourism as well as for the cut- and dried-flower trades, thatching reeds, and medicinal plants. Apart from the income streams that flow from these stocks, this vegetation protects the acid, sandy soils of the rugged mountains, ensuring a steady flow of clean water to the large human population of the Western Cape region. *Invasive alien plants* (IAPs), particularly trees and shrubs (Richardson and Van Wilgen 2004), threaten natural capital in this region. Research on the ecology and impacts of the principal invader-plant species provided the rationale for the *Working for Water* program (WfW), initiated in 1995 (Van Wilgen et al. 1996; Van Wilgen 2004). This government-sponsored program seeks to protect water supplies and restore productive and conservation land, primarily through labor-intensive IAP clearance operations at a national scale.

In this chapter, we address two questions: (1) Are the currently applied methods for clearing IAPs adequate for restoring natural capital following invasion? and (2) Do the benefits of restoring natural capital outweigh the costs? We consider the mechanisms of invasion and the opportunities for intervention that arise at each stage. We also introduce a framework that combines issues of scale (biome, watershed, plant community) and invasion history, to conceptualize the opportunities and actions required for restoring natural capital, as well as the costs and benefits.

Priorities and Actions Required for Restoring Natural Capital in Fynbos Ecosystems

IAP species affect various natural capital stocks (indigenous plants) and flows (particularly water and aesthetic values), depending on the species invading, the ecosystem being invaded, the history of invasion, and the scale being assessed. We present the main priorities for restoration at the different scales, under recent and long invasion histories, in a conceptual framework (table 22.1).

TABLE 22.1

Conceptual framework to identify the main priorities for restoration by removal of invasive alien plants

Scale	Main priorities	Actions
Biome: 90,000 km^2	Prevent entry of new alien plant invaders.	• Screen entry points for new potential plant invaders. • Collaborate with enterprises relying on alien plant species.
	Prevent spread of emerging alien plant invaders.	• Implement rapid clearing and follow-up actions on emerging plant invaders. • Initiate biological control research for emerging plant invaders.
	Strategic plans for control of transformer alien plant species.	• Identify and map transformer plant species. • Prioritize watersheds for clearance based on (1) potential for further spread of plant invaders; (2) magnitude of impacts on biodiversity, ecosystem functioning, and productivity; and (3) importance of watershed for conservation, agriculture, and water production. • Initiate biological control research for transformer species not yet under substantive control.
Watershed: 50–250 km^2	Prioritize clearance to minimize ecological impacts of alien plant invasion.	• Clear low-density stands and outliers first. • Clear from top of watershed downward. • Clear sites with high biodiversity and consumptive-use value.
	Restore natural capital.	• Restore direct consumptive-use value (e.g., harvesting potential). • Restore indirect consumptive-use value (e.g., clean water yields). • Restore nonconsumptive-use value (e.g., diversity of habitats, tourism potential).
Plant community: 0–10 km^2	Restore natural capital.	• Control invasion by alien plants by recommended methods ensuring adequate provision for follow-up control.
	1. Medium-dense alien stands (25%–75% canopy cover)	• No further action required beyond alien plant control; community on a trajectory of natural recovery.
	2. Recent closed stands (>75% canopy cover)	• Reintroduction of serotinous overstorey species, if local species important in flower-harvesting trade.
	3. Long-term closed stands (two or more cycles)	• Reintroduction of serotinous overstorey species and other long-lived species.
	4. Long-term closed stands + at risk of soil erosion (granite and shale substrata)	• Post-fire physical stabilization of unstable slopes. • Scarification or soil turnover in heat scars. • Reintroduction of fast-growing, short-lived species for soil stabilization.

Note: Actions are for mountain fynbos; adjustments should be made for other vegetation types (e.g., earlier seeding intervention where seed banks are short lived).

Fynbos typically is a fire-prone, sclerophyllous, shrubland vegetation growing on nutrient-poor soils under a Mediterranean-type climate (Cowling and Holmes 1992). Although species-rich, fynbos has low forage value. Biodiversity underpins the total economic value of fynbos ecosystems by providing the basic structural components (i.e., "stocks") for ecosystem functioning and, ultimately, productivity (i.e., "flows") (Turpie 2004). Fynbos biodiversity directly supports the flower harvesting, horticulture, and other specialist consumptive-use industries. An analysis of fynbos products indicated that the average value of harvests for all fynbos ecosystems is US$4 ha^{-1} y^{-1} at farm gate level (Turpie et al. 2003) (The exchange rate used for all economic values is ZAR6.58=US$1[31 December 2003].) However, the total economic value of fynbos natural capital greatly exceeds this (Turpie 2004) because of the value of the services it provides. Direct nonconsumptive use (tourism), indirect consumptive use (beekeeping and water production), and nonuse (future use and existence) values were estimated to exceed $192 ha^{-1} y^{-1} for mountain fynbos. No comparable information is available for lowland sand plain fynbos.

Biome-Scale Requirements

It was the demonstrated negative effects of IAPs on water resources that prompted government action in the form of the WfW program (Van Wilgen et al. 1996). Economic analyses of the hydrological impacts of IAPs show that large-scale clearance programs are justifiable on the basis of the water benefits alone, despite the high costs involved (Van Wilgen et al. 1997; Higgins et al. 1997; Turpie and Heydenrych 2000; Turpie 2004). Nevertheless, these high costs could have been avoided had potentially invasive plant species been prevented from entering the region and becoming established. Thus, at national and biome scales, the top priority is to implement a screening system and risk analysis, in collaboration with enterprises reliant on IAPs (agricultural, forestry, and horticultural), to ensure that only low-risk species are imported (Richardson and Van Wilgen 2004) and that new species are closely monitored. The second priority is to prevent the spread of emerging invader-plant species, which have a high local proliferation and impact but have not yet spread extensively. The third priority is to plan strategically to control the well-established invasive species, especially *transformers* that replace indigenous vegetation (Wells et al. 1986). Transformer species should be identified and mapped and watersheds prioritized for clearance based on the criteria outlined (table 22.1). Any transformer species not under substantive biological control should be targeted for further research (Zimmermann et al. 2004).

Watershed-Scale Requirements

At the watershed (or catchment) scale, IAPs have the potential to alter landscapes and reduce water availability, thus impacting nonconsumptive-use values of fynbos ecosystems and downstream users, respectively. Consumptive-use values may also be reduced in watersheds that are heavily invaded or support extensive areas of closed alien stands.

Strategic planning at this scale of operation should involve IAP species mapping and the prioritization of management units to minimize further impact. For example, to halt the spread of IAPs in a watershed it is essential that isolated plants and low-density, alien plant populations are cleared first to reduce the invasive front. As some invasive alien species can be dispersed in water, riparian habitat clearance should start from the top of the watershed.

Of the economic and ecological benefits sought, clean water yields will be obtained most quickly following IAP clearance and subsequent restoration of riparian and adjacent plant communities (Scott et al. 1999). If water is the most important natural resource from the watershed, then riparian restoration should be a priority. However, time-consuming riparian clearance of alien plants should not precede the removal of outlying low-density IAP stands, in order to prevent expansion into pristine watershed areas.

Plant Community–Scale Requirements

Mountain fynbos vegetation is fairly resilient to the impacts of alien plant invasion provided that clearance treatments do not damage the long-lived, indigenous soil seed-banks on which post-fire regeneration depends (Holmes and Cowling 1997a). Vegetation cover and structure rapidly recover following IAP clearance, but species richness remains lower and is the most sensitive disturbance indicator at the plant community scale (Holmes and Cowling 1997a; Holmes and Marais 2000; Euston-Brown et al. 2002).

Healthy ecosystems mend themselves, but degradation may approach a threshold below which recovery fails (Whisenant 1999). Although there are some conflicting studies (e.g., Grime 1997; Symstad and Tilman 2001), the level of remaining biodiversity is probably a major factor influencing ecosystem functioning and recovery potential.

Fynbos research indicates potential thresholds for different vegetation and substratum types beyond which ecosystems do not recover sufficiently following invasion to reinstate ecosystem structure and functioning (figure 22.1). For example, following a *fell-and-burn* treatment for clearing hakea (*Hakea sericea*), of Australian origin, indigenous species richness and cover were at 40% of noninvaded fynbos on granitic soil, and severe soil erosion followed (Richardson and Van Wilgen 1986). On a sandstone soil, however, no soil erosion resulted from a similar treatment to clear pine (*Pinus radiata*) and acacia (*Acacia mearnsii*), though recovery of fynbos richness (13%) and cover (45%) values were even lower relative to intact fynbos (Holmes and Foden 2001).

Where there is a low recovery probability above the self-repair threshold, additional interventions will be required to prevent further environmental degradation by potential secondary IAPs and other biophysical changes. For example, fynbos on moderate slopes with fine-textured soils may require physical stabilization following IAP clearance and burning. Indigenous seed-banks are depleted during severe slash fires so that a range of native species, including faster-growing herbaceous and longer-lived shrub species, should be sown prior to the rainy season.

Further actions may be required to restore the direct, consumptive-use values that in fynbos vegetation are strongly linked to biodiversity. Reseeding is necessary for biodiversity recovery in areas exceeding five hectares that have been invaded by closed alien stands for two or more fire cycles, because seed dispersal distances are short (meters rather than kilometers) for most fynbos species (Holmes and Richardson 1999). In areas smaller than five hectares, natural recolonization from surrounding intact vegetation may be sufficient. In all cases, recovery should be monitored closely and additional actions implemented where required.

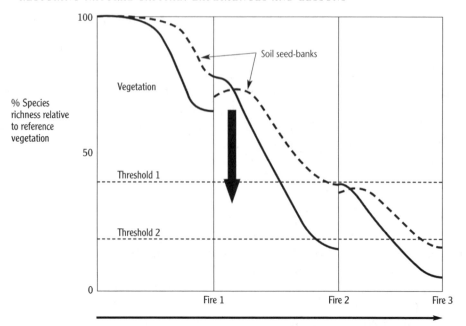

FIGURE 22.1. Schematic diagram illustrating fynbos restoration potential following dense, alien tree invasion in relation to noninvaded fynbos. The black arrow indicates the impact of an unnaturally severe fire on seed-banks and recovery potential. Thresholds 1 and 2 indicate the richness levels below which ecosystem structure and functioning may not adequately recover, on granite and sandstone substrata, respectively. Data from Holmes and Cowling (1997a,1997b); Holmes and Marais (2000); Holmes and Foden (2001); Holmes (2002); and Holmes et al. (2000).

One of the first impacts of closed alien stands on biodiversity is to eliminate the overstory shrub component that is dependent on canopy-stored seeds for post-fire regeneration (Holmes and Cowling 1997a). This guild containing members of the Proteaceae and Bruniaceae, important in the flower-harvesting trade (Coetzee and Middelmann 1997; Rebelo and Holmes 1988), should be reintroduced to enhance natural capital, even where ecosystem functioning and a high proportion of biodiversity have been restored through IAP clearance alone. This is essential as overstory Proteaceae are considered keystone species in fynbos (Vlok and Yeaton 1999).

Finally, we consider that the currently applied methods for clearing invasive alien plants are not adequate for natural capital restoration, where closed alien stands have been present for over two fire cycles (twenty to forty years), or intense fires have occurred over large areas (>5 ha). In most other cases, appropriate IAP control operations alone will result in natural recovery. Fortunately, closed stands represent only 1.2% of alien invasions in mountain fynbos (Rouget, Richardson, et al. 2003), and a high proportion of those may be recent stands with good restoration potential. The situation is less positive for lowland sand plain fynbos, where occupation by closed alien stands is similar (Rouget, Richardson, et al. 2003). In such sites, restoration potential falls rapidly during the first cycle of dense invasion by leguminous trees (Holmes 2002) and may require additional interventions such as soil nitrogen reduction and species reintroductions to minimize secondary invasions.

Costs and Benefits of Restoring Natural Capital in the Cape Floristic Region

Invasive alien species result in significant costs through losses and damage to natural and agricultural systems, estimated for the United States alone at $137 billion per year (Pimentel et al. 2000). Such a comprehensive figure for South Africa is not yet available, but it must be substantial, despite the positive benefits of some invasive species (Van Wilgen et al. 2001). For example, the net present cost of watershed invasion by one tree species, *Acacia mearnsii*, is estimated at $1.4 billion in water lost through transpiration (De Wit et al. 2001).

Biome-Scale Costs and Benefits of Restoring Natural Capital

Negative economic impacts of invasive alien trees and shrubs are likely to be incurred when they attain medium to high densities (>25% canopy cover). The total costs of IAPs in relation to reduced water yields in the fynbos biome amount to $34.7 million per annum, based on alien stand extent (Rouget, Richardson, et al. 2003) and estimated values of water yield loss in medium and high densities (Turpie et al. 2003). Additional economic losses, using average values generated for fynbos vegetation in Turpie et al. (2003), include potentially $0.9 million per annum of florists' material and $0.7 million per annum for pollination services. The negative impact of IAPs on tourism, future use, and existence values are probably profound but extremely difficult to measure. Nevertheless, the total estimate for losses of the more tangible flows from the natural vegetation of the Cape Floristic Region amounts to $36.3 million per annum as a result of current levels of alien plant invasion.

The Working for Water IAP-clearing program spent $219.4 million nationally between 1995 and 2000 (Van Wilgen et al. 2002), and projected costs for controlling IAPs in the fynbos biome average $45.3 million per annum (table 22.2). Although slightly higher than the estimated natural capital losses due to current invasion levels, this expenditure is justifiable in terms of securing the future natural capital benefits. Furthermore, there is an opportunity cost to be considered in not implementing alien plant control, as IAPs will continue to spread exponentially and occupy hitherto noninvaded land (Higgins et al. 1997). Given that the

TABLE 22.2

Costs of the Working for Water (WfW) invasive alien plant control program in the Western Cape Province.

Scale		Costs (US$ million yr^{-1})	
		2002–03 financial year costs	Projected annual requirement
Biome	Implementation	13.63	42.43
	Planning	0.08	0.26
	Research	0.63	1.95
	Biological control	0.23	0.71
	Total	14.57	45.35
Watershed (George: 80 km^2)	Total	0.21	0.12 maximum 0.41 minimum 0.032

Source: WfW unpublished data.
Note: Costs are for a 30-year and 25-year clearing scenario for biome and watershed scales, respectively. The Western Cape Province is used to represent the fynbos biome.

Western Cape has a rapidly growing human population, particularly in the metropolitan areas, and limited freshwater resources, uncontrolled expansion of IAPs into natural vegetation of the surrounding mountain watersheds would have negative consequences for economic development and, potentially, human health in the province.

It has been demonstrated that biological control (research and implementation) offers large investment returns. For example, the calculated benefit:cost ratio of 104:1 was based on a reduction in the extent and spreading rate of long-leafed wattle (*Acacia longifolia*) and related declines in negative economic impacts (Van Wilgen et al. 2004). Should insufficient future funding be secured to ensure the required mechanical and chemical control of IAPs, effective biological control would prevent targeted species from rapidly spreading to the detriment of natural capital.

Watershed-Scale Costs and Benefits of Natural Capital Restoration

A modeling study on the economic viability of IAP removal in the watershed supplying the town of George compared a twenty-five-year clearance and no-clearance scenarios (Larsen and Marais 2001). Water production was perceived to be the most valuable product from the watershed area and was the easiest to quantify. The method used calculated the net present value of the costs (capital, management, and maintenance) of bulk water supply schemes in the watershed over a forty-five-year investment period, with a discount rate of 8%. Results indicated that the average discounted cost of bulk water supply was $0.05 m^{-3} higher for a no-clearance scenario than a twenty-five-year clearance scenario (Larsen and Marais 2001). This suggests that for water supply alone, the benefits of clearing IAPs at the watershed scale clearly outweigh the costs.

Plant Community–Scale Costs and Benefits of Restoring Natural Capital

IAP clearance costs depend on a number of factors besides alien stand density. The IAP species determines whether herbicides are required to prevent resprouting after felling, and the nature of difficult terrain can increase clearance costs. In addition, depending on the sensitivity of the habitat, management costs will vary for handling felled material. For instance, in riparian areas it may be advisable to remove biomass above the twenty-year flood level to minimize the risk of logjams downstream. Where extraction costs are minimal, biomass could even be sold to offset some of the clearance costs.

In mountain fynbos, the total cost of restoring natural capital by clearing low-density (5%–25% canopy cover) IAPs yields a benefit to cost ratio of 25.6 for *Pinus pinaster* or *P. radiata* (4% discount rate over fifty years) (table 22.3). However, as the canopy cover of the alien stands increases, clearing and restoration costs rise, while benefit to cost ratios decrease (to 2.9 and 2.4 at a 4% discount rate for recently closed stands of pines and acacias [*Acacia mearnsii* or *A. saligna*], respectively). These figures indicate the importance of early intervention in minimizing restoration costs and maximizing the long-term benefits. Early clearance of the low-density alien stands and outliers is a cost-effective investment for securing natural capital, as fynbos recovery potential is high at this stage.

Although several studies have shown that fynbos may recover well following dense invasion in terms of plant cover and structure, species richness appears to remain lower than ref-

TABLE 22.3

Costs and benefits of alien plant clearance at community scale in mountain fynbos

Alien stand	Species	Clearance schedule	Average costs of restoration (US$ ha^{-1})		Benefit of restoring natural capital ($ ha^{-1} yr^{-1})		
			Alien control	Additional restoration	Consumptive Use		Benefit:cost ratio 4% (8%)
					Direct	Indirect	
Light	Pine	Initial	86	• None required	0.2	31.6	25.6 (10.6)
Medium dense	Pine	Initial + 1–3FU	318	• None required	1.4	109.1	7.9 (3.4)
Closed, recently	Pine	Initial + 3–4FU	746	• None required	2.7	191.1	3.2 (1.3)
Closed, long term	Pine	Initial + 3–4FU	746	• Reintroduce serotinous overstorey species[a]: $76 ha^{-1}	2.7	191.1	2.9 (1.2)
Closed, long term, and soil damage	Pine	Initial + 3–4FU	746	• Stabilize slopes using physical measures[b]: $532 ha^{-1} • Sow seed mix of fast-growing and long-lived species[c]: $486 ha^{-1}	2.7	191.1	1.3 (0.5)
Light	Acacia	Initial	93	• None required	0.2	31.6	23.8 (9.9)
Medium dense	Acacia	Initial + 2–3FU	395	• None required	1.4	109.1	6.8 (3.0)
Closed, recently	Acacia	Initial + 3–4FU	1,058	• None required	2.7	191.1	2.6 (1.1)
Closed, long term	Acacia	Initial + 3–4FU	1,058	• Serotinous overstorey species: $76 ha^{-1}	2.7	191.1	2.4 (1.0)
Closed, long-term, and soil damage	Acacia	Initial + 3–4FU	1,058	• Stabilize slopes using physical measures: $532 ha^{-1} • Sow seed mix: $486 ha^{-1}	2.7	191.1	1.3 (0.5)

Sources: Restoration costs from field trials (Holmes 2001); benefits calculated from Turpie et al. (2003).
Note: Benefit:cost ratios calculated using both 4% and 8% discount rates over a fifty-year period. Data are actual clearing costs Working for Water (WfW) averaged over the canopy-cover classes (light 5%–25%, medium dense 25%–75%, closed >75%), incorporating initial and up to four follow-up (FU) clearing treatments per stand.
[a] A seed density sowing trial indicated that five seeds/m^{-2} is sufficient to reinstate an overstorey shrub layer of one to two mature plants/m^{-2} (P. Holmes, unpublished data). This is equivalent to 5,000 Proteaceae cones/ha^{-1} (mixed species), collected and sown at a cost of R500 ha^{-1}, or approximately six person days ha^{-1}. Costs are lower than other seed mixes, as cones are quick to collect and may be strewn directly onto the soil to release their seeds in situ (P. Holmes, personal observations).
[b] Commercial installation costs are $2,050 ha^{-1} for logging at 3-m intervals; material costs depend on whether local, alien tree logs are used or treated timber droppers bought (R0.5/m^{-1}) and transported to site (D. van Eeden, personal communication). Comparable installation costs for job-creation program such as WfW are $532 ha^{-1} (C. Marais, unpublished data).
[c] P. Holmes, unpublished data. (The exchange rate used for all economic values is ZAR6.58=US$1[31 December 2003].)

erence noninvaded vegetation due to local extinctions. Plant species with canopy-stored seed-banks, such as overstory Proteaceae and Bruniaceae, are among the first to be eliminated by dense stands of IAPs. As these taxa form the basis for the flower-harvesting industry, it may be beneficial to reintroduce them in areas where this forms part of the local economy. However, species reintroduction costs are high ($76 ha^{-1}) relative to average annual flower-harvesting values ($3 ha^{-1}) in mountain fynbos (table 22.3). A less costly outcome may be achieved without compromising the restoration of the site by reseeding part of the area to establish a plant population that would accumulate seed for natural dispersal following fires.

Vegetation that fulfills some ecological function may recover following the clearance of dense old stands (i.e., >40 years or >one fire cycle) of IAPs. However, fynbos seed-banks and hence species and functional guild richness will be poorer than at sites cleared after shorter invasion periods (figure 22.1). Provided that there are no immediate threats to ecosystem functioning, such as soil instability or lack of indigenous cover to counter secondary alien plant invasions, the additional costs of sowing a comprehensive fynbos seed mix ($486 ha^{-1}) may not be warranted. Species reintroduction via propagated material is an order of magnitude more costly (P. Holmes, unpublished data). Before a final decision is made on whether to reintroduce species, the plant community should be assessed in its landscape context. If the invaded area is extensive, or no indigenous seed sources occur within 0.5 km, then natural recolonization will be slow (Holmes and Richardson 1999) and species reintroduction should be considered to speed up and enrich fynbos recovery. On state-owned land it may even be possible to reduce seed collecting and sowing costs through initiating government-funded, labor-intensive programs to relieve poverty.

Slope-soil stabilization and plant reintroduction will be required where major soil damage has occurred (table 22.3) (Euston-Brown et al. 2002), particularly those vulnerable sites in granitic areas subjected to high-intensity fires used for removing standing or felled invasive alien shrubs. Indeed, soil stabilization should take place before the onset of winter rains to avert severe erosion (Richardson and Van Wilgen 1986; Euston-Brown 2000), though the commercial costs are extremely high, owing to the purchase and transport of treated timbers and labor. If logs or branches of the felled vegetation can be used and labor supplied through the Working for Water program rather than at market prices, costs may be markedly reduced (table 22.3). These costs should be weighed against the in situ benefits of natural capital restoration, as well as the avoidance of potentially expensive soil-erosion impacts downslope. A locally indigenous fynbos seed mix should be sown that includes both fast-growing species and longer-lived components. If the fast-growing species are in short supply owing to recent fires, commercially available, noninvasive, annual grass seed can be used to provide initial soil-surface stability. The costs of such restoration actions generally will outweigh the short-term benefits to a local landowner, while without the restoration of natural capital, the long-term environmental and social costs could be unacceptably high.

Contribution

Natural capital in fynbos vegetation is closely aligned to its biodiversity, because fynbos products, such as tea, flowers, and thatch, are derived from its rich floral diversity rather than its forage value. Invasion by alien trees and shrubs threaten direct, as well as indirect, use and nonconsumptive values.

We consider that for mountain fynbos ecosystems, current IAP-control programs are adequate for restoring natural capital in most situations. Fynbos seed-banks persist in the soil, and, provided that alien plant clearance does not destroy this, sufficient biodiversity regenerates to sustain ecosystem functioning and services, including the supply of clean water, pollination, and tourism and existence values. An exception may be the regeneration of species that underpin the flower-harvesting industry, many of which do not have soil-stored seed-banks. In a few local situations after prolonged invasion or intense fires, indigenous seed-banks may be depleted to the extent that basic ecosystem structure and functioning does not readily recover. In these cases, additional restoration actions such as species reintroductions and physical soil stabilization are required following IAP control.

At biome and watershed scales, alien plant control programs are justified in terms of the benefits for securing future water supplies. Furthermore, the high opportunity costs of delaying clearing underscores the urgency for initiating and sustaining control programs and motivates for investment in biological control as a safety net against future escalations in alien plant invasions. It is more economical to prevent the entry and spread of new invader species, through screening protocols and early eradication, than to control them at a later stage.

At the plant community scale, the first priority is to clear the low-density alien stands (<25% cover) in order to secure natural capital. The benefit-cost ratio for total clearance (initial and follow-ups) of low-density alien invasion averages 25:1 compared to 2.7:1 (at 4% discount rate) for closed alien stands. Not only does it make ecological sense to begin by controlling the lower-density alien stands, it also makes economic sense to remove them before they form dense stands. Additional investment in restoring natural capital, such as species reintroductions, may be warranted in areas that are important in the flower-harvesting industry and at sites where recovery following clearance is unlikely to perform basic ecosystem functions such as soil stabilization. However, costs are high relative to the annual benefits, and only from a long-term and national perspective will the expenditure be justifiable.

Acknowledgements

David Richardson acknowledges support from the DST Center for Excellence for Invasion Biology.

Chapter 23

Return of Natural, Social, and Financial Capital to the Hole Left by Mining

J. Deon van Eeden, Roy A. Lubke, and Pippa Haarhoff

Opencast mining under arid conditions is notoriously slow and costly to repair (Lubke and van Eeden 2001; Burke 2001; Milton 2001). Little attempt was made to revegetate or restore mined surfaces in arid parts of South Africa prior to 1998, when an increase in mining and other environmentally damaging activities led to the promulgation of legislation compelling remediation. The natural capital of the area is the vegetation and landscape from which flows a surprising array of opportunities. This chapter deals with an unusual synergy between a mining company, a museum, and a commercial landscaping enterprise that have funded and implemented restoration of coastal semidesert vegetation at an abandoned phosphate mine and provided sustainable employment for local inhabitants.

The Chemfos phosphate mine is situated in the Anyskop area of the west coast of the Western Cape Province of South Africa (figure 23.1). This harsh, arid environment (mean annual rainfall ca. 200 mm) has supported a succession of land uses over the past 10,000 years, including hunting, pastoralism, agriculture, and mining (Burman and Levin 1974). Despite leaving a legacy of discarded machinery and vast tracts of disturbed and barren land invaded by alien plants, the mining site had two major assets: a vast, partly exposed deposit of fossil mammals and a village with people who could restore and develop these natural resources into a *fossil park*.

Site Background

The post-mining landscape reflects the approximate paleotopography of the early Pliocene. Phosphate deposits were formed in a marine environment when the sea levels were approximately 26 m higher than today (Hendey 1981a). Subsequent sea-level changes and climate patterns have molded the land, which has supported a succession of fauna and flora at different stages in its evolution. Large-scale opencast mining consumed the rich phosphate deposits, destroying the coastal scrub habitat, while on adjacent properties agriculture fragmented and degraded the natural vegetation.

The discovery of one of the richest Pliocene fossil deposits known in 1958 (Hendey 1981a, 1981b, 1982) led to rewarding paleontological research and an exciting post-closure possibility for the site. A mining company (Samancor, now part of BHP Billiton) funded the

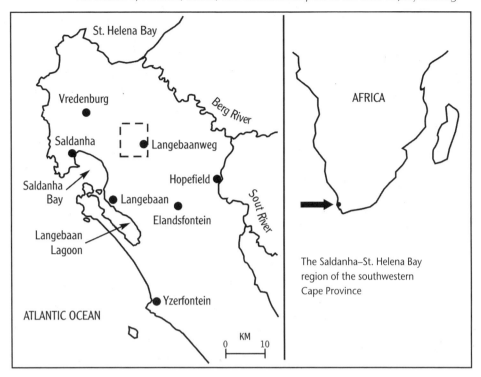

Figure 23.1. The Saldanha–St. Helena Bay region of the Western Cape, showing the position of the Chemfos mine site (reproduced with permission from Hendey 1982).

initial proposal to restore the mining area and maintain the fossil deposits as a national heritage site within a restored buffer of coastal vegetation.

This chapter describes some of the techniques and approaches used in the restoration of the old mine within the context of the new land use adopted for the 650-hectare site. In adapting methods for redressing the environmental degradation, some of the previous employees received training completely unrelated to their previous positions. Thus, this led to the regeneration of livelihoods and dignity for many people by providing socioeconomic benefits not usually associated with the abandonment of mining.

Historical Land Uses of the Natural Capital in the Anyskop Area

For thousands of years hunter-gatherers used the Anyskop area, as is evident from the stone tools recovered (Conard 2003). Subsequently, early accounts by Europeans noted that the local inhabitants exploited the natural resources for food, shelter, and trading (Axelson 1977). For example, in 1648, French explorers recorded trading tortoises and ostrich feathers with the locals while also commenting on the soil fertility and the presence of elephants and other large mammals in the Saldanha Bay area (Burman and Levin 1974).

By 1825, European land use for winter livestock grazing was formalized, which even at an apparently low-impact level rapidly caused the extermination of predators and large herbivores (chapter 7). Since 1907, Anyskop and other regional farms were used for both grain production and livestock grazing (Du Preez 1988; Dirk Loubser, Langebaanweg, Vredenburg District, 2004, personal communication). However, the grazing potential of the natural vegetation is low because dry biomass production averages only 486 kg ha^{-1} yr^{-1} (Craven 1994). Unfortunately, plowing and grazing have contributed to a marked decline in renewable natural capital through destruction of natural vegetation and exposure of the sandy soil to erosion by the strong winds characteristic of this coastal region.

The consumption of nonrenewable natural capital began in 1941 with the establishment of the Chemfos mine. Phosphate extraction rates increased during the 1960s when the small, independent mine was taken over by Samancor. However, phosphate production never exceeded 130,000 tons per year (CES 1996a), and operations were scarcely viable due to low prices and fluctuating demand for phosphate fertilizers (caused largely by droughts) in the maize-growing areas of South Africa. In 1993, economic factors related to aging technology, increasingly deeper deposits, and growing competition from cheaper chemical phosphate forced Samancor to suspend operations.

During the mining period, unsustainable practices—which included the minimum handling of materials—were favored because of the prevailing economic constraints and nonexistent legal requirements. Topsoil and overburden were often dumped injudiciously, with little consideration for the resulting land-form, let alone separating valuable topsoil from subsoil and rock. As a result, natural vegetation recovery from the seed-bank within the fresh topsoil was hampered, and almost 650 hectares of vegetation were destroyed. Practical considerations, such as dust suppression and land stabilization, led to the introduction of invasive alien plants, including *Acacia cyclops*, *A. saligna*, and *Eucalyptus* spp., to recolonize mobile sands, especially on the large slimes dams. Indeed, using seed-bearing branches of these species to brush-pack bare soil resulted in establishing dense stands of invasive alien plants, which also hindered natural vegetation recovery.

No restoration budget was available during the mining era, and Samancor was therefore left with a legacy that required much strategic planning before a closure certificate could be granted by the Department of Mineral and Energy Affairs (DME 1992).

Restoration of Natural Capital as an Exit Strategy

In 1995, Samancor approached an environmental consultant to compile what was called a *rehabilitation plan* and henceforth called as such for a post-mining environment that was self-sustaining and aesthetically compatible with the surrounding area (CES 1996a). It soon became obvious that the major priority should be ecological restoration around the valuable fossil site (Hendey 1981b, 1982), where research has already generated over seventy scientific publications. Important extinct vertebrate finds include *Agriotherium africanum*, the first bear discovered south of the Sahara, and *Homiphoca capensis*, a true seal, as well as four penguin species (Hendey 1981b).

In 1996, the fossil-rich area of the mine was declared a National Heritage Site, and therefore restoration had to provide for future paleontological research, while making the site suitable for use by the Iziko Museums in Cape Town. Indeed, the exit strategy was driven by a vision for a post-closure mine use that exceeded the minimum requirements set by legislation (CES 2001). Although it receives no direct benefits, BHP Billiton funded and continues to fund the closure process and is responsible for natural capital restoration.

Conceptual Rehabilitation Plan

A rehabilitation plan was prepared (CES 1996a) and followed by a detailed scheme for revegetation strategies (CES 1996b). Action was required on a broad front that included decommissioning the plant and associated buildings, while upgrading useful infrastructure at both the mine and the village. Natural vegetation restoration and setting up a fossil park provided a major focus for regional development (BCD 2000; CES 2001).

Mining infrastructure with no potential future use was demolished at a cost of US$486,300. However, farmers and merchants bought reusable steel structures and scrap metal, while fertilizer distributors purchased unsold phosphate. This provided employment for six months to ten local men, though the majority of work was completed by outside contractors.

The old office buildings were upgraded, and a tourist and administration facility was created for the Iziko Museums. Negotiations are currently under way for creating a joint management venture, with the Iziko Museums as the new owner and supported financially by BHP Billiton until the project becomes self-funding. As of 1998, the post-mining land-use and development support amounted to $395,000.

Socioeconomic realignment of the mine village was carried out at a cost of $378,000 through June 2004. This included the transfer of assets such as houses, a church, a community hall, and fifty hectares of land to a newly formed management company comprising residents and representatives of the mining company and the fossil park. The mining company accessed state subsidies for partial upgrading of facilities, and contributed the balance, as well as financing the consultation and management required during the five-year process (2000–2005).

Vegetation restoration activities, undertaken at a cost of $898,846 over ten years (1996–2005), supported about twenty people annually. These included sand stabilization, reintroducing the diverse array of plants and animals, and generally improving the aesthetics, which, combined, provided a suitable setting for generating maximum renewable flows (e.g., income, education) from the site's fossil resources.

In the initial study by CES (1996a), the vegetation and soils of the mine and nearby reference sites were surveyed (Boucher and Jarman 1977) to select species and provide background information of the original habitats (Lubke et al. 1998). The rehabilitation plan implemented a series of thirty-two trials on different soils resulting from mining operations (CES 1996b; CES 1996a). These included replicated evaluation of different topsoil combinations, soil amelioration, brush packing, and wood chip mulching. Seeding and planting at different densities were also assessed, as was the response of different species to the soil types (CES 1996b; Lubke et al. 1998). Using an adaptive management approach, the monitoring and evaluation results were adapted and applied on a larger scale and further refined for implementation into new areas (CES 1997), with all the spatial information recorded on geographic information systems (GIS) for future reference (CES 1997, 1999).

Species Evaluation and Selected Techniques for Restoring Natural Capital

Establishing natural vegetation cover was considered essential due to the degraded nature of the post-mining environment and the soil exposure that resulted from invasive alien vegetation clearance. There were few native pioneer plants available for seed harvesting on adjacent farms, because the vegetation had not recently been burnt. By contrast, vegetation alongside local roads provided an important seed source for pioneer plants because these areas were recovering from the impacts of construction and pioneer species were abundant (Lubke and van Eeden 2001).

As seed collection and processing were costly, it was beneficial to use appropriate species for the prevailing soil conditions. In fact, 90% of the seeds were collected by hand and processed separately, with limited use made of vacuum harvesting. In addition, as the phenology of native species was not well known, and influenced by the fluctuations in rainfall and temperature, continuous revision of the timing of seed collection was necessary to attain greatest results and flexibility.

Post-harvest seed handling contributed significantly to the total seed cost. Traditional techniques were adapted to deal with the large volumes of seed required for restoration on a scale of hundreds of hectares. It was decided not to clean the seed, but rather to process the collected material to allow seed release from the fruits. Thus, to determine the required quantity, a *clean seed equivalent* was calculated per unit volume of mixed seeds and plant detritus.

Plants that readily released their seeds (ca. 65% of species collected) were dried under cover, with the material turned regularly to allow rapid desiccation and seed release. Seeds of the remaining 35% of species were processed using a hammer mill. The seeds were then washed from fruits (pods, capsules) and allowed to dry. As the mill damages some large seeds, a motorized handheld vacuum with a plastic impeller was used as an alternative. The sucking impeller provided the correct impact level for the material passing through to shred the dry fruits without excessive seed damage (Lubke and van Eeden 2001).

Seeding took place in autumn (April–May), and germination during the first winter season (June–August) in 1997 was good. However, warm winds during spring (September) resulted in high mortality rates, particularly of short-lived perennials that colonize disturbed sites. The initial seed mixtures used were as diverse as possible and included fifteen annuals, seven grasses, twenty-nine shrubs, and five trees.

In addition to sowing seed in the winter season of 1997, transplanted seedlings were collected from adjacent areas and established in trial sites. They survived well, as they were a year older than seedlings established from seed and had stronger root systems and greater reserves. As no local indigenous plants were commercially available, propagating plants was necessary. An experimental nursery was established and comparisons were made between the success rates of propagation from seeds and from cuttings of local species (Lubke and van Eeden 2001).

As rooted cuttings were required for planting out at the end of May, when rainfall patterns were reliable, cuttings were collected during February, since commercial cutting production normally allows for six to eight weeks. Initial results were poor, with few cuttings producing roots in the heated propagation fog bed, due to the dormancy of most of the plant species during summer. Experimental propagation during spring yielded much better results and thus rooted cuttings were grown in trays and left dormant for the five summer months, November–March, before being transplanted.

Hydroseeding of a selected seed mixture (10 kg ha^{-1}) suitable for the various soils was carried out in April prior to the first rains, with different seed combinations used for sandy and calcareous soils. Additives included microbial inoculants, FireGrow seed-germination stimulant, and Hydropam soil binder, to reduce dust. FireGrow, a plant-smoke derivative that stimulates seed germination, was developed at Chemfos (De Lange and Boucher 1990), and the optimum concentration of this additive in the hydroseed mixture was determined by bioassay (Meets and Boucher 2004).

By 2002, the rehabilitation plan was achieved with good vegetation cover on the mine site. Invasive alien plant removal began in 1997, and by 2004 all mature alien vegetation, except that maintained in two research sites, had been eradicated. In addition, the Plant Protection Research Institute agreed to use the Chemfos mine as an experimental site for the release of seed-feeding weevils in the biological control of *Acacia cyclops* (Donnelly 1992, 1997); this ongoing program has been partially successful (CES 2001).

During 2003 and 2004, the restoration emphasis was on further diversification in the cultivation techniques. The production of slow-growing geophytes and climax shrubs in the nursery was undertaken for reintroduction once the plants were well established. During the last maintenance phase (until August 2006), this activity was continued.

Community Involvement in Restoring the Natural Capital

Adaptability through diversification enabled the local community to reinvent themselves socioeconomically. Utilizing different layers of natural capital, the community transformed itself, ensuring survival in the short term while exploring opportunities to ensure independence from mining and move toward long-term sustainability.

This process of exploring their own abilities, and alternative ways of generating income streams, could not have come about without a change in mindset of all those involved. The resulting increase in self-worth may well contribute to the continued refinement of sustainable consumption of flows, ensuring not merely survival but also long-term prosperity and well-being.

Emerging Wood-Selling Contractors

Removal of the invasive tree species, *Acacia cyclops* and *A. saligna*, provided firewood for local use. Wood sellers were given free access to managed woodlots to exploit this resource, and revenue from sales supported fifteen families for two years. In total, from 1996 to 1998, three contractors sold 4,500 m^3 of firewood with a value of $19,000. If this practice could be sustained on the 650 hectares of land, the resulting flow from firewood would support five families indefinitely, but at the cost of the natural vegetation being replaced by alien woodlots.

The removal of invasive alien plants with a commercial value was followed by those with no firewood value, as well as the disposal of seed-bearing brush by burning. Subsequent removal of germinating seedlings required a minimum of five years. The total cost of the alien plant-eradication program was $395,137 or $608 ha^{-1}. By comparison, the flow generated by the consumption of the stock represents only 0.045% of the cost of eradicating the problem, which was largely due to the high percentage of young plants yielding no firewood.

Communal Vegetation Restoration

Samancor considered that the local mining community needed retraining to carry out the natural vegetation restoration. Creating a local skill base in horticultural techniques would allow the restoration techniques, developed at Chemfos, to be applied to a much wider area, generating a self-sustaining enterprise (Lubke and van Eeden 2001). Additionally, it has resulted in other activities such as commercial plant propagation and landscaping that have supported up to 120 people since 2004.

Seasonality of Restoration Work

The seasonal nature of the restoration work, under arid conditions, resulted in a fluctuating labor demand and uncertain employment opportunities. The seasonal trend in manpower requirements is reflected in the monthly salaries from 1996 until 2003/04. In autumn and winter (April to August), the monthly salary total (i.e., the total wages earned by all the people employed for the month) for the mine project averaged $6,000, when more people were working, compared with only $2,900 in the summer months (figure 23.2).

Unstable income can lead to an array of problems among workers and their families, as well as the occasional loss of skilled and semiskilled horticultural workers to other industries. To address this problem, other opportunities were sought to expand the income base of workers so as to ensure a more consistent cash flow and economic sustainability. It was important to retain the restoration skills while broadening the base of the enterprise, so that work could continue throughout the year for all of the permanent and most of the seasonal staff.

The seed collection and processing skills, developed during the Chemfos mine restoration project, now provide employment for more than twenty laborers. Since 1997, new South African regulations obliging industries to restore damage caused by their activities have created an opportunity to supply seed to developers. Indeed, the range of plants propagated for

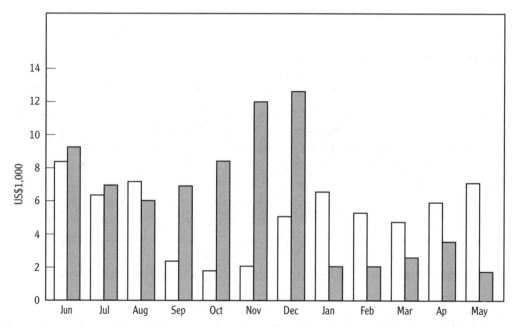

FIGURE 23.2. Seasonal trends in income flows from Chemfos project (open bars) versus landscape gardening and outside restoration contracts (shaded bars). Y-axis shows total combined monthly income for employees.

the project became increasingly popular in other restoration projects, as well as for domestic gardens and the business premises, because their hardiness and low water requirements made them ideal for landscaping in semiarid environments. In 2005, approximately 300,000 plants were propagated for other projects.

The Life Span of the Natural Capital Restoration Project

Initially, the project had a proposed life span of five years, based on the anticipated time to clear alien plants and introduce stable, self-sustaining natural vegetation at the site. Due to climatic constraints (below average rainfall for three years) and higher than predicted maintenance requirements, the project life span was extended. The ten-year plan included two more seasons to diversify vegetation by introducing propagated geophytes and climax vegetation, and two further years of maintenance with continued control of invasive alien plants. It is considered that after this period the old mine site can be easily managed as a conservation area, without risk of it reverting to a degraded landscape and without specialized maintenance.

It is unlikely that the restoration of the mine would have been as successful if this project was the sole activity for the team. The overheads and management costs of the business were partly covered by other commercial projects in the region, thus reducing this component of the mine restoration cost to 50% in years four to six, and to below 15% in years seven to eight. Seeds collected from local plants provided a valuable resource not easily obtainable in the region, and selling the surplus to other projects allowed the restoration budget costs to be reduced. The contractor also paid rent for the use of the facilities, and the combined contribution of income from rent and seed sales amounted to $30,000 between 1997 and 2002.

Between 2002 and 2005, the amount invested in restoring vegetation at Chemfos declined as the process drew to an end. This influenced project viability, as the initial agreement between the client and contractor allowed for a fixed cost, plus 14% for overheads and profit, with a limit on management expenses. For the mining company this was very cost-effective but had only limited appeal for restoration contractors. The unusual nature of the project and the contractors' commitment to success, however, led to exploring other avenues of ensuring sustainability. Additional work was generated, which increased the income and profitability for Vula Environmental Services, while safeguarding the retention of the local skill base for related projects (figure 23.3).

It is essential to consider the financial viability of a restoration project for the duration of the process, that is, before the environmental damage, such as that brought about by mining, has been carried out. Restoration costs should form a central part of any mining feasibility study. Hence, if restoration is undertaken during the life span of an environmentally detrimental industrial activity, cost fluctuations and viability can be offset. This becomes economically more problematic following cessation, such as at mine closure, if these additional expenses have not been anticipated and budgeted for correctly.

Establishment of the West Coast Fossil Park

The Iziko Museums saw the establishment of a fossil park as an opportunity to add *public paleontology* to their traditional education and academic roles. The well-curated collection

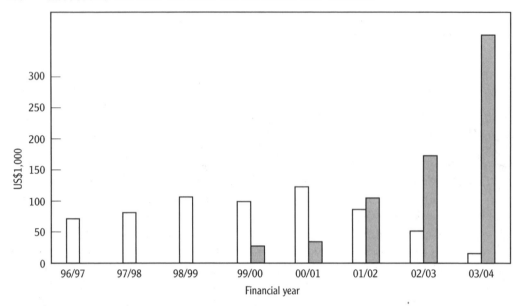

FIGURE 23.3. Comparison of wages earned from the Chemfos project (open bars) versus other activities, such as landscape gardening and outside restoration contracts (shaded bars), demonstrates the declining flows generated by the Chemfos project toward the completion phase, and the growth in alternative, restoration business activities.

housed in Cape Town (Hendey 1981b, 1982) is largely reserved for scientific purposes, with few specimens on display. Establishing a new facility where the public could see fossils in situ made it possible to unlock sustainable nonconsumptive flows from a previously unexploited stock.

Maximizing the potential benefits from the fossils, however, requires capital that could not be generated by the current project. The park operating costs during 2003 were $127,660, while the earnings were only $37,900. The shortfall was funded mainly by the mining company (as part of the restoration process) and by other donors. A business plan was formulated to make provision for new display and educational facilities, which would accommodate increased numbers of tourists and students. The enhanced visitor experience should prolong the visit and create greater opportunities for contributing to the income and educational flows.

Implementing the business plan required an investment of $1,368,000 over six years. The cumulative interest from this investment on the South African financial market was expected to yield the $567,000 required for the development of the facilities. Subsequently, after six years the mining company could withdraw its investment and the fossil park would thereafter be self-funding. The mining company has pledged funds to support the development.

Contribution

Restoration projects situated in remote areas can contribute greatly to the socioeconomic well-being of local communities. However, the limited life span of the restoration work could merely delay the inevitable hardship for those remaining if these projects do not develop and diversify into long-term sustainable activities. Though this is not the primary concern of mining companies, the human capital lost when skilled people fail to find alternative employment is as tragic as the natural capital loss. For example, social decay can lead to environmental damage through overexploitation of natural resources such as vegetation and wildlife by uncontrolled harvesting and poaching.

Natural capital restoration at the old Chemfos mine will never be truly completed because much of the capital has been consumed. Nevertheless, the establishment and sustainable use of the restored environment can contribute greatly to mitigating the damage caused by mining and has generated many benefits. The people living in the former mining village, now called Green Village, have employment opportunities connected with the fossil park, which employs eight villagers on a full-time basis. The restoration project has also created job opportunities for Green Village and the broader local community. In addition, schools and universities benefit as they have access to an exceptional paleontological site for educational and research purposes, which is preserved for posterity. On a broader scale, the fossil park and associated restoration site is a novel tourism attraction, which should generate long-term economic returns.

Costs and timeframes of restoration projects can be underestimated during the conceptualization and lifetime of environmentally damaging projects, such as mining. Continued reevaluation and adequate flexibility are required to ensure the successful restoration of natural capital to generate flows in a severely modified environment, with consideration given to both natural and human capital. Hence, one should be evaluating appropriately the potential natural capital "hole" left behind *before* an environmentally damaging activity, such as mining, begins.

This case is unusual in that few sites can fall back on the value of their fossil resources for educational and research outreach or have the viable option of retraining the local community to ensure sustainable alternative employment in situ. A more general lesson is that the values that people place on natural capital, such as coastal vegetation, change and increase over time as new sustainable flows (e.g., use of plants in horticulture and landscaping) are recognized and marketed.

Acknowledgements

The authors wish to thank BHP Billiton for their vision and continued financial support of the project and commend the enthusiasm and determination of many people in developing the West Coast Fossil Park, notably our colleagues from the Iziko Museums of Cape Town. This chapter is dedicated to the workers and members of the local community who made the restoration process a reality.

Chapter 24

Protecting and Restoring Natural Capital in New York City's Watersheds to Safeguard Water

CHRISTOPHER ELLIMAN AND NATHAN BERRY

In the late nineteenth century, New York City, realizing that local water sources could not satisfy the long-term needs of its burgeoning population, decided to tap into the rivers and streams of the upstate Catskill and Delaware River (Cat-Del) watersheds (figure 24.1). From 1907 to 1965, the City built a much heralded—but locally controversial—engineering behemoth that today channels through a series of reservoirs and tunnels more than one billion gallons (4.4 billion liters) of clean, unfiltered water to nine million New Yorkers every day, up to 125 miles (200 km) away.

For decades, while the City focused on the reliability of its engineering infrastructure, there was less concern for the natural health or future of the Cat-Del watersheds, which provide 90% of the City's daily water supply. In the late 1980s, however, the United States Environmental Protection Agency (USEPA) promulgated stricter rules governing the management of unfiltered surface-water supplies. The new regulations ended a long period of relatively passive watershed management and coincided with the realization that water quality was increasingly threatened in many of the City's reservoirs. The City faced a choice: either devise a detailed watershed-protection plan to safeguard the water supply or build a filtration facility to remove impurities. The capital costs of filtration alone would fall in the $6–10 billion range and require annual operating costs in excess $300 million. The City estimated that for $1 billion, it could restore and preserve its upstate resource and avoid the crippling cost of filtration. With the financial benefit of avoiding filtration through pollution prevention so clear, the City chose to follow a watershed-protection program that many consider to be the premier example of the economic rationale for protecting and restoring natural capital.

To date, the City's watershed-protection efforts have sustained and arguably improved water quality. Its long-term strategy for watershed protection is, however, not clearly defined; in fact, it is currently under review and debate, which in many ways is timely. The watershed today confronts more intense threats and challenges from those faced at the establishment of the protection plan nine years ago. Addressing these threats requires a full appreciation of the value of natural capital.

This chapter summarizes the steps taken thus far to secure and restore the natural capital of the City's watersheds to maintain and enhance ecosystems services and will argue that the initiative has more political and conceptual hurdles to surmount in order to ensure the long-term protection of this most valuable asset.

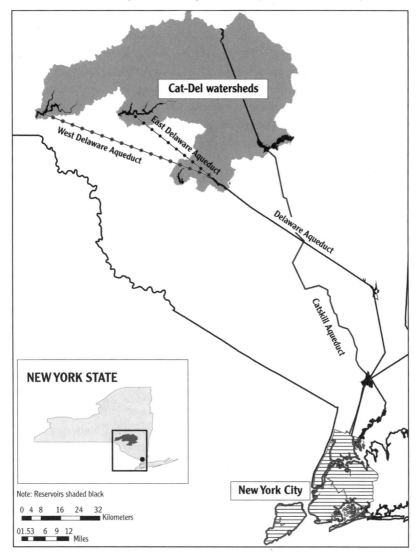

FIGURE 24.1. Water from the Cat-Del watersheds is transported by two underground aqueducts to three holding reservoirs located closer to New York City, wherein the water from the Cat-Del reservoirs is mixed and allowed to settle, then brought to the City's water distribution facility. Map courtesy of Gillian Weber (Source: Open Space Institute). For a more detailed map, go to DEP website (http://www.ci.nyc.ny.us/html/dep/html/wsmaps.html).

Building a Water System

Following the American Civil War (1860s), the eastern forests were widely clear-cut to fuel the industrialization of a booming nation. Approximately 80%–90% of the Cat-Del region was logged intensively (Kudish 2000), first for white pines, then hemlock for local tanneries, and then cleared for agriculture. The widespread landscape devastation and siltation of the streams prompted concern for water quality and led to a rare New York State Constitutional Convention in 1894 that established the Catskill Forest Preserve and, later, the Catskill Park.

The state mandated that its lands within the park would remain "forever wild" and would never be developed or harvested for their natural resources. This level of protection had never before been granted, in New York or elsewhere, and it inspired a nascent wilderness movement that influenced Theodore Roosevelt, Gifford Pinchot, John Muir, Bob Marshall, and Aldo Leopold—the group that first articulated the value of natural capital and the public goal of not just protecting, but restoring, nature's assets. The full flowering of this intellectual movement was the 1964 federal Wilderness Act, established seventy years after the Catskill Forest Preserve.

Initially it was New York State that restored upstate watershed lands, usually by acquiring abandoned land and allowing reforestation. New York City, with the concurrence of state government, saw the Cat-Del watersheds as its principal water source and, in 1907, began work there on its new water system. Armed with land use oversight and powers of eminent domain, the City condemned lands and evicted and flooded entire villages. The theme of an overbearing urban interest eclipsing local wants and needs persists as a real and recurrent perception, afflicting the political dynamics in the Cat-Del region to this day. By 1927, the Ashokan and Schoharie reservoirs, providing 148 billion gallons of storage, were added to the City's water system. By 1965, the Cat-Del system comprised six reservoirs with 475 billion gallons (560 billion liters) of storage, fed by a watershed that covers over 1,600 square miles (4,200 km^2) (figure 24.1).

The Federal Surface Water Treatment Rule

In 1989, the USEPA promulgated the *Surface Water Treatment Rule* (SWTR), which requires filtration of all public surface-water supplies unless water system operators meet certain requirements. In addition to satisfying numerical water quality criteria, public water system operators must maintain a watershed-management program with a goal to maximize their land ownership and/or controlled land use. The rule stipulates that a "public water system must demonstrate, through ownership and/or written agreements with landowners within the watershed, that it can control all human activities which may have an adverse impact on the microbiological quality of the source water" (National Primary Drinking Water Regulations 2002).

Most other unfiltered urban water systems in the United States owned a majority of their watersheds, and none owned less than 40%. New York City owned only 36,000 acres (14,400 ha), less than 4% of Cat-Del watershed lands, and New York State owned another 20%. With so little watershed land controlled by public agencies and specifically managed for water purposes, the Cal-Del system was in danger of not complying with the SWTR.

Understandably, the New York City Department of Environmental Protection (DEP), with its major construction projects and challenges, has historically filled its large staff with engineers, most of whom innately prefer a technological approach, such as filtration, to water-quality issues. For the past thirty-five years, DEP has been working to build a mammoth backup water-delivery system, the City's Water Tunnel Number Three, at enormous cost and engineering challenges. This and other projects have dominated the budget of the agency, with watershed-protection initiatives receiving less attention. This internal priority setting had been reinforced by a disinclination to address upstate enmity toward City watershed oversight. A filtration plant at the vast scale required by the City was in many respects a more discrete effort than alternatives such as land acquisition and regulation. The staggering cost of filtration, however, compelled the City to reach out to watershed communities and landowners in order to negotiate an agreement to protect water quality.

The City's primary programs fell into two broad areas: improving water quality–related infrastructure and strengthening land management and acquisition. Watershed landowners, aware of potentially intensified City regulation and intrusion, and unwilling to have their local control further compromised, resisted through political influence at the state level. To defuse the conflict and avoid political gridlock, representatives of New York State, New York City, watershed communities, and certain environmental groups forged the historic New York City Watershed Memorandum of Agreement (MOA). Signed in January 1997, the MOA brought about the first material changes to watershed regulations since 1953. The overarching goal was to protect the watershed without compromising the lives of its residents. It adopted land-acquisition targets, procedures, and funding levels; introduced more modern watershed regulations; established requirements and funding for improving water-related infrastructure; and committed technical and financial resources for environmentally sound economic development in watershed towns.

Improving Water-Related Infrastructure

The USEPA and the MOA identified a number of infrastructure initiatives the City had to advance. Chief among them were the repair or replacement of septic and other subsurface systems, wastewater-treatment-facility upgrades, and storm-water containment. Infrastructure improvements were imperative to limiting source-point pollution in the Cat-Del region, thereby preserving and restoring its natural attributes, particularly water quality.

At the establishment of the protection plan, up to 50% of the watersheds' 30,000 septic systems were either improperly sited, not functioning properly, or not working at all (USEPA 2000). The proximity of these systems to waterways presented a great, yet largely remediable, threat to the watershed. In keeping with the premise of the MOA, that upstate residents should not shoulder the cost of watershed protection by themselves, the City agreed to contribute 100% of the septic repair or replacement expenses, provided the eligible participant's primary residence was in the watershed. For homeowners whose primary residence was outside the watershed, a growing demographic, and whose septic system needed repair or replacement, the City would contribute 60% (USEPA 2000). As a result, the City has, since 1997, spent more than $20 million and has allocated substantial additional funds to continue the program. While not curing septic problems, the program has shown that septic waste at current population levels can be controlled.

Similarly, over forty wastewater treatment plants (WWTPs), which process both community sewage and storm water, were discharging, prior to the agreement, effluent levels incompatible with watershed health. Since the signing of the MOA, the City has committed more than $270 million and upgraded, or is in the process of upgrading, all WWTPs in the watersheds (New York City Government 2006). The City has also made, and continues to make, substantial infrastructure improvements on roadways, sewer lines, and sand-and-salt storage facilities. In total, the City has already spent or committed roughly $700 million to improve water quality–related infrastructure in the Cat-Del watersheds (DEP 2006, personal communication).

Land Management and Acquisition

The USEPA notes that owning the land in a water system operator's watershed "is the best means of protecting water quality" (USEPA 2006). To that end, the City has designated priority regions and land parcels in the watersheds (on a scale from 1 to 4) and allocated $250 million to purchase land in fee or easement over a ten-year period (with the option to spend an additional $50 million if needed). Land is purchased from willing sellers at the land's fair market value (FMV) and then managed by the City. Where the City purchases easements, the land continues to be owned and managed by the seller, but the development rights are held and extinguished by the City. The easements are often valued at around 60% of FMV; given their flexibility, landowners have been willing to sell easements, and this appears to be a win-win situation for both parties.

Traditional rural land uses, such as farming and, in particular, forestry, are generally compatible with watershed management, if best management practices are adopted and watercourses are adequately buffered. In 1992, the City established and funded the not-for-profit Watershed Agricultural Council (WAC) to focus on means to support local agriculture while protecting its water supply. Once a farmer adopts a WAC-approved farm management plan, the farmer's structural costs of implementing the plan will be paid. Of commercial (large) watershed farms, comprising over 80,000 acres (32,400 hectares), 92% have signed up for an approved plan. These farmers are also eligible to sell agricultural easements to WAC. According to WAC executive director Tom O'Brien, farmers accept easement offers about 60%–70% of the time. To date, WAC easements and pending purchases protect fifty-six farms encompassing 14,500 acres (5,900 hectares). Additional farms have applied for easement sales that cover another 22,000 acres (8,900 hectares), although there is no guarantee that WAC will extend offers to all these farm owners (T. O'Brien, director of WAC, 2006, personal communication; A. Olney, WAC easement program manager, 2006, personal communication).

Since 1997, the City, through DEP or WAC, has approached Cat-Del landowners representing approximately 380,000 acres (153,800 hectares) and acquired over 72,000 acres (29,140 hectares) from willing sellers through fee (48,000 acres, 19,400 hectares) and easement (24,000 acres, 9,700 hectares) at a cost of roughly $180 million (D. Tobias, head of DEP's land acquisition program, 2006, personal communication). Including state lands, just over 30% (up from 24%) of the Cat-Del lands are now protected.

Land Use Today

The Cat-Del region, from the rugged east to the rolling landscape farther west, is home to countless creeks and rivers, and human use generally occurs in close proximity to these watercourses. A USEPA assessment of the region noted that "as a result of topographic constraints, the majority (90%) of urban development and agricultural land use is located near streams" (Mehaffey et al. 2001). At the establishment of the watershed-protection program, the Cat-Del region had experienced relatively little change in the type and extent of land use over much of the past century. The Cat-Del region was long home to small-scale agriculture and silviculture businesses, which peaked in the decade after World War II and have since been in decline, as farming and forestry have become increasingly globalized industries. Even with agriculture programs such as WAC's, the region continues to lose farmers and farmland, often to development. Between 1992 and 2001, forest cover on private land in the region dropped from 86% to 79%, representing a net loss of close to 123,000 acres (50,000 hectares) of land classified as "forest" (Tyrell et al. 2004).

Over the past decade, the United States has witnessed a booming real estate market, in both housing starts (new construction) and prices, and from 2001 to 2004, New York State ranked high (18th out of 50) in the number of new housing units. Of the four Cat-Del counties, three placed in the top quartile in percentage increase in housing starts among New York State counties (United States Census Bureau 2005). Real estate prices have risen dramatically—in many areas by multiples of two or three over just a few years—and continue to rise in the Cat-Del region, creating incentives to convert more working lands to residential or commercial properties. Five casinos and one other mega-resort project have recently been proposed (and two are still under consideration) for the Cat-Del region, threatening further potential pollution sources.

These figures suggest that New York City and New York State must do more to safeguard the watershed from construction and development activities. The City has regulated WWTPs in terms of treatment levels and quantity of daily flow, but it has not, with the exception of phosphorus- and fecal coliform–restricted basins, been able to place a limit on the number of WWTPs allowed or the density of septic systems in specific regions, let alone the watershed as a whole (MOA). To do so would firmly limit development and the number of people permitted to settle, and as such would antagonize the watershed population; hence, this option remains politically unpalatable. Under the circumstances, this lack of regulation is not surprising, but it nonetheless poses a long-term threat to the health and security of the downstate water supply.

Furthermore, critical regulations the City has promulgated may not sufficiently address pollution vectors. To control the quantity and quality of storm-water runoff from large-scale development, the City requires the submission of a quantitative analysis demonstrating that the new construction will have no effect, or a positive one, on storm-water quality (NRC 2000). Not only is this a difficult analysis to perform, it is optimistic to believe that storm-water quality will improve or be unaffected by most new development. And while the City has imposed setback distances and buffer zone regulations on new construction of impervious surfaces and the storage of petroleum and other hazardous materials, these distances may be insufficient to protect waterways. Similarly, the City's watershed rules do not preclude development in areas with steep slopes, despite the inherent dangers (NRC 2000). Such weaknesses in the MOA allow development around some of the most vulnerable reservoirs and waterways.

Since the City is unable to sufficiently regulate the land it does not control, it has established the goal of buying, either in fee or easement, the most vulnerable land parcels. While the city has successfully protected at least 7% of the Cat-Del watersheds since 1997, limitations hamper its effectiveness and efficiency. First, the City is currently prohibited from purchasing any land with a habitable structure. Therefore, if a landowner wants to sell his land and house, the City is essentially eliminated as a bidder, no matter how crucial the property is to watershed protection. Moreover, the City's use of retrospective FMV appraisals, which tend to lag market valuations, and protracted review process lasting up to eighteen months or more, puts it at a competitive disadvantage when competing against developers for choice land. Additionally, the land that is most dramatically appreciating in value is that closest to waterways, yet the City has been unable to respond to economic principles that would suggest it should allocate acquisition dollars as quickly as possible. As land prices escalate, selling to developers becomes more attractive for watershed property owners than walking through the City's lengthy land acquisition processes. The City must acquire land without bureaucratic rigidity. It should increasingly look to DEP and WAC easements over full-fee acquisition as a means of protecting the natural capital of the land. Many landowners who are unwilling to transfer their land to the City would be willing to sell the development interests in their properties, if the terms were attractive. Most important, the City needs to allocate more funds to the program and become more flexible with the price it is willing pay for land. With dramatic increases in land prices, the City naturally needs to commit funding to maintain a meaningful acquisition program. Finally, as the City is, in fact, buying water quality, not merely land, a retrospective FMV appraisal should not apply, but rather the City's offer should reflect the land's contribution to water quality and quantity—its natural capital value. For example, the City might offer a "water quality premium" over FMV for crucial land parcels (priority levels 1 and 2).

Challenges and Potential Solutions for Watershed Protection

The historic *New York City Watershed Memorandum of Agreement* (MOA) was driven by a federal mandate and is the product of political negotiation. In promulgating the SWTR, the USEPA set limits on pollutant levels in the City's water supply, effectively imposing a floor on the City's protection efforts (the lowest legal level of protection). Similarly, federal filtration rules have had the effect of placing a ceiling on the highest level of watershed protection; that is, the City should be willing to fund filtration avoidance up to the cost of filtration. There is an optimal level of protection along the continuum, from ceiling to floor. Calculating that level is speculative and beyond the scope of this chapter, but it is likely that the MOA's protection prescriptions constitute the floor. The watershed protection program was born of negotiation, in which an important City objective was to limit its financial commitment, and the watershed communities' objective was to keep the City out of watershed matters altogether. In many respects, then, these two parties had common goals—that is, to meet the lower threshold of USEPA standards. This lower threshold is not fixed, but rather moves over time as water security necessitates. And with unprecedented conversion of land from open space now under way, protection efforts should rise accordingly.

The City could not anticipate the move by downstate residents to acquire property upstate following the September 11, 2001, attacks on the World Trade Center. In fact, the tragedy only intensified the existing trend in real estate appreciation in the watershed region, which started, ironically, just as negotiations for the MOA were concluding in 1997. Arguably, the City's most effective and controversial tools—aggressive regulation of land use and selective use of eminent domain—were essentially negotiated away in 1997, leaving the City hobbled in its efforts to protect water quality and quantity for half the population of New York State. To be sure, the City has done quite well in restoring water quality by upgrading local infrastructure and addressing agricultural runoff and nonpoint-source pollution. But where the

City has had less success, and where the MOA is most dated, is in averting the real and escalating threat of residential and commercial development.

It is unclear what steps the City will elect to take to confront the adverse land use trends in the region. There is an inverse relationship between the City's land use regulatory powers and the cost of watershed protection, particularly acquiring interests in land. The powers the City ceded in the MOA have undoubtedly made watershed protection more difficult and costly, as it now must acquire the rights and interests that it could once stipulate. The scope and scale of Cat-Del protection require that multiple levels of government work in concert to protect the region's natural resources. Despite the City's recent success in safeguarding its water supply, more support from the state is needed to ensure robust protection. In addition to requesting the state to take a more active role in watershed protection in general, the City should press the state for broader land use oversight and targeted regulations. Without more state involvement in safeguarding the Cat-Del water supply, its protection will be incomplete and insecure.

Contribution

This case highlights two major challenges to restoring natural capital in a watershed with multiple owners and private stakeholders. The first is navigating complex upstate-downstate (or upstream-downstream) relationships where one party's actions and goals conflict with those of the other. The second is gaining widespread acceptance that ecosystem processes in intact natural landscapes can filter large quantities of water more cost effectively and reliably than human-made filtration plants. In turn, realistic values need to be placed on natural capital, as the use of land for ecological services competes in the economic market with alternative uses such as property development.

Chapter 25

Making the Restoration of Natural Capital Profitable on Private Land: Koa Forestry on Hawaii Island

LIBA PEJCHAR, JOSHUA H. GOLDSTEIN, AND GRETCHEN C. DAILY

Hawaii Island is a mosaic of public forest and private land, and its natural capital includes more endangered species than any other state in the United States (Czech et al. 2000). This necessitates innovative restoration, especially on private land, to protect and expand biodiversity conservation. The privately owned properties are predominantly ranches or farms that often retain some native forest, but pressure to develop or diversify agricultural lands is increasing with the rise of tourism and the decline of historically dominant economic ventures, such as sugarcane and cattle (Simmons 1999; Bowen et al. 2000). Alternative scenarios of future land use could be sustainable forestry of native species and the implementation of conservation areas at one extreme, and a proliferation of subdivisions and ranchettes at the other.

Koa (*Acacia koa*, Fabaceae), a fine timber tree and one of Hawaii's many natural capital assets, has potential for sustainable harvesting that is compatible with restoration of other biodiversity assets. In this chapter we present the results of semistructured interviews (Pejchar and Press 2006), economic models (Goldstein et al. 2006), and a literature review, which together suggest that koa forestry could be a promising means of aligning conservation and economic incentives on private lands.

Endemic to Hawaii, koa is both an ecologically and culturally important species and one of the most valuable hardwoods in the world (Scowcroft and Jeffrey 1999; Wilkinson and Elevitch 2003). Ancient Hawaiians used koa for homes, oceangoing canoes, surfboards, paddles, and spears. Today it is locally cut and carved into high value–added products such as musical instruments, fine furniture, and crafts. In addition to producing wood products, koa reforestation could provide a diversity of ecosystem services, such as carbon sequestration, soil conservation, water storage, and fire prevention.

This native tree has also proved useful in forest restoration (Whitesell 1990; Scowcroft and Jeffrey 1999) and constitutes excellent foraging and breeding habitat for several threatened and endangered species (Menard 2001; Pejchar et al. 2005). Because koa wood sells for a high premium (Winkler 1997), selective harvesting may be economically feasible (Rice 1997) and could have minimal effects on native plant and animal communities in production forests.

Koa is currently in high demand but in low supply (Dudley 1997). Because of strict regulations in state forestlands, much of the koa available is salvaged from ancient dead or dying trees scattered throughout private pastures. There is little koa regeneration in these areas because seedlings cannot compete with exotic pasture grasses, or they are eaten by cattle and other introduced ungulates (Scowcroft and Jeffrey 1999). Indeed, koa is in such high demand that there have been several recent cases of poaching on state land (Leone 2002).

Approach: Interviews and Models Evaluating the Prospects of Natural Capital Restoration

Responding to the increasing scarcity of koa, a shifting economy of land use, and continued threats to wildlife, private landowners, public agencies, and nonprofit conservation organizations in Hawaii are all the more interested in koa forestry (Loudat and Kanter 1997). Several are pursuing conservation measures related to koa or using the funds from salvage koa to fund regeneration. Others, however, are hesitant to switch to koa forestry for a variety of reasons.

We conducted twenty semistructured interviews from September to November 2003. Ten interviews were with practitioners, that is, persons responsible for managing land with koa or that could have koa, or persons in some way dependent on this species for their livelihood. These were large private landowners (ranchers), public land managers (state and federal), wood millers and craftspeople, and forestry consultants. The other ten were classified as scientists, that is, academic, agency, and nonprofit researchers and regulators.

The respondents we interviewed cumulatively own, or work for those who own, more than 90% of all public and private land on the Island of Hawaii, including the top five landowners, in terms of hectares. As we approached twenty interviews, all respondents were recommending that we talk to people or organizations that we had already interviewed. Given this result, and considering that Hawaii is a small state and koa forestry a young industry, we believe that a sufficient number of appropriate people were interviewed to draw informed conclusions (Pejchar and Press 2006).

Each interview lasted one to two hours and was recorded, unless requested otherwise (five cases) or conducted over the phone (two cases). We assured all respondents anonymity, and therefore quotes will be attributed only to the type of respondent (e.g., landowner) throughout. We first asked respondents to describe their association with koa, and what potential (if any) they saw for koa forestry/reforestation as a means of restoring both economic and ecological value to land. Next, we asked all respondents to address the primary question: What are the most important factors limiting koa forestry for conservation and/or profit in Hawaii? We encouraged respondents to think about limiting factors from a range of arenas—silviculture, science, economics, policy, and culture. The remainder of the interview was loosely structured, with no preset questions, to allow each person to fully discuss the issues based on his or her particular expertise and experience.

We transcribed all interviews and analyzed the data in two steps. First, we tabulated the most commonly discussed limiting factors and reported how frequently they were cited. Then, using this table along with secondary data from published reports, we discussed the problems and potential solutions associated with each major issue that emerged from the interviews. The statements in our discussion thus reflect the opinions of the respondents.

We complemented the largely qualitative results of these interviews with a quantitative analysis. Using financial models, we examined the economic potential of koa forestry with a focus on identifying practical business strategies, which draw upon timber and nontimber conservation revenue streams, to address the key economic barriers facing landowners. Our goal with both interviews and models was to investigate a promising tool for sustaining liveli-hoods and biodiversity in Hawaii, and to use the results to draw conclusions applicable to the broader community of environmental scientists and policymakers engaged in conservation on private land.

Results: Risks and Opportunities of Restoring Natural Capital on Hawaii's Private Land

Twenty-five factors currently limiting koa forestry as an economically viable tool for restora-tion were cited by at least two respondents (table 25.1). These factors and the results of our fi-nancial models were grouped into five general categories: (1) science and silviculture, (2) en-dangered species and public opinion, (3) culture and changing land use, (4) economic investment and risk, and (5) providing incentives.

Science and Silviculture

Techniques for growing koa for restoration projects are well developed, though there is al-most no knowledge on how to manage native stands such as koa for forestry. This information

TABLE 25.1

Major limiting factors for koa forestry on the island of Hawaii

Limiting Factor	Frequency cited by respondents N = 20
Lack of silviculture information	10
How to manage for profit and wildlife	10
Landowners wary of federal government	10
Endangered species could hold up harvest	9
Large initial investment	9
Lack of information on how to grow koa at different elevations/climate regimes	7
Soil and water requirements	7
Lack of information on relevant wildlife	7
Koa forestry not "on the radar screen" for most landowners	7
Need a demonstration that koa forestry works	7
Disease causes high tree mortality	6
Lack of information on how to produce high-quality wood	6
Need for seed selection research	6
What other crops could be planted with koa to get financial return earlier?	6
Public opposition to cutting trees could interfere with harvest	6
High economic risks associated with new industry	6
Long-term investment—must be thinking long term and be able to afford it	5
Lack of land, land going to development instead	4
Competition from pasture grasses limits establishment and growth	4
Frost—causes tree mortality at high elevations	4
Incentive programs: not enough funds	3
Incentive programs: only accessible to few large landowners	3
Tax benefits/disincentives	3
State unwilling or unable to harvest/plant koa (thus not contributing to industry)	3
Infrastructure (processing facilities) lacking	2

Source: Authors' analysis.

gap was one of the three most important concerns cited by landowners. As a consequence, because silvicultural techniques for exotic species are well developed, landowners are inclined to plant eucalyptus or pine, which have little or no biological value (Lamb 1998; Simmons 1999). Similarly, there are clear prescriptions for alternative land uses, such as ranching, coffee, and housing developments, compared to native forestry. These land uses have little conservation value but are attractive because they are established practices. Nevertheless, several prominent ranches are now growing koa for harvest and restoration purposes. Their experiments with various management techniques will be vital for evaluating effective management options for future.

Endangered Species and Public Opinion Regarding Restoration

The politics of restoration, however, could hinder these efforts. Nearly half of the respondents thought that a concern that endangered species could prevent harvest in the long term was keeping people from reforestation. Some landowners maintained that *Safe Harbor Agreements* were inadequate: they believed these agreements would not protect them because laws can change. They were concerned about the costs and risks associated with surveys and about privacy. In the past five years, however, three landowners have entered into Safe Harbor Agreements for the nene, Hawaii's endangered goose (*Branta sandwicensis*). Thus, tough political issues such as endangered species protection are being tackled on private lands. Finally, gaining public approval for sustainable forestry is difficult, as there is a small but quite vocal segment of the public that opposes tree harvest anywhere for any purpose. Increased understanding of working koa forests and the alternatives (rangeland, exotic plantations, development, etc.) via demonstration projects could alleviate this pressure. For instance, the Institute for Pacific Island Forestry is currently establishing two experimental forests with goals that explicitly include public access and education (Lingle 2006).

Culture and Changing Land Use Affecting Natural Capital

Although gaining public support for these projects is important, garnering landowner interest and enthusiasm is essential. Most landowners interviewed in this case study are ranchers or investors, not farmers or foresters. Thus most do not have the mindset, lifestyle, or expertise to grow trees. Paniolo (Hawaiian cowboy) culture is deeply embedded in Hawaii. Even though ranching may not be the best economic choice, it is an attractive way of living. However, the newest generation of landowners, now inheriting ranches, is less interested in ranching. At the extremes, these landowners have the choice of a diversity of conservation-friendly land uses, including koa forestry, or dividing the land into small parcels, stripping it of all marketable koa, and selling it to developers. There is an important window of opportunity for building silvicultural knowledge and expanding incentives for koa forestry during this current period of land transition.

Economic Investment and Risk Concerning Natural Capital Restoration

Private landowners' interest in koa has expanded since the 1980s due to strong growth in koa prices (Winkler 1997). This interest, however, is counterbalanced by a few specific factors making investments in koa forestry financially risky. Koa forestry, like all high-value hard-wood cultivation, involves large upfront costs for forest establishment, along with a waiting period of three to five decades (estimates vary depending on location) (Elevitch et al. 2006) prior to obtaining revenue from the timber harvest (Friday et al. 2000). Information limita-tions, as highlighted by respondents, compound the financial risk of koa investments by mak-ing it difficult for landowners to tailor management practices to fit their economic and eco-logical conditions. Furthermore, koa forestry investments must compete with forestry projects focused on exotic species, which often have shorter rotation periods (as little as five to eight years for *Eucalyptus* plantations in Hawaii) (Whitesell et al. 1992). Landowners that grow koa show some combination of the following characteristics:

1. They grow koa because they enjoy their land and value native Hawaiian forest for its own sake.
2. They have other sources of income.
3. Their land is zoned for conservation, and other, more intensive, land use practices are restricted.

Our financial models showed that koa forestry, in principle, can align conservation and economic incentives (Goldstein et al. 2006). A forestry venture based solely upon timber rev-enue covering 202 hectares (500 acres), for example, has an attractive net present value of $1,119/ha. The cash flow trajectory, however, is problematic. In terms of actual cash outlay, total project costs are approximately $1.04 million with $0.38 million (37%) of this coming in the first year for forest establishment. Cumulative, current-year timber revenue over the ten harvest years is projected to be $26.3 million, which corresponds approximately to a 25:1 to-tal revenue to total cost ratio. While this ratio is attractive in isolation, the large time lag be-tween when costs are incurred and revenue received is potentially untenable for a landowner with no other major income sources. Respondents cited concern over the magnitude of the initial investment (table 25.1), and the results of our models support this concern. The bar-rier of upfront costs is most formidable for the initial conversion from ranching to forestry, since, if successful, future rotations could be financed by profits from earlier rotations. Iden-tifying ways to finance upfront costs through landowner-specific arrangements, combined with additional and earlier nontimber revenue streams, would be particularly beneficial for landowners. Expanding the set of conservation revenue streams, public and private, could greatly contribute toward making koa and other restoration activities economically attractive on private lands.

Providing Incentives and Creating Markets for Natural Capital Restoration

Respondents cited the need for a range of programs that serve the diverse requirements of landowners. Existing incentive programs for sustaining and restoring forest cover, such as the *Forest Stewardship Program* and *Forest Legacy*, are sometimes underfunded or sporadically funded (State of Hawaii 2001), yet they have tremendous potential to tip the balance in favor of restoration.

Hawaii is presently exploring opportunities to launch a *Conservation Reserve Enhance-ment Program* (CREP), which is a federal government landowner-assistance program, target-ing watershed and wildlife benefits through retirement of environmentally sensitive agricultural lands. Koa forestry projects are one of CREP's eligible land uses and, by participating, landowners would receive a mix of land rental and cost-share payments. Combining participation in CREP with future timber harvests results in a large increase in net present value to $4,104/ha. Perhaps even more important, CREP's payment structure greatly improves the cash flow prior to timber harvest by providing a revenue stream from the project's start, as well as offsetting approximately 42% of total management costs.

There are also efforts under way to harness existing, and create new, nongovernment incentives for restoration. Existing incentives not yet flowing into Hawaii include the sale of carbon credits and access to new markets from forest certification. While interest in carbon credit–based forestry projects is increasing, our economic results show that a koa forestry venture based solely upon carbon credits has a strongly negative net present value of –$1,206/ha and is not viable, given current market conditions. Carbon credits can be part of a profitable strategy, though, if the landowner is able to also participate in CREP; this strategy yields a net present value of $1,443/ha. This further highlights the value of government assistance programs in increasing opportunities for advancing win-win conservation practices on private lands (Goldstein et al. 2006).

New incentives under consideration include potential payments for other ecosystem services that are provided by koa forests. The focus is on those that relate intimately to human well-being and are central to imminent policy decisions, such as hydrological services and soil conservation (provision of irrigation and drinking water and flood control) (Gutrich and Donovan 2001); provision of biodiversity and recreational values; and generation of other cultural values that flow from the land and the people who live there. Ecotourism is another tool that is underutilized in Hawaii. The unique wildlife in upland koa forests has potential for generating ecotourism dollars for landowners and mobilizing political will for conservation.

Creating the financial mechanisms and institutions to support these potential incentives requires considerable scientific and social understanding, which is just now being developed. The scientific underpinnings of this work will include maps showing (1) the levels, types, and value of services supplied by land under koa (or other native) forest; (2) the degree of spatial congruence (or separation) in the supply of different services; and (3) the forecasted changes in both services and societal needs for them, under alternative scenarios of demographic, land use, and climatic change. The aim is to characterize the tradeoffs—in cost-benefit terms familiar to decisionmakers—of alternative futures and alternative conservation investments.

Koa forestry presents one such enticing investment for restoring conservation value on private lands. The information that we have generated through our interviews and economic models has contributed to an increasing knowledge base for koa forestry. Stakeholders are now mapping out a variety of pathways for expanding native forestry ventures on private lands. As noted earlier, a few landowners are already managing koa. Learning from these front-runners will benefit other landowners who are interested in koa but have not yet decided to invest. For koa forestry, the biological, economic, institutional, and cultural barriers have now largely been identified. The time is ripe for future work to push further with developing novel policies, incentives, and markets to create opportunities where barriers still exist. Toward this end, some of Hawaii's largest landowners are partnering with non-profit organizations and business entrepreneurs to pursue carbon credits, forest certification, and new political initiatives to support koa forestry and other conservation land uses.

Contribution

Conservation biologists, policymakers, and land managers are well aware of the problems facing biodiversity. Public, nonprofit, and private partnerships that encourage economically viable restoration on private land provide an opportunity to construct innovative solutions beneficial to all parties. The following recommendations are based on this case study but are broadly applicable to other innovative restoration projects on private lands.

Research needs to be focused on the needs of practitioners in Hawaii and elsewhere (Langholz et al. 2000). For example, our interviews revealed a gap between what practitioners and scientists deemed important for making koa forestry work for conservation (Pejchar and Press 2006). Practitioners were primarily concerned with federal government involvement and endangered species issues, investment costs and risks, and the fact that koa is often not on the radar screen as a viable alternative to other land uses. Scientists, in contrast, were concerned with balancing profit and wildlife, and the technical aspects of silvicultural research. Practitioners must also be brought into the decision-making process. For example, private landowners are rarely on endangered species recovery-plan teams even though much critical habitat in Hawaii is on private land.

Restoration on private land in Hawaii demands the integration of ecology and economics (Shogren et al. 2003; Mascia et al. 2003). Lack of information on land use economics, management prescriptions, and the biology of relevant organisms has increased risk to the landowner. Scientists and economists can help landowners make informed choices by answering the following questions: What are the economic/ecological tradeoffs of various land use alternatives? How can the landowner maximize both profit and habitat? And, how sensitive are the profit margin and wildlife to different aspects of the management plan? Ecologists can contribute with cross-disciplinary research on how biota respond to various management regimes, and how habitat value can be maximized without losing economic value.

Developing financial strategies that meet the diverse needs of private landowners, while also achieving restoration goals, is another key step in broadening opportunities for restoration on private land. This is particularly important for decreasing the burden of upfront costs borne by the landowner. In this context, government incentive programs can be especially useful in strengthening economic incentives to adopt conservation practices, such as native forest restoration on private land. Because of this, incentive programs are likely to be most successful if funds are reliably available. When programs are widely advertised and then cut or perennially underfunded, landowners lose trust in public-private partnerships. In fact, most landowners who adopt conservation-oriented, land use practices do so because they value nature. Voluntary incentive programs help them leave legacies without losing livelihoods; thus, the importance of financial incentives, regulatory relief, fair tax, and strategic zoning laws cannot be overemphasized.

Finally, this study demonstrates that working with landowners and investors to embark on innovative flagship projects that restore natural capital is crucial (table 25.1) (Wilcove and Lee 2004). With promising but controversial projects like native tree forestry, the public and often even concerned scientists need convincing. Projects such as koa forestry have the potential for fostering open and trusting relationships between public, private, academic, and nonprofit partners and reducing the restoration risks for landowners. Overall, the case of koa forestry suggests that, if driven by economics and informed by ecology, restoration efforts can succeed on private land in Hawaii.

Acknowledgements

Mahalo to all of the people interviewed for their valuable time and information, and for their commitment to Hawaii's forests. We are grateful to J. Scheuer for several pivotal discussions that shaped this study. D. Press, K. D. Holl, J. L. Lockwood, and T. K. Pratt provided useful comments. T. Male, J. B. Friday, and The Nature Conservancy of Hawaii provided key data for the financial models. We thank Esevier for permission to reprint portions of a previously published article (Pejchar and Press 2006). This research was funded by the Department of Environmental Studies, UC Santa Cruz; the Stanford Institute for the Environment; a National Science Foundation Graduate Research Fellowship to J. Goldstein; a Hawaii Conservation Alliance fellowship to L. Pejchar; and the State of Hawaii Division of Forestry and Wildlife.

PART III

Restoring Natural Capital: Tactics and Strategies

One of the key issues in relation to natural capital and its restoration is *valuation*, that is, what value we should assign to ecosystem goods and services and the stocks that assure their flow, if any. We have shown in chapters 1 to 3 that valuation can, at best, be only partial. This raises an additional issue, namely, how to rank both the monetary and nonmonetary values of natural capital to give effect to the right decision. We therefore shall give special attention to this issue and, accordingly, the first two chapters in part 3 examine the various valuation and decision-making techniques available, from two very different perspectives. The first is an ecological economics approach, while the second is a decision-making framework for community groups, government departments, and private organizations that need to make decisions about whether or not to invest in the restoration of natural capital. Such decisions require a major and long-term commitment, and they typically involve multiple stakeholders and objectives combined with considerable uncertainty. Taken together, these two chapters should aid in the decision-making process regardless of the scale involved.

The next two chapters of part 3 offer a clear-eyed look at tactics and strategies aimed at overcoming the formidable physical and biological as well as the socioeconomic obstacles to the restoration of natural capital that exist at *local and landscape levels*. The following two chapters are then devoted to tactics and strategies at overcoming the socioeconomic, institutional, and political obstacles at the *global level*, including those of an economic, legal, institutional, and governance nature, as well as those related to culture, education, media, and marketing. The final two chapters are devoted to tactics and strategies aimed at financial and nonfinancial mechanisms that help make the restoration of natural capital work. They focus on *policies and institutions*—two aspects that qualify for the ancient phrase sine qua non.

One of the brighter aspects of globalization is that people everywhere are more aware of our collective ecological footprint and its current and projected consequences at the global level (see chapters 1 and 5). The linkages between and among scales are also increasingly understood, even if human and market behavior—as usual—lags behind. What is now needed is for that awareness to be translated into new policies, plans, and behavior—locally, regionally, and internationally. As stated often in the book, the goal presented here is nothing less than jointly forging a new future, a new trajectory, for people managing habitats and ecosystems as if nature matters. Radical changes in investment and management are envisioned in a firm and collective move away from dynamics-as-usual toward achieving the fulfilled relationships envisioned in chapter 2.

The obstacles to restoring natural capital mentioned are also responsible for ecological and economic degradation or *rot*. In simplistic terms, to improve human well-being and quality of life, and—to mention yet another sine qua non—to generate greater hope for our future, it is vital to find ways to stop both the economic and ecological rot caused by the mismanagement and waste of biological resources, and the long-ingrained failure of people to even imagine investing in replenishing natural capital stocks when they begin to dwindle.

Chapter 26

Valuing Natural Capital and the Costs and Benefits of Restoration

WILLIAM E. REES, JOSHUA FARLEY, ÉVA-TERÉZIA VESELY,
AND RUDOLF DE GROOT

Economics is frequently defined as the study of the *efficient* allocation of scarce resources among alternative desirable ends. Natural capital is, of course, the ultimate potentially scarce resource meeting an ultimately desirable end. Not only do healthy ecosystems improve quality of life in many ways, they sustain life itself. Moreover, without natural capital, no production whatsoever is possible (chapters 1, 3, and 31). While many economists argue that human ingenuity can develop substitutes for natural capital, this belief ignores the laws of thermodynamics (Georgescu-Roegen 1971). It can also lead to patently absurd assertions, such as that of Schelling (the 2005 Nobel laureate for economics) who argued that losing one-third of agricultural output to climate change would be only a minor setback, since agriculture accounts for only 3% of the gross national product (GNP) (Schelling 1992). Maintaining and improving the quality of life for present and future generations is clearly a high-level desirable end. Thus, natural capital is extremely important as both a means and an end in economics.

Conventional economic analysis strives to balance the marginal benefits of an activity (assumed to be declining) with the marginal costs (assumed to be increasing). Since natural capital is a prerequisite for life, its total value is actually infinite. However, since some natural capital must be depleted or altered to produce manufactured capital (chapters 3 and 4), the human enterprise necessarily encroaches on pristine nature (or natural capital). The questions then become, How much natural capital should be left intact to provide life support functions and other ecosystem services, and how much can safely be converted to manufactured capital?

If we could accurately determine all the costs and benefits of development and economic growth, and ecological changes were smooth and predictable, this question could theoretically be answered by marginal analysis. We would reach an optimal level for both natural and manufactured capital when the value of the next increment of manufactured capital just equals the value of natural capital sacrificed in the process. Conceptually, we should therefore allocate resources to restoring natural capital whenever the marginal benefits of doing so exceed the marginal costs. The practical problem is that we do not understand all the benefits the restoration of natural capital provides, and many of these are public goods that markets ignore and whose values may be incommensurable with market values. This makes it extremely challenging to balance costs and benefits. This chapter considers a variety of approaches to value natural capital and its restoration.

Measuring the Economic Benefits and Costs of Restoring Natural Capital

Monetary valuation of nonmarket benefits of natural capital is increasingly used to measure the economic benefits and costs of restoring natural capital, albeit with considerable controversy. On the positive side, monetization theoretically provides a common metric or currency to compare costs and benefits (though the comparison of present and future values remains problematic, as will be discussed later), and the results are easily communicable to decisionmakers. Indeed, decisionmakers often request monetary valuations as part of their deliberations, and proponents argue that decisionmakers will simply ignore the benefits of restoring natural capital if the latter are not quantified and priced. Even when methods are controversial and values imprecise, a list of the many ecosystem services, with their imputed values, can draw attention to the full range of benefits provided by restoration (see table 26.1 for an overview of monetary valuation methods).

Most direct and indirect restoration costs lend themselves to monetary valuation, as do certain market and near-market benefits of restoring natural capital. In some cases, the value of restoration can be inferred from calculable avoided-damage costs, as in the case of the well-known New York City–Catskill Mountain restoration program (chapter 24). Similarly, the restoration of floodplain and riparian zone vegetation can reduce flood damage, and restoring wetlands can diminish storm damage to built capital. Again, the money value of the avoided costs can readily be quantified.

Unfortunately, it is not possible to evaluate many of the other benefits of natural capital restoration by reference to market prices only. Healthy ecosystems provide many life support functions that are not marketed, yet whose loss would have unacceptable impacts on human well-being (MA 2005f). Restoring natural capital increases ecosystem resilience, making it less likely that we will lose these functions. So valuable are these services that many believe that access to them is a human right—this is actually enshrined in some national constitutions (e.g., in Costa Rica and Brazil; see also chapter 29).

When economists try to value intangible ecosystem services, they confront two serious obstacles characteristic of public goods. First, the services are generally nonexcludable by nature, meaning that there is no feasible institution or technology that can prevent someone from using the resource at will (see chapter 3). Any payments for their use must be voluntary or coerced (see chapter 33). Second, most ecosystem services are nonrival as well as nonexcludable, which means that use by one person does not preclude beneficial use by others (the resource is not scarce in the conventional sense). For example, no one living in a coastal area protected from storm surges by healthy wetlands can be excluded from this benefit and, if one person benefits, this does not leave less storm protection for others.

To value nonexcludable resources, economists must construct a social-demand curve. A common approach is to create a hypothetical market, typically achieved by asking people how much they would be willing to pay to preserve or provide the service in question (see "contingent valuation" in table 26.1). Alternatively, one could estimate the impact of the service on the value of market goods, for example, by comparing housing prices in protected versus unprotected areas (see "hedonic pricing" in table 26.1). Nonrivalness presents another challenge. Since each additional unit of service benefits everyone, the value to society of a marginal change in the service is the summation of marginal values across all the individuals who benefit.

TABLE 26.1

Monetary valuation methods of natural capital's services

Method		Description	Application examples
1. Direct market valuation	Market price	The exchange value (based on marginal productivity cost) that ecosystem services have in trade.	Mainly applicable to the goods (e.g., fish), but also some cultural (e.g., recreation) and regulating services (e.g., pollination).
	Factor income or production factor method	Measured effect of ecosystem services on losses (or gains) in earnings and/or productivity).	Natural water quality improvements, which increase commercial fisheries' catch and thereby the incomes of fishermen.
	Public pricing	Public investments, e.g., land purchases, or monetary incentives (taxes/subsidies).	Investments in watershed protection to provide drinking water or conservation measures.
2. Indirect market valuation	Avoided (damage) cost method	Services that allow society to avoid costs that would have been incurred in the absence of those services.	The value of the flood-control service can be derived from the estimated damage if flooding would occur.
	Replacement cost and substitution cost	Some services could be replaced with human-made systems.	The value of groundwater recharge can be estimated from the costs of obtaining water from another source (substitution costs).
	Mitigation or restoration cost	Cost of moderating effects of lost functions (or of their restoration).	Cost of preventive expenditures in absence of wetland service (such as flood barriers) or relocation.
	Travel-cost method	Use of ecosystem services may require travel, and the associated costs can be seen as a reflection of the implied value.	Part of the recreational value of a site is reflected in the amount of time and money that people spend traveling to the site.
	Hedonic-pricing method	Reflection of service demand in the prices people pay for associated marketed goods.	Clean air, presence of water, and aesthetic views will increase the price of surrounding real estate.
3. Surveys	Contingent-valuation method (CVM)	Asks people how much they would be willing to pay (or accept as compensation) for specific services through questionnaires or interviews.	Often the only way to estimate nonuse values. For example, a survey questionnaire might ask respondents to express their willingness to increase the level of water quality in a stream, lake, or river in order to enjoy activities like swimming, boating, or fishing.
	Group valuation	Same as contingent-valuation method but as interactive group process.	
4. Benefit transfer		Uses results from other, similar areas to estimate the value of a given service in the study site.	When time is limited for original research and/or data are unavailable, benefit transfers can be used (with caution).

Source: Compiled after various sources in De Groot et al. (2006).
Note: For more information and examples on monetary valuation, see http://biodiversityeconomics.org; http://www.naturevaluation.org; or www.ecosystemvaluation.org.
de Groot, R., M. Stuip, M. Finlayson, and N. Davidson 2006. Valuing wetlands: Guidance for valuing the benefits derived from wetland ecosystem services. Ramsar Technical Report No. 3, CBD Technical Series No. 27.

There are further formidable ethical, theoretical, technical, and practical obstacles to monetary valuation of ecosystem services. For example, as noted earlier, it is questionable whether it is ethical to place monetary values on services considered to be human rights. A second ethical issue is that demand curves are determined by preferences weighted by income (see also chapter 3). This approach makes the (un)ethical assumption that an individual's "vote" on the value of nonmarket benefits, freely provided by natural capital, should be weighted by his or her success in the market economy. While it would be fairly simple to elect politicians in the same manner, most people would find this approach to politics morally reprehensible.

Vatn and Bromley (1994) describe three major theoretical-technical shortcomings of conventional economic theory that impede monetary valuation of natural capital and thus confound conventional cost-benefit analysis. The first of these is the *Cognition Problem*: Valid evaluation of life support goods and services assumes that individuals have perfect knowledge about all the benefits of those goods or services and the underlying ecosystem functions providing them. In the real world, however, perfect knowledge is unattainable, so some valuable attributes may be disregarded out of ignorance. Perhaps most significant in this context is that many functional contributions of species and ecosystems are essentially beneath perception, that is, they are cognitively "invisible" when working normally. Vatn and Bromley (1994,133) describe such "functional transparency" to mean that "the precise contribution of a functional element in the ecosystem is not known—indeed is probably unknowable—until it ceases to function." Obviously, we cannot value what we cannot know. Compounding the ignorance facing even the best informed scientists, most valuation research seeks information from lay people after a superficial briefing on the topic that lasts only a few minutes and is inadequate to convey even the limited amount we do know.

A second obstacle is the *Incongruity Problem*: If the different attributes of natural capital assets are "incongruous" or fundamentally at odds with each other in the assessors' minds (e.g., the timber value of a cypress swamp versus its habitat value for sustaining the last remaining ivory-billed woodpeckers), then a single measure such as hypothetical price will not reflect all the important information. Incommensurate values simply cannot be placed upon a common scale. In these circumstances, "social norms restrict or reject the commodity fiction" (Vatn and Bromley 1994,135), so that restoration and conservation decisions have to be based on something other than the perceived money value of the natural capital stocks in question.

Vatn and Bromley's third obstacle to monetary valuation is the *Composition Problem*: In a complex dynamic ecosystem, the whole may actually be dependent on each of its fundamental parts so that the value of any single component cannot be understood independent of the value of the whole. Thus, the value of individual ecosystem components should not be derived from their perceived utility to humans but rather from their functional contribution to maintaining the integrity of the whole system. Vatn and Bromley (1994,137) argue that, in such circumstances, "the commoditization of environmental goods (as reflected in contingent valuation studies, for example) can be looked upon as a product of the felt need to value them. It is not immediately obvious to many—other than economists—why it is necessary to characterize environmental attributes this way." Some ecological goods and services remain technically *impossible* to price (see also chapter 2).

Yet another problem is the fact, already emphasized, that monetary values are marginal values (see chapters 3 and 4). However, nonlinear changes, surprises, discontinuities, and irreversible thresholds characterize ecosystems. In other words, a marginal increase in certain activities such as commercial fishing or logging may lead to nonmarginal, even "catastrophic," outcomes such as stock collapse or regional climate change, respectively. Under such circumstances, marginal valuation can be disastrously misleading.

Modifications of, and Alternatives to, Cost-Benefit Analysis

Although none is ideal, there are alternatives to conventional cost-benefit analysis. For example, a cost-effectiveness approach would typically begin by setting a specific restoration target. A community might decide, based on the best available science, how much restoration of natural capital is required to provide an adequate amount of ecosystem services to maintain or enhance human security and well-being. The community would assess the costs of various options to achieve this goal, including the opportunity (forgone benefits) and transaction costs, and then choose the least costly option.

More sophisticated alternatives might use one of several multicriteria analysis tools, all of which follow a similar logic. First, working with affected parties, define the desired goals for a project or program, along with different (often incommensurable) criteria by which to evaluate success in achieving them. Second, generate alternative strategies for achieving the goals, assess how well each strategy meets our evaluation criteria, and discard those strategies inferior to others for all evaluation criteria. To rank remaining options and facilitate the final strategy choice, we would develop weights for the different evaluation criteria, again with input from affected parties. While assigning weights is similar in principle to monetary valuation, the process is far more transparent and amenable to stakeholder participation (Farley et al. 2005).

Human Time Preference: The Discounting Problem

Both the monetary valuation of ecosystem services and its alternatives have another serious limitation: how to compare costs and benefits occurring at different times. The standard approach is to use discounting (as has been done in chapter 21, for example), which represents the present value of costs and benefits occurring in the future as being worth exponentially less than the same nominal benefits/costs today.

There are several justifications for this temporal discounting, the major one based on the opportunity cost of money. Money today can be invested and grown to some projected future value—discounting is essentially this process in reverse—but this opportunity is lost to money acquired only in the future. Problematically, discounting assumes the economy will continue to grow indefinitely (see chapter 4), which is unrealistic. In fact, economic growth is now consuming critical natural capital instead of just the "natural interest" (see chapters 5 and 6), so the future economy is likely to shrink. Economic decline justifies a *negative* discount rate (see also chapter 29) because a unit of economic output in the future will provide greater net benefits than an equivalent unit today. Similarly if natural capital is declining, its marginal utility is presumably increasing, which would also justify a negative discount rate. In addition, estimated returns on monetary investments are often too high because the negative impacts of the investments are ignored. If we justify ignoring negative impacts that occur in the future because their discounted present value is so low, we are guilty of circular reasoning.

Curiously, economists often further justify discounting natural capital because technology may develop a superior substitute for a given resource, making it less valuable in the future. While this can happen and should be supported, society actually often develops new uses for resources that increase their scarcity—consider petroleum over the past 150 years. This makes resources more, not less, valuable over time. If anything, this would again justify a negative discount rate. Ultimately, technology is a complement to natural resources, not a substitute (Daly and Farley 2004).

A second general justification for discounting is personal time preference. People prefer benefits today to benefits in the future due to impatience, greed, and uncertainty. Conventional economists assert that we must respect this preference despite two major counterarguments. First, if the economists' argument is valid, how can we then argue that society should prefer benefits for people alive today to those of future generations? Second, society has a longer life span and less uncertainty than individuals. Social discount rates should therefore be lower than private ones.

There are many other problems with discounting, and there is a vast literature proposing modifications and alternatives (see Henderson and Bateman 1995; Weitzmann 1998; Frederick et al. 2002; Young and Hatton McDonald 2006). However, even this cursory assessment makes it clear that discounting the future benefits of natural capital restoration may not be justified.

Who Benefits from, and Who Should Pay for, the Restoration of Natural Capital?

Identifying all the parties benefiting from the restoration of natural capital is complicated by the variety of benefits and their diffuse spatial and temporal distribution of restoration. Those investing in restoration may gain in a range of ways: for example, a company from regulatory compliance (chapters 9 and 10); a local community from improved water quality (chapter 24); a corporate sponsor from the "green" image (chapter 23); or an environmental nongovernmental organization from payments received for carbon sequestration services (chapter 19).

With potentially many beneficiaries, who should pay? The costs include the direct expenditure of conducting on-the-ground restoration activities (which may be minimal, if natural capital can restore itself), and the opportunity costs incurred from not exploiting alternative uses of the restored land and resources. Traditionally, two main principles apply in exploring who should pay for ecosystem maintenance and improvement. The *impacter pays* principle—an extended version of the *polluter pays* principle—requires individuals or entities that damage ecosystems to meet the costs of restoration activities (chapters 9 and 10). The *beneficiary pays* principle requires anyone who benefits from restoration to contribute to the costs of undertaking it (chapters 12, 24, and 25). The application of these principles might result in cost-sharing arrangements among or between impacters and beneficiaries. Both principles are explicitly normative, stating an ethical position on who ought to bear the costs of achieving desired environmental goals. The final decision is necessarily made in the political arena, based on views about rights and fairness, seasoned by both community expectations and power politics (Pannell 2004).

Impacter Pays

Legislating the impacter pays principle is equivalent to introducing a new obligation for restoration under existing property rights. The impacters might restore the ecological damage themselves or mitigate it by buying damage credits. These credits would then go toward paying for the restoration of equivalent ecosystem services—for example, carbon sequestration services—in some other location. This mechanism depends on an established institutional framework for credit trading and strong capacity for regulation and compliance monitoring.

Individuals or businesses that face the legal responsibility for restoration will not necessarily bear the economic costs of its provision. These costs can be passed on to consumers through higher prices or passed back to employees and shareholders through lower wages and dividends (Dodds 2004). Adoption of the impacter pays principle may thus have social consequences, as it will add to the costs faced by resource users or consumers. This may be important, especially if the commercial viability of resource users is uncertain. However, such problems should be addressed by access to temporary credit rather than through release from responsibility (Marshall 1998).

To apply the impacter pays principle for historical damages, a retrospective liability framework is needed. This is a common approach in contaminated site legislation (Caunter et al. 2005). However, retrospective application of the impacter pays principle gives rise to certain practical problems, such as the difficulty in identifying and locating the person(s) originally responsible for the natural capital damage. Moreover, since monitoring and enforcement can result in considerable implementation costs, this can reduce the benefits of applying the impacter pays principle and may compromise the achievement of restoration outcomes (Aretino et al. 2001). Some even consider it unfair to penalize impacters retroactively for complying with the accepted legal frameworks and policies of the past. From this perspective, financing the repair of past degradation should be based on the beneficiary pays principle, which is likely to imply some funding from taxpayers (Tilton 1995).

Beneficiary Pays

The beneficiary pays principle means that anyone who benefits from the restoration of damaged ecosystems should contribute to the cost. The principle has two components—*user pays* and *beneficiary compensates*—that should be used together. Under the user pays principle, individuals and groups contribute to the costs of undertaking restoration activities that directly benefit them. Under the beneficiary compensates principle, governments typically contribute to the restoration costs on behalf of the general community, if restoration generates public (nonexcludable, nonrival) benefits (Young 1992; chapters 21, 22, and 32).

If specific beneficiaries are lumped into the category of general community, and their private benefits are described as public benefits, the general community may be burdened with costs that individuals or groups should meet under the user pays component. This potential misapplication may occur because it avoids the problem of having to identify specific beneficiaries and the associated compliance costs of collecting contributions.

Since adoption of the beneficiary pays principle may require beneficiaries to pay for restoration services that they have not paid for previously, it will have social implications. To help individuals adjust, governments may consider providing assistance or introducing new arrangements gradually.

To summarize, who benefits from and who should pay for restoration will be context dependent and contingent upon ecological, economic, social, and legal circumstances. Economics can inform this debate by quantifying the distributional effects of alternative policies and the impacts of alternative policies on efficiency. The ultimate decisions, however, will flow from community perceptions of the problem, public expectations, legal norms, and the distribution of political power.

Behavioral and Sociocultural Barriers to Restoring Natural Capital

Some of the barriers to natural capital restoration are deeply rooted in human nature and prevailing cultural beliefs. At bottom, people are not natural conservationists. On the contrary, like other species, *Homo sapiens* has an innate predisposition to occupy all accessible habitats and exploit all available resources (Rees 2004). The human expression of these tendencies, however, differs from that of other species in two important ways. First, human progress has been marked by efforts to reduce or eliminate systemic negative feedback that would limit the scale of human enterprise (chapter 1). Second, our technological capacity to exploit natural capital is cumulative and constantly improving. One result of these evolutionary forces is the "remarkable consistency in the history of resource exploitation: Resources are inevitably overexploited, often to the point of collapse or extinction" (Ludwig et al. 1993).

Spurred by such observations, Fowler and Hobbs (2003) tested and rejected the null hypothesis that *H. sapiens* is "ecologically normal," that is, that humans fall within the normal range of natural variation observed among similar species for a variety of ecologically relevant measures. In terms of population size, energy use, carbon dioxide emissions, biomass consumption, and geographical range, the human impact is greater than that of other species by orders of magnitude. For example, human biomass consumption is almost two orders of magnitude above the upper 95% confidence limit, in comparison with ninety-six other mammalian species. These data indicate that *H. sapiens* may be the most destructively formidable predatory and herbivorous vertebrate ever to have existed. Ironically, the overuse of natural capital witnessed today is the result of humanity's extraordinary evolutionary success (Rees 2004).

There are many other dimensions to humanity's bias against conservation/restoration. For example, humans are naturally self-interested as is reflected in the ethical framework of neoliberal economics. This framework is "*utilitarian, anthropocentric* and *instrumentalist* in the way it treats [natural capital]. It is utilitarian in that things count to the extent that people want them; anthropocentric, in that humans are assigning the values; and instrumental, in that biota is regarded as an instrument for human satisfaction" (Randall 1988, original italics). This value framework, with its emphasis on short-term, individual, utility maximization, is clearly hostile to the long-term, common-pool values that would accrue from restoring natural capital, particularly given the difficulties in quantifying and pricing the latter.

Societies, of course, are highly complex, and short-term individual utility maximization is only one facet of human nature; we are also capable of cooperative and altruistic behavior and long-term planning. However, certain aspects of our biological predisposition against sustainable behavior are currently being reinforced by purely cultural factors. Perhaps most important has been the purposeful "social construction" of our modern, growth-based consumer society, abetted by the creation of a multibillion-dollar advertising industry dedicated to making a secular religion of consumption. The attendant risks to natural capital are heightened by the fact that our education systems generally fail to instill in nonscience students even a rudimentary understanding of natural capital's role in providing and maintaining global life-support functions. Consequently, the public remains almost totally ignorant of humanity's ultimate dependence on adequate stocks of functional natural capital.

Meanwhile, technological advances and globalization increase the average person's alienation from natural capital (chapter 2). Consider the accelerating process of global urbanization. The migration of hundreds of millions of people from the land to cities separates people both spatially and psychologically from the ecosystems that support them. Most critically, with urbanization, people shift from valuing natural capital for subsistence (exploiting the direct *use* value of natural capital) to seeing natural capital as a means of making money through trade, that is, through the commodification of natural capital and the exploitation of its *exchange* value. This has the potential to lead to overexploitation for short-term economic gain as ecological ignorance and human time preference (discounting) combine to underestimate the value of the future income stream from essential natural capital stocks.

Worse still, as people become increasingly dependent on distant stocks and flows of natural capital (e.g., on imported food), they both place less value on their own natural assets—they may pave over local farmland—and are blinded by distance to the effects of unsustainable natural capital exploitation in the exporting regions. Not only do such factors undermine incentives for ecosystem maintenance and restoration, they accelerate the depletion of remaining stocks of natural capital everywhere.

There is a final concern. Scientists' warnings about the need for natural capital restoration and conservation are undermined by the growth ethic and society's arrogant confidence in the power of technology to compensate for the destruction of ecosystems and the loss of natural life-support functions. To quote the late professor Julian Simon, "Technology exists now to produce in virtually inexhaustible quantities just about all the products made by nature" and "We have in our hands . . . the technology to feed, clothe, and supply energy to an ever-growing population for the next seven billion years" (cited in Bartlett 1996, 342). This comforting, almost defiant belief now pervades neoliberal economics as "the principle of near-perfect substitution" (see also chapter 2).

Such powerful cultural myths serve as a kind of intellectual armor on the body politic, often deflecting the sharpest barbs of reality and, in this case, reinforcing the growth-bound, natural capital–depleting status quo. The substitution myth persists despite contrary econometric analyses showing that, because of the hidden costs of shifting resources from consumption to investment, "it is not possible to substitute capital for environmental life support and maintain material well-being" (Kaufman 1995, 77). When prevailing cultural mythology reinforces innate human behaviors that have become destructively maladaptive, it is doubly difficult to muster the economic resources and political support necessary for society to properly value, conserve, and restore natural capital stocks.

Contribution

Uncertainty, narrow self-interest, a spreading consumer ethic, and the naturally conservative behavior of human societies confound efforts to restore natural capital. This is increasingly problematic. Systems science tells us that we face irreversible thresholds beyond which continued depletion of natural capital may be catastrophic to global civilization, yet we have no idea where many of these thresholds lie. Indeed, we face the distinct possibility that we are nearing critical "tipping-points" and that, without active restoration, the loss of natural capital and global life support will soon be irreversible.

Given irreducible uncertainty and unprecedented high stakes, the prudent course for humanity would be to focus its efforts on generating the resources and developing the institutions necessary to restore natural capital. It could even be argued that, since adequate stocks of critical natural capital are a prerequisite to human existence and the aggregate economy is already in a state of overshoot (WWF 2006), additional research into valuing natural capital is a waste of valuable time. We should simply get on with the needed restoration. Regrettably, most of the people with the authority and influence to bring about natural capital restoration on the scale required are not yet convinced of the need, or they remain in the sway of the aligned powerful forces maintaining the status quo. In these circumstances, monetary valuation of natural capital, whatever its limitations, may yet prove to be a useful tool for convincing the public and policymakers of the need to act now. And fortunately, scientific evidence is mounting that sustainable use (and thus restoration) of ecosystems is economically more profitable than unsustainable use (Balmford et al. 2002).

Whatever the outcome of the restoration debate, it is clear that, despite the prevailing substitution myth, humans cannot survive without natural capital. Restoration is our only real option. The tragedy is that the power of our growth-with-substitution myth may prevail over the warnings of our best science (see also chapter 2). In these circumstances, those with at least a hint of biophilia in their psyches may take some small comfort in knowing that natural capital can flourish without humans.

Chapter 27

A Decision-Analysis Framework for Proposal Evaluation of Natural Capital Restoration

MIKE. D. YOUNG, STEFAN HAJKOWICZ, ERICA J. BROWN GADDIS,
AND RUDOLF DE GROOT

Community groups, government departments, and private organizations frequently need to make decisions about whether to invest in the restoration of natural capital (Hajkowicz et al. 2000). When these decisions require a major commitment, they typically involve multiple stakeholders and objectives, and considerable uncertainty (Gough and Ward 1996). Since public scrutiny and expert review are essential for transparency and public participation, the process can be facilitated by rigorous, repeatable, structured methods to resolve the question, To restore or not to restore? This chapter provides such a decision-making support framework.

Deciding to Restore Natural Capital

When deciding whether to invest in the restoration of natural capital, it is important to consider the nature of the decision being made.

Decision Characteristics

Gough and Ward (1996) indicate that one of the major characteristics of natural capital restoration decision making involving ecosystem services is the existence of considerable uncertainty and the presence of multiple stakeholders with conflicting objectives, coupled with a need to assess intangible outcomes (see also chapter 26). In fact, probably the most important decision is what the restoration objective should be. In many cases it may be more cost-effective to prevent degradation rather than to wait until restoration is necessary. Such decisions must consider landscape ecology and the relative importance to the community and landholders of ecosystem services. Unfortunately, private property ownership often conflicts with community interests (see chapter 21).

Once a potential restoration target is identified, several steps are required for the restoration of the system, while decisions must be made throughout the process. A first step might be the negotiated removal of the causes (or drivers) of ecosystem service loss and then determining whether the system can restore itself (see chapter 28). The second step may involve decisions to restore ecosystem processes in general (see chapter 9), whatever they may be at a chosen site. Both of these stages facilitate the self-organization of a system to provide more ecosystem services. If desired species do not colonize naturally, the third step may involve the reintroduction of certain species to restore a particular ecosystem structure.

Classification of the Decision to Restore Natural Capital

Decisions associated with restoration can be approached in many ways. The purpose of the classification presented here is to enable assessment of the decision-support techniques discussed throughout the remaining chapters. Three major types of restoration decisions can be identified:

1. *Prioritization decisions* involve appraising many alternatives, with decisions being either cardinal or ordinal. *Cardinal prioritization* requires a quantitative performance index to be assigned for each alternative, while *ordinal prioritization* requires only information on the relative performance of the alternatives. For example, prioritizing a large number of projects competing for limited funds is a prioritization decision.
2. *Allocative decisions* involve the distribution of resources among competing stakeholder groups, with the apportionment of pollution permits providing an example of allocative decision making. If these permits are made tradable, the result can be the efficient restoration of natural capital (chapter 32).
3. *Threshold decisions* typically involve a step change in managerial approach. When making threshold decisions, it is necessary first to identify an appropriate threshold and second to identify the alternatives associated with the threshold. Determining the minimum, acceptable water quality standards for environmentally significant streams is an example of a threshold decision. Restoration of natural capital is required, typically, when water quality falls below this level. Once the threshold is set, restoration becomes mandatory when water quality falls below that level—irrespective of the cost of doing so.

Classification of the Decision Support for Restoring Natural Capital

Figure 27.1 classifies decision support into two main categories:

- *Descriptive and predictive decision-support techniques* are explanatory in nature. They can be used to illustrate how complex environmental or social processes function. Typically, they try to avoid incorporating human value systems to identify a rational or appropriate decision, while tending to describe or predict what *will* happen rather than what *should* happen. Predictive models can help decisionmakers explore the *realm of the possible*, that is, assess the possibility of the maximum expected performance, which can provide a comparative benchmark for alternatives.
- *Prescriptive decision-support techniques* combine social, economic, and biophysical data with preference statements to prescribe a rational, optimal, or *best* decision (i.e., what should occur). Tradeoffs among prestated objectives, such as economic efficiency, job creation, and environmental preferences, are common. Nevertheless, the prescribed decision may not necessarily be what is adopted in practice, once further discussion and evaluation of recommendations occur. Prescriptive restoration decision-support techniques provide abstractions or simplifications of reality, and as with any type of model, they always contain inaccuracies.

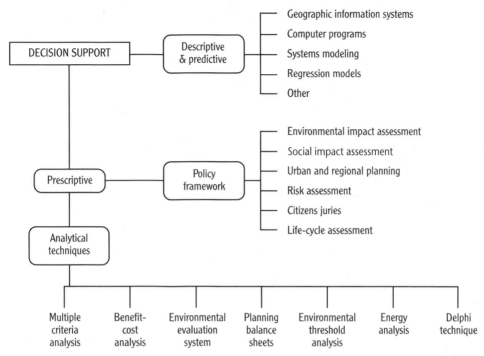

FIGURE 27.1. Classification of dominant restoration of natural capital decision-support techniques and frameworks. (Descriptive and predictive approaches are not discussed in this chapter. For more detailed information on each technique, see Hajkowicz et al. 2000).

Despite their limitations, prescriptive models can be extremely useful in helping decisionmakers understand a complex decision problem, while reducing the issues to be resolved before making a final decision. Typically, prescriptive models focus on benefits and costs, broadly speaking.

Prescriptive decision-support models are broken up into two subcategories of analytical techniques and policy frameworks. Analytical techniques involve a highly specialized, repeatable, and structured process for identifying an optimal or best decision alternative. Cost-benefit analysis (CBA) and multiple criteria analysis (MCA) provide examples. Typically, analytical techniques, like these, follow well-developed methodological protocols. As professional protocols are well understood, it is important to document any deviations from standard analytical procedures.

In contrast, *policy frameworks* for decision support are more loosely structured and capable of handling a much broader range of issues and input. Most policy frameworks represent the legal and institutional processes by which decisions are made, and they need to mesh with other laws and cultural practices of a society. They need to be able to accommodate input from the many stakeholders, community groups, industry, and government, and consequently they cannot be as rigidly structured as decision-support analytical techniques.

Often these analytical techniques are applied as part of, or within, a policy framework. For example, MCA might be used to assess the performance of alternative development scenarios during an environmental impact assessment (EIA). However, it is most unlikely that restoration practitioners will need to choose between application of a policy framework and an analytical technique. Generally, the questions will be which policy framework(s) should be adopted, and which analytical techniques are required.

Decision Uncertainty Associated with Natural Capital Restoration

The state of knowledge guiding decision making is regarded by Wynne (1992) to be at one of four levels: (1) under conditions of risk, (2) under conditions of uncertainty, (3) under conditions of ignorance, and (4) under conditions of indeterminacy. Within this framework, a decision involves a risk when the distribution of probabilities is known and the causes of change fully understood; uncertainty involves situations where the causes of change are not fully understood but can be explained after a change has occurred. Indeterminacy applies to situations where the state of a system changes and the causes of that change cannot be explained.

Under conditions of ignorance, the outcomes or the probability of their occurrence are not fully known. The extent to which each of these is important depends in part at least on the goals of the restoration project. For example, restoration of hydrologic functions may have significantly less uncertainty associated with it than the introduction of a nonnative species to control another nonnative invasive species (see McNeeley 2001; Gobster 2006).

Depending upon the preferences of those involved in the analytical processes, uncertainty can cause decisionmakers to exhibit preference for the status quo until more information becomes available and/or more alternatives emerge.

In the case of an indeterminate decision, the cause of the issue being managed is not known. In this situation, very careful experimentation should be undertaken.

Conflicting Objectives in Restoring Natural Capital Decisions

Smith (1993) indicates that traditional environmental and natural resource decision making has concentrated narrowly on the following three questions: (1) Is it technically feasible? (2) Is it financially viable? (3) Is it legally permissible?

Contemporary understanding of rational, restoration decision making introduces a requirement for a broader range of factors to be considered. Typically, restoration decisions are characterized by multiple objectives. For example, it is common to try to satisfy economic, ecological, social, and cultural objectives, which in some cases demands a choice between restoration of selected ecosystem services and full restoration of ecosystem structure. In other examples, where irreversible losses have occurred, while full restoration is not possible, many of the ecosystem services sometimes can still be recovered.

The Restoring Natural Capital Decision-making Processes

While this chapter focuses on so-called rational decision-making processes, it needs to be recognized that many decision-making processes do not necessarily follow rational principles, as described by decision theory. The violation of these rational-decision principles can be attributed to the nature of subjective values and probability functions, combined with the failure to consider all factors that influence a process, or to the specific shortcuts used by decisionmakers when making judgments. Avoidance of rational processes can also be used to produce outcomes that favor one group over another.

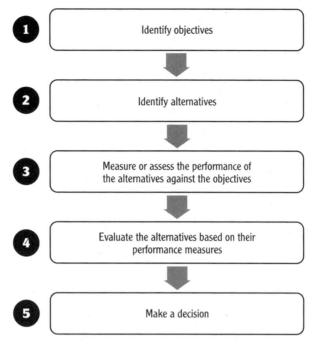

FIGURE 27.2. The rational decision-making process (Hajkowicz et al. 2000).

Rational Model of Decision Making

Exposed to many unforeseen and unpredictable factors, decisions associated with natural capital restoration rarely follow the rational model (figure 27.2). Nevertheless, this provides the foundations for most objective decision making (Smith 1993), and is unidirectional, rigid, and highly structured. However, there are few opportunities for revision or consultation, and Lindbolm (1959, 82), in discussing the rational model, observed that

> as to whether the attempt to clarify objectives in advance of policy selection is more or less rational than the choice of inter-twinning marginal evaluation and empirical analysis, the principle difference is that for complex problems the first is impossible and irrelevant, and the second is possible and relevant.

A requirement for a less rigid or "intertwinning" decision procedure makes it possible to consider new alternatives as the process advances and to allow objectives to be respecified. Typically, decisionmakers will need to revisit the stage of specifying alternatives and objectives many times.

The Process of Decision Making About Restoring Natural Capital

Figure 27.3 presents a generalized version of the restoration decision-making process that builds upon the rational model. Its circular form emphasizes the cyclical nature of most restoration decision-making processes, with the gradual change from light to dark gray representing progression through the procedure. In practice, however, restoration decision-making processes are (very) messy, because external factors, such as political interventions,

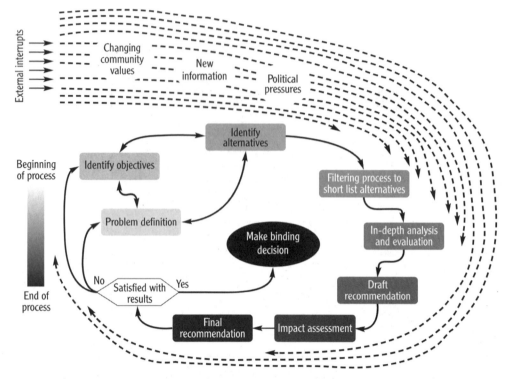

FIGURE 27.3. A generalized restoration decision procedure (Hajkowicz et al. 2000).

the emergence of new information, and changing community values, continuously influence the procedure.

Another important aspect of the process is legitimacy, because as the procedure develops, different actors and stakeholders become involved. Professionally trained analysts, for example, play a major role during the filtering processes and in-depth analysis, as shown in figure 27.3, while impact assessment typically involves extensive stakeholder engagement.

The feedback loops between the stages of problem definition, identification of objectives, and identification of alternatives associated with ecosystem service provision, show the interrelatedness of these tasks. In practice, this stage of the process is fuzzy, and it can be difficult to define the problems or opportunities without having objectives. Similarly, developing objectives can be complicated without knowledge of the alternatives (and vice versa).

Once the restoration alternatives have been identified they are usually subject to some form of screening process that removes those that are clearly inferior and disliked by all stakeholder groups. If, for example, two alternatives offer similar outcomes but one costs much more, then the more expensive option can quickly be eliminated. Indeed, screening allows the available resources to be concentrated on more detailed analysis.

Once a smaller set of restoration alternatives has been identified (typically two to four options), these are subject to in-depth analysis and full specification. In most cases, one of the alternatives needs to include the no-restoration option. A classic set of alternatives would be to consider (1) leaving a stream as it is, accepting the consequences of increased sediment load, and accepting the status quo; (2) restoring upstream wetlands; (3) restoring a riparian corridor; or (4) dealing with the farmland sediment source. Once a short list of alternatives has been identified, use of structured decision-support techniques can begin.

After the completion of an in-depth analysis, a draft recommendation is made, with high-level decisionmakers becoming more involved in the process. Typically, the level of analysis and involvement of decisionmakers will depend on the magnitude of the impacts of the considered project; small-scale or low-impact projects may require less in-depth analysis as well as less high-level oversight. The most common approach at this stage involves various forms of impact assessment. Sometimes, high-level decisionmakers sense that the subsequent stages may fail or cause serious political problems, and, as a result, they seek to delay the decision by calling for more information and/or consideration of other alternatives. Political problems are common when a key person has expressed a view or position that is not supported by detailed analysis. Procrastination and filibustering occurs until a way can be found to make it appear that the views of the key person are not in conflict with the view supported by the analysis and engagement processes used.

On completion of the impact assessment phase, a final recommendation is made. If deemed unacceptable, the procedure may be repeated and the stages cycled through again. Otherwise, restoration of the natural capital in question begins, subject, of course, to available resources, budget authorization, the absence of court injunctions, or protesters. However, not all restoration decisions will follow all these stages.

Community-based Natural Resource Management

Many studies have found that major restoration decisions not involving the community and key stakeholders in the decision-making process fail. This has led to the emergence of *community-based natural resource management* (CBNRM) (chapter 29), which is defined as "a process by which the people themselves are given the opportunity and/or responsibility to manage their own resources, define their needs, goals and aspirations and make decisions affecting their well-being" (Experts Workshop 1991). Nevertheless, it needs to be recognized that communities are not homogenous and people within them often hold conflicting views. Central governments have a responsibility for ensuring that the views of all stakeholders have, at least, been considered.

CBNRM involves a shift in autonomy from government institutions to frameworks that empower the broader community (Matsumara 1994). This is leading to requirements for decision-support techniques that are accessible to the community and the use of information derived from community sources.

With specific reference to CBNRM, Fellizar (1994, 205) defines a community as "a group of people living in a geographically defined area, with a common history and definite pattern of relationships." Its members have a common interest in how the natural resources are managed. In other terms, the community members will all be stakeholders standing to incur some loss or gain from a decision.

For example, the community in a catchment could comprise agricultural producers, environmental conservationists, and tourists, who derive economic benefits and ecological and recreational values, respectively. These and other groups will incur loss or gain as a result of how the catchment is managed and therefore should be considered part of the community. This understanding implies that the community, as defined for purposes of restoration decisions, may have members who do not reside within the local region. Unfortunately, the development of mechanisms to engage the global community in a cost-effective manner remains problematic.

Numerous authors have argued that CBNRM is a more effective means of achieving sustainable development goals compared to traditional top-down approaches (Renard 1991; Campbell and Siepen 1994; Matsumara 1994; chapters 15, 16, and 19), as it is generally considered to foster a sense of increased stewardship over natural resources and promote inter- and intragenerational equity.

Transparency in Restoring Natural Capital Decisions

The shift toward CBNRM has led to calls for increased transparency in government (and private sector) decision making. Transparency means that the basis for a particular decision is clear to all stakeholders. While it is simple to argue that nothing should be ambiguous or hidden from the public eye, in practice this ignores the possibility of legitimate protection of privacy and confidentiality. Failure to maintain privacy and confidentiality detracts from many of the values central to democratic government.

Decision-support methods can sometimes be too complex or used in such a way that the reasons behind a decision are unclear; this can occur either incidentally or intentionally. Ideally a decision-support framework will provide a structure that makes explicit the reasons behind a particular decision and intentional "clouding" difficult or impossible.

A common problem with decision support is that the procedures are not fully understood by the decisionmakers or the stakeholders being impacted. The end product of such decision-support techniques can be a single index representing the performance of alternative options. However, decisionmakers will be unlikely to trust an index without knowledge of its underpinning assumptions, and hence McAllister (1980, 265) cautions against over reliance of the "grand index" in decision making:

> Although grand index schemes are appealing as elegant technical solutions to the evaluation dilemma, they are neither valid nor acceptable. There is no simple shortcut to the time-consuming personal task of reviewing the many consequences of proposed actions until a holistic impression of their significance forms, which can be used to judge the preferred action.

Implementing Decision Support

Most, if not all, decision-support techniques and processes applied to the restoration of natural capital have the common purpose of seeking to improve restoration decisions by increasing transparency, multiple stakeholder participation, fairness, and decisionmaker learning and understanding. To this list, one can add comparability, comprehensiveness, accuracy, timeliness, cost, and so forth. The key challenge for the practitioner is to decide which approach best suits the stage at which a decision-making process stands. In practice it is rare that any single technique or framework will be best suited to a particular problem. Typically, the techniques are not mutually exclusive, and therefore different approaches can shed light on different aspects of what are often complex decision choices.

Administrative Arrangements

These have a profound impact on restoration decision processes and outcomes. To a large extent, administrative structures determine which stakeholders are involved and the nature of alternatives and issues considered. Regarding administrative arrangements, Dovers (1999, 81) identifies the following opportunities:

1. To improve information capacities (gathering, manipulation, and communication)
2. To improve policy and management coordination and integration across sectors
3. To increase longevity and persistence in policy processes and initiatives
4. To enhance policy learning across space and time
5. To improve capacities and techniques for policy instrument choice and comparative policy analysis
6. To provide clearer policy and statutory mandates (more direction, less direction) to improve institutional capacities
7. To enhance and institutionalize community participation in policy and management

In countries such as Australia, the emergence of community groups is facilitating a significant devolution of decision-making responsibility from government to community members. This devolution of responsibility is confronting community groups with complex decisions that they are not well equipped to make. While many of the techniques currently available appear too complicated or time consuming, in practice if one wants to make a decision in the confidence that it can be relied upon, their use cannot be avoided. Hence, community groups are increasingly required to apply structured and transparent decision-making processes. In South Australia, for example, the Natural Heritage Trust (a major source of funding for community-based projects) requires cost-benefit analysis (CBA) for projects that exceed AUS$150,000 (US$105,000) in any year.

Applying Decision Support at Various Stages of the Decision-making Process

The type of decision support that is most relevant will largely depend on what stage has been reached in the decision-making process. Figure 27.4 shows a simplified form of the process and identifies where the various support techniques have most pertinence, as well as the community involvement stages of greatest importance. CBNRM plays a significant role in the decision process, even though it is not classified as a support technique or policy framework; it supplies legitimacy to the procedure.

While each of the decision-support techniques could be useful at every stage, there are some points where certain approaches are likely to have more (or less) relevance. Multiple criteria analysis, for example, is extremely useful for filtering alternatives, because it can be structured to appraise quickly a large number of options against a set of criteria. While there are many exceptions, at the beginning, analytical techniques tend to dominate, which, toward the end, are replaced with policy frameworks.

Supporting Decisions in the Early Stages of the Process

Decisions that occur in the filtering stages of the procedure involve prioritization. Here the main task is to rank a large set of alternatives against a set of objectives. In such cases, formal

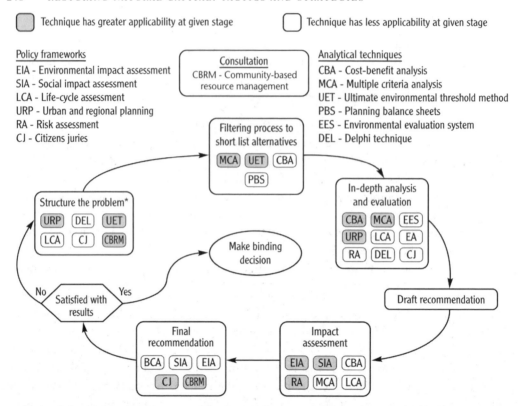

Structuring the problem involves the stages of identifying objectives, identifying alternatives, and defining the problem as identified in the natural resource management decision process in chapter 1.

FIGURE 27.4. Application of decision-support techniques and policy frameworks in the natural resource management-decision procedure. (For a fuller description of these techniques, see Hajkowicz et al. 2000.)

or informal multiple criteria analysis (MCA) and its many variations have most applicability. Typically, budgetary constraints will not permit a wide range of policy frameworks or CBA to be used.

Through the construction of an effects table, MCA can provide a useful way of evaluating many alternatives against a set of multiple and conflicting objectives. Effects tables are normally prepared by subgroups of professional analysts and stakeholders, with the data being at either a quantitative or an ordinal measurement level. If a performance index is not required for each alternative, then ordinal-level data can be relied on more heavily and the requirements for quantification relaxed.

Conversely, CBA is a more demanding and more structured form of MCA, which can be used to prioritize decisions if sufficient resources are available. However, the MCA procedure is more flexible than CBA and therefore provides opportunities for time and cost saving. Detailed CBA is usually reserved for a smaller number of proposals, which, following screening, are deemed to have significant potential.

Another drawback of CBA is that priority is often defined by noneconomic (and unmonetizable) criteria (see also chapter 26). While nonmarket valuation techniques are improving, they struggle, faced with evaluating social and cultural considerations. Some people consider that MCA can handle such criteria better and more transparently than CBA, though the contrasting view is that CBA is more rigorous and less prone to subjectivity. Once a priority list has been obtained through MCA, more in-depth analysis can be conducted by applying other policy frameworks (e.g., risk assessment, environmental impact assessments [EIA], social impact assessments [SIA], citizens juries) or CBA (see figure 27.4).

Supporting Decisions Near the Conclusion of the Process

Decisionmakers are faced increasingly with binary decisions near the end of the process; that is, the range of natural capital restoration alternatives has been subject to analysis and investigation such that only two remain (generally to accept or reject a proposal of some type). Major binary decisions are generally made by democratically elected or appointed community leaders who rely on, among other things, the above techniques to act as a surrogate for a community referendum on the final choice. As such, it needs to be recognized that such considerations extend well beyond traditional CBA.

Decisionmakers will primarily be interested in assessing whether a proposal is ecologically, economically, and socially sound. This is achieved through EIA, SIA, and other forms of impact assessment, with a CBA possibly having relevance as an economic check of the proposal's financial soundness. The citizens jury approach may also have value to incorporate community attitudes into the decision-making process.

In binary decisions, the focus is on loosely structured, value-laden, political issues, and it is unwise to make a binary decision by placing total reliance on the results from a single analytical technique, especially when that technique is poorly understood by the community. In this regard, MCA has significant shortcomings, as it involves methods understood by only a minority of technical experts. Its use in the lead-up to a final binary decision choice that is being closely watched by many stakeholders can lead to perceptions of unfairness or concealed factors influencing a decision. The calculation of performance indices by methods such as MCA can also detract from its role at this stage.

Contribution

It is evident that major restoration decisions require the use of more than one technique and, in a world freed from budgetary and time constraints, it would be desirable to apply as many relevant techniques as exist. However, in a resource-constrained world only a few techniques can be used even though methods that combine several techniques are emerging. For example, MCA can be used within an EIA to assess the environmental merits of alternative development proposals, as it often includes socioeconomic variables and thus has broader application than simply assessing proposals on their "environmental merits." Similarly, combinations of CBA and MCA are being tested.

When there is no agreement, as is often the case, Rawls (1987) advises the use of techniques that reveal the extent to which a decision is characterized by an "overlapping consensus." He recommends that a search for a consensus affirmed by opposing theoretical, religious, philosophic, and moral doctrines is likely to be a just one and, in a resilient fashion, likely to thrive over generations. While appealing, this approach is challenging since it is not based on any internal sense of mathematical logic. Nevertheless, a decision, affirmed by many different techniques and consistent with alternative sets of value systems and value weights, has the elusive attraction of apparent rigor, because it has passed all tests.

At the beginning of this chapter it was noted that the objective of decision support is to improve the rationality of the decision-making processes. A rational decision was defined as being consistent with the values of the decisionmaker and the information available. Nevertheless, there will always be a requirement for decision-making processes that occur above and beyond a decision model or policy framework. Hence, other factors that cannot be incorporated into a neat model must be given consideration. These include political considerations, social concerns, or a potentially limitless range of issues important to individuals in a society.

There will always be factors that cannot be properly measured or do not fit neatly within a policy framework, and, as such, there will be no quick or simple means by which they can be taken into account. The lengthy and complex process of learning about restoration decision alternatives and reflecting on community values cannot and should not be sidestepped.

Acknowledgement

This chapter is an adapted version of Hajkowicz et al. (2000).

Chapter 28

Overcoming Physical and Biological Obstacles to Restoring Natural Capital

Karen D. Holl, Liba Pejchar, and Steve G. Whisenant

Human activities over the past century have resulted in dramatic habitat transformation and reduced species numbers, which in turn cause loss of ecosystem services that would otherwise benefit humans (MA 2005a; chapter 5). For example, in many areas of the United States less than 20% of the original wetland cover remains (Zedler 2004), which negatively affects flood control, sediment uptake, water purification, recreation, habitat, and water supply services (NRC 1992). Mitsch et al. (2005) estimate that restoring wetlands along <1% of the Mississippi River would reduce nitrogen inputs into severely hypoxic parts of the Gulf of Mexico by 40%. Although this is a relatively small wetland area, it is still four times the total amount conserved and restored under the U.S. Department of Agriculture's Wetland Reserve Program. Likewise, Zhao et al. (2004) estimate that the loss of 71% of tidal wetlands between 1990 and 2000 on Chongming Island, China, the largest alluvial island in the world, resulted in a concomitant 62% reduction of ecosystem services worth between US$855 million and $911 million.

Restoration should never be considered a substitute for conserving the remaining natural ecosystems (Aronson, Clewell, et al. 2006), given the many obstacles to restoration and its variable success. Nevertheless, the extensive loss of natural ecosystems means that restoring ecosystems is essential to replenish natural capital (see chapters 1, 3, and 26). In this chapter we provide an overview of the physical and biological obstacles to restoring natural capital at both local and landscape scales (table 28.1). Rather than discussing the full list of biophysical obstacles for each ecosystem type, we classify them into four categories (figure 28.1): (1) the inability to restore ecosystem processes due to ongoing stresses; (2) the case that an ecosystem has crossed a threshold beyond which it cannot recover without human intervention (at least during an acceptable human timeframe); (3) the lack of a source of propagules; and (4) a poor understanding of ecosystem processes and/or restoration methods. We note that many degraded ecosystems suffer from obstacles in multiple categories. For each general obstacle, we provide specific examples from a range of terrestrial and aquatic ecosystems worldwide, drawing on case study chapters from this book and our experience in other ecosystems. We briefly suggest potential approaches for overcoming these obstacles, referring to strategies outlined in other chapters (figure 28.1).

We use the Society for Ecological Restoration (SER) definition of ecological restoration as "the process of assisting the recovery of an ecosystem that has been degraded, damaged, or

TABLE 28.1

Examples of physical and biological obstacles to restoration at multiple spatial scales

Soil conditions
    Low nutrient availability
    Altered texture
    Altered structure (too compacted or loose)
    Low organic matter
    Too wet (hypoxic) or dry
    Excessive temperature extremes
    Secondary salinization
Microclimatic conditions
    Unsheltered soil surfaces that affect temperature and evaporation
    Altered wind patterns
    Low humidity
Water quality
    High nutrients—eutrophication
    Excessive water temperature extremes
    Low dissolved oxygen
    High levels of chemical contaminants
    High sediment loads
Disturbance regimes—changes in frequency, intensity, and periodicity
    Fire
    Flooding
    River-flow regimes
    Large grazers
Altered topography and geomorphology
    Altered slope and runoff patterns
    Increase or decrease in elevation (wetlands)
    Altered channel meandering patterns (rivers)
Changes in population processes
    Lack of dispersal and colonization of propagules
    Low genetic diversity and connectivity with other populations
Community processes
    Lack of seed dispersers
    Lack of pollinators
    Altered herbivory
    Lack of mutualists
    High levels of exotic species

Source: Authors' analysis

destroyed" (SER 2002). We highlight the fact that the goal of restoration is to set an ecosystem on a trajectory toward recovery. Given historical contingencies and the dynamic nature of ecosystems, it is impossible to achieve a highly fixed endpoint (SER 2002). Therefore, restoration efforts aim to facilitate the recovery of species and ecosystem functions, recognizing extensive natural variability.

Ongoing Stresses Impeding the Restoration of Natural Capital

A primary obstacle to restoring natural capital is ongoing stresses (or drivers of ecosystem degradation) to a particular ecosystem that impede the recovery and restoration of natural capital. These stresses often result from the small size of the project, which allows the impacts of surrounding land uses to continue to influence the system. Ongoing stresses make it impossible to restore a trajectory of recovery, so continuing inputs are required to maintain the site in a restored state. In other words, the symptoms rather than the cause of the degra-

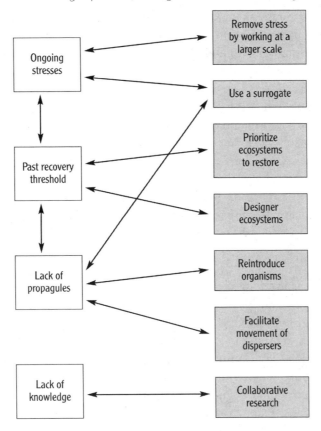

FIGURE 28.1. Conceptual framework for general categories of biophysical obstacles to ecological restoration (open boxes) and strategies for overcoming those obstacles (shaded boxes). Source: own analysis.

dation are being addressed. For example, altered hydrological regimes are a common stress to rivers; in 2000, there were over 45,000 large dams worldwide (WCD 2000). Along the Sacramento River, the largest river in California, much money is being spent planting trees and shrubs to restore natural capital in the form of riparian forests (chapter 17). However, flow regulation by Shasta Dam limits the recruitment of plant species requiring high flows for establishment and reduces the channel scouring and meandering essential for proper river functioning. Therefore, recovery of the full complement of species and ecosystem processes is not possible. Likewise, many lake restoration efforts, such as dredging, phosphorus inactivation, and artificial circulation, are actually ongoing lake-management methods rather than strategies that will alter ecosystem trajectory. Unless high inputs of nutrients and chemical contaminants are reduced, such interventions will be required ad infinitum (NRC 1992).

Certainly, the logical strategy to overcome such obstacles would be to remove the stress or cause of the problem. Unfortunately, in many cases, this is not politically, economically, or ecologically feasible, or not possible at the small scale of most restoration projects. One strategy therefore is to move the restoration effort to a larger scale, both ecologically and jurisdictionally. For example, by restoring riparian buffer strips upstream it may be possible to reduce nutrient inputs resulting in recovery of lakes or estuaries (Schoonover et al. 2006). Along the

Sacramento River, restorationists are buying extensive farmlands to restore riparian forests where there is high flooding risk, and it is politically feasible (Golet et al. 2006). This will allow more channel meandering, thereby restoring riverine processes, rather than just investing in localized tree-planting efforts.

Another strategy for overcoming ongoing stresses is to use a proxy or surrogate for the process that cannot be restored. This will likely require ongoing management, and often such efforts result in only partial restoration. For example, along the Colorado River controlled releases are being tested as a way to compensate for the lack of high flows that deposit river sediments; to date these efforts have met with limited success (Stevens et al. 2001). In ecosystems where large grazers such as elk or bison previously created patchy resource distribution, domestic grazers, such as cows or sheep, can sometimes provide a substitute management tool for maintaining habitat heterogeneity (Weiss et al. 1999; Hayes and Holl 2003).

Crossing Thresholds of Natural Capital Recovery

Restoration ecologists have long recognized the existence of threshold barriers to recovery (Bradshaw 1984; Hobbs and Norton 1996; Whisenant 1999; chapters 3 and 9). All ecosystems have the capacity for natural recovery following some moderate level of disturbance, although a greater level of disturbance may preclude recovery in human timeframes without significant intervention. For example, post-mining substrates often contain significant amounts of acid-forming materials that continue to release H^+ ions, lowering the soil pH to levels that inhibit vegetation development (Hossner et al. 1997), if there is no human intervention to facilitate revegetation. Similarly, irreparable degradation has occurred in Western Australia, where decades of conversion from native shrubs and trees to agriculture has reduced landscape-scale transpiration, as the annual crops cannot access the deeper water table (Cramer and Hobbs 2002; see also chapter 9). These changes have brought saline groundwater toward the surface, killing vegetation and creating serious damage that is unlikely to be reversed by any practical treatment (Hobbs et al. 2003).

To maximize the resources available for natural capital restoration, funds should be allocated toward protecting and restoring systems that have yet to pass over a threshold. For example, Moody and Mack (1988) argue that to be most effective in limiting the spread of invasive alien plants, removal efforts should focus on small, nascent populations, rather than on dense, well-established populations, a fact highlighted in chapter 22. In Hawaii, where many birds in the lowlands have gone extinct due to avian malaria, bird conservation efforts should prioritize habitat restoration in disease-free, high-elevation forests that could eventually serve as a source population to recolonize disease-ridden areas, if birds evolve resistance (Scott et al. 1986; chapter 25).

An unpopular, pragmatic approach may be to focus on those systems where one can get the "most bang for the buck," although this means writing off certain areas as beyond recovery during the short term. This is a serious concern for restoring natural capital, particularly when land degradation has negative impacts on human health and well-being and may raise environmental ethic and justice issues; this is a particular concern if such areas are predominantly among lower-income communities. For example, after the Exxon-Valdez spill in Alaska, there were immediate efforts to clean the oil off beaches. The heavy tidal scouring in this area would have decomposed the oil within a few years, but not intervening was not an option, given that many beaches were important for Alaska Natives' rituals (D. Kelso, California, 2002, personal communication). Decisions to not restore certain highly degraded lands will be controversial, but we argue that given the severe degradation of natural ecosystems and the immense amount of funding needed for restoration, there will necessarily be difficult decisions regarding resource allocation.

Another option for ecosystems past a recovery threshold is to create *designer ecosystems* that restore certain functions or species (McMahon and Holl 2001). For example, in China, deforestation has resulted in 360 million hectares of highly eroded land, comprising 38% of the land area and resulting in a soil loss of five billion tons annually (Li 2004; see also chapters 19 and 29). True restoration at this scale, and with an escalating demand for wood products, is impractical. As a result the government has undertaken an aggressive campaign of replanting, primarily with a few stress-resistant, alien tree species (afforestation), as well as some endemic species with high value; this effort is aimed largely at slowing desertification and providing firewood for China's immense population (Li 2004; Ma 2004). Similarly, Janzen (2002) suggests planting *Gmelina arborea*, an alien pulpwood tree, in abandoned pastures in northwestern Costa Rica. This species facilitates establishment of native tree seedlings by enhancing seed dispersal and shading out pasture grasses, while providing an income for landowners six to eight years later. What is not clear is whether *G. arborea* will have long-term negative effects on soil conditions. Deciding to create a new ecosystem on a site may seem like "playing God," but all human decisions about ecosystems, including degradation, destruction, management, or restoration, require some degree of value judgment. The decision to reintroduce certain species or to rehabilitate certain functions is often preferable to not intervening. Nevertheless, restorationists should be clear from the outset about the goals of any restoration effort (Clewell and Aronson 2006).

Restoration practitioners may elect to restore natural capital that has passed a threshold. This may be feasible with sufficient intervention, such as at mining sites, with a combination of topsoil restoration, removing invasive alien species, and seeding and planting (Ward and Koch 1996; chapter 22). However, practitioners must recognize that attempting to restore an ecosystem that has crossed one or more thresholds will usually be extremely labor and capital intensive and may require long-term maintenance.

Lack of a Source of Propagules for Recolonization

A third major obstacle to recovery of natural capital is a lack of organisms to recolonize a disturbed area once the stress is removed. In some cases, particularly with large fauna, these animals may be extinct or in extremely low numbers. For example, some large grazers were extirpated at the time of European colonization in South Africa (chapter 7), while elsewhere many seed-dispersing and pollinating birds in Hawaii have gone extinct due to avian malaria (Cox and Elmquist 2000; Sodhi et al. 2004). Often the organisms are still present, but the remaining populations are sufficiently distant from the site being restored that natural colonization is impossible or extremely unlikely. Given the severe fragmentation of many ecosystems (an issue highlighted in chapter 8), lack of plant and animal dispersal has been noted as a major factor limiting recovery in a range of terrestrial ecosystems, such as tropical forests, grasslands, and temperate forests (reviewed in Bakker et al. 1996; Holl 2002; Honnay et al. 2002). Simply put, improving site conditions is irrelevant if the intended organism(s) cannot arrive at the site.

In the case where an organism has gone extinct or is regionally extirpated, it may be necessary to introduce a surrogate. For example, in California coastal grasslands, introduced cattle (Hayes and Holl 2003) seem to play a role similar to native elk (Johnson and Cushman 2006) in maintaining the diversity of rare annual plants, but they may degrade riparian vegetation if not properly managed. Similarly, in Hawaii, many plant species have gone extinct due to the loss of native bird pollinators, although a few plant species have persisted due to the introduction of the Japanese white-eye (Cox 1983).

If propagule sources are present but distant, several strategies can facilitate recolonization. Both plants and animals are often intentionally reintroduced to disturbed sites as a part of restoration efforts. Reintroducing species may raise complex issues about the genetics of the source population, in particular whether it is locally adapted or sufficiently diverse to avoid inbreeding depression (Morrison 2002; Hufford and Mazer 2003; chapter 10). If the organism is rare, it is also important to consider potential impacts of removing individuals or seeds from the remaining populations to recolonize disturbed sites.

Restoration efforts often focus on enhancing natural recolonization by making the habitat more attractive to fauna, particularly seed dispersers, by providing canopy architecture, perching structures, food sources, and shade (Morrison 2002). Mills et al. (chapter 21) found that planting shrubs increases seed dispersal into degraded South African thickets. Some restoration and conservation efforts also work at the landscape scale to provide corridors, stepping stones, or a more favorable surrounding habitat matrix to facilitate faunal movement through the landscape (Holl et al. 2003; chapter 8). For example, birds are more likely to move through shade coffee plantations than across more intensive agriculture land uses (Raman and Mudapa 2003).

Incomplete Knowledge of Natural Capital Function

Despite an exponential growth in our understanding of ecosystem recovery, lack of knowledge of natural capital functions and of the most effective restoration methods remains a substantial barrier to restoring natural capital (see also chapter 5). In addition, often ecosystem degradation happened so long ago that we lack clear knowledge of the predisturbance ecosystem (chapter 7). A classic example of poorly understanding the intricate connections within ecosystems led, in part, to the failure of a restoration project to provide salt marsh habitat for the endangered light-footed clapper rail in southern California (Boyer and Zedler 1996; Zedler and Adam 2002). Even after a number of years, the constructed marshes failed to support a cordgrass canopy that was sufficiently tall to provide nesting habitat for the clapper rail. Only with ongoing monitoring and manipulative studies was it discovered that poor cordgrass growth resulted from nitrogen limitation, and that the soils used for construction were inappropriately sandy and did not include enough fine sediment. As a consequence, the short cordgrass did not support a predatory beetle, which consumes scale insects that eat cordgrass. Hence, the wrong choice of soils resulted in an outbreak of scale insects, causing the cordgrass to senesce prematurely, or die (Boyer and Zedler 1996). While certain aspects of this system might have been predicted (e.g., lack of fine sediments affecting nutrient cycling), the complex interactions were not appreciated until after the mitigation failure. Like Humpty Dumpty, it is a lot easier to take ecosystems apart than to put them back together again (Bradshaw 1987; Lockwood and Pimm 1999).

Lack of knowledge of natural capital function and restoration is probably the easiest obstacle to overcome as long as restoration ecologists document and share the results of both successes and failures. We strongly recommend that scientists and restoration practitioners collaborate to design, monitor, and adaptively manage projects in order to maximize the potential for learning more about ecosystems and to guide future restoration efforts (Bradshaw 1987; Holl et al. 2003). Restoration efforts are some of the largest ecological experiments being undertaken, although they are underutilized as such and may require using nontraditional analytical techniques (Holl et al. 2003). It is also important to recognize the substantial amount of local (or traditional) ecological knowledge that exists among nonscientists, which can be critical in designing successful restoration projects (chapters 15 and 16).

Contribution

We have briefly discussed four major categories of physical and biological obstacles to restoring natural capital, which include a long list of site-specific factors. Despite these, we contend that the social, economic, and cultural obstacles, discussed later in this book, are more restrictive than the biophysical ones. Most physical and biological obstacles can be overcome with sufficient funding and/or a long-term commitment, if ongoing demands from humans allow the possibility of working at a spatial scale sufficiently large to recreate important ecosystem processes. Examples of such success stories are cited in other chapters in this book and elsewhere, but they are relatively small in number compared to the numerous restoration failures. To overcome physical and biological obstacles, restoration plans should be designed in collaboration with local human communities, if the projects are to have any chance of succeeding.

Given that there will never be enough money to undertake all the needed restoration projects, hard decisions will have to be made about which sites to restore and the extent of resources to be allocated. Being clear on the physical and biological obstacles to recovery in each case and the relative potential benefits to humans in the form of restored natural capital will help in prioritizing these efforts.

Acknowledgements

We appreciate discussions on this chapter with many participants at the Restoring Natural Capital workshop in St. Louis, Missouri, in August 2005. We thank Bob Scholes for providing the initial framework for the chapter.

Chapter 29

Overcoming Socioeconomic Obstacles to Restoring Natural Capital

Christo Marais, Paddy Woodworth, Martin de Wit, John Craig, Karen D. Holl, and Jennifer Gouza

The previous chapter dealt with the physical and biological obstacles to the restoration of natural capital. This chapter focuses on overcoming the socioeconomic obstacles, grouped into three categories: (1) economic; (2) legal, institutional, and governance; and (3) cultural, educational, media, and marketing. In his seminal paper, *The Tragedy of the Commons*, Hardin (1968) recognized all three categories. He discussed the human population and its demand for natural resources as a result of a growth in numbers, as well as society's demand for privileges (economics); access to, and the competition for, resources (legal); and conscience, guilt, and anxiety (culture and education). We discuss these three categories of obstacles, followed by some recommendations for overcoming them.

Economic Factors Impeding the Restoration of Natural Capital

Historically the motivation for restoring natural capital was economic, aimed at making degraded areas productive. Little emphasis was placed on the long-term ecosystem benefits resulting from such restoration (Mentis 1999; Milton 2001). This meant that little research was carried out on the monetary quantification of ecosystem goods and services, the flows resulting from natural capital functions; although in most cases the restoration costs were known, it was more difficult to argue the economic benefits. In fact, most species and some services do not have clear economic values associated with them, as these are not traded in the marketplace; yet, as indicated in chapter 2, this does not imply that they have no intrinsic value. In recent decades, environmental and ecological economists have begun to develop methods to quantify some of the benefits that restored ecosystems supply, though this is far from being routinely applied (see chapters 3 and 26).

Community-based Natural Resource Management

Increasing economic functions and land scarcity often renders it impossible for natural capital to compete with short-term, commercial, land-use options. This is further complicated by the degradation drivers often being misinterpreted, resulting in unsuccessful restoration efforts. Resources are therefore misallocated with the result that management efforts and funding becomes fruitless (Homewood 2005; chapter 28). For example, the degradation drivers of the woodlands around Lake Malawi National Park were misinterpreted during the 1970s. It was thought that deforestation was caused by an increase in the local populations "that depend on forest products for cooking and building" (Abbot 2005). However, in 1993/94, it was discovered that it was not increased domestic fuelwood demand having a detrimental impact, but commercial smoking of fish caught in Lake Malawi. Therefore, restoring the natural capital of the park would be futile if the focus was on reducing the domestic wood consumption. This example indicates that *community-based natural resource management* (CBNRM) projects can potentially fail if they lack shared values and understanding between conservation bodies, aid agencies, and the rural communities involved (see also chapters 15, 16, and 27). The key to success is to ensure the buy-in and ownership of the project by communities, irrespective of whether it takes place in a developed or developing country. Society has to experience and understand the benefits tangibly, and therefore restoration projects have to be mainstreamed into local economies. However, care should be taken to ensure that the potential benefits of restoration projects are not overstated, and, for example, in the context of CBNRM, the emphasis could be changed from *community-based* (local economy) to *natural resource-based* community development, as was illustrated in chapter 19.

Discount Rates

Another obstacle to the competitiveness of restoration projects is the analytical method used to calculate their costs and benefits. Generally, restoration projects have high initial costs, while the benefits accrue later (see, for example, chapter 25). Which discount rate to use is therefore pertinent, because the higher the rate, the smaller the future benefits, and vice versa. Hence, there is an argument in favor of using low, and even negative, discount rates when future benefits are likely to be high or increasing due to resource scarcity (OECD 1994, 194–95; see also the extensive discussion in chapter 26).

Gross Domestic Product and Practical Evaluation of Ecosystem Services

In addition, the *gross domestic product* does not include the value of goods and services rendered by natural capital (see also chapter 6). When provision is not made for changes in the quantity and quality of natural resources and ecosystem services, it is assumed ecosystem services do not contribute to welfare but only to costs. Consequently, it is in the short-term and commercial interest of society to keep the value of resources as low as possible (Blignaut 2004b). This convention is changing, albeit slowly, with the introduction of natural resource accounts, and in cases where the benefits of ecosystem goods and services have been internalized, such as the New York City–Catskill example (chapter 24). Watershed services are, however, very localized, and benefits derived from prudent management cannot necessarily be borne elsewhere.

In general, the practical valuation of ecosystem services has not materialized, and much more technological intervention is needed to "unlock" the value of nature's services (Sagoff 2002). Issues such as the tragedy of the commons (Hardin 1968; chapters 3, 26, and 31) and property rights, or the lack thereof, complicate matters (Landell-Mills and Porras 2002). As a result, natural resources are mostly undervalued, and once they are valued, their worth cannot be translated into economic gains in a straightforward manner (chapters 32 and 33).

Transaction Costs

The transaction cost of initiating payments for ecosystem services can also impede the implementation of restoration projects. According to the *Clean Development Mechanism* (CDM) under the Kyoto protocol, carbon emissions can be reduced in two ways: physically lowering emissions, and sequestering carbon through forest restoration, while providing an income to rural people (Landell-Mills and Porras 2002; Jenkins et al. 2004). The latter example is, hypothetically, an ideal opportunity to restore both social and natural capital simultaneously. Unfortunately, these transactions are extremely costly. The CDM executive board, for example, has two conditions for the validation and certification of projects for its adaptation fund. The first is equal to 2% of the certified emission reductions, while the second is determined by the size of the project. The costs vary from US$5,000 for a project of less than 15,000 tons CO_2 sequestered per crediting period ($0.33 per ton) to $30,000 for a project of more than 200,000 tons ($0.15 per ton). (See chapters 21, 31, and 32 for further discussions on CDM.) Smaller projects are therefore being disadvantaged, making it more difficult for poorer communities to access the market (Knowles 2005). These costs exclude brokerage fees, which is an additional transaction cost. Hence, the upfront costs of restoration are high, and the economies of developing countries cannot afford them, often causing dependence on international funding, with the associated increased transaction costs and inevitable delays.

Examples of Restoring Natural Capital

In many respects Costa Rica is leading the way, especially among developing countries, in restoration of natural capital. In 1997 the country initiated a program called *Pago por Servicios Ambientales* (Payments for Environmental Services Program, or PSA) aimed at the protection and restoration of its natural resources, based on the services from natural ecosystems (Nicholls 2004). The PSA is based on five types of services (table 29.1).

The PSA is very small when compared to China's *Sloping Land Conversions Program* (SLCP) (also known as the *Grain for Green* program). With a budget of $40 billion, aimed at converting 14.67 million hectares to forests, SLCP has already reforested 7.2 million hectares after four years, while benefiting 15 million farming households (Xu et al. 2005; chapters 19 and 28). The stated goals of China's central government are to conserve soil and water in fragile landscapes and also to restructure the rural economy. This can allow farmers to gradually shift into more environmentally and economically sustainable activities, such as livestock and off-farm income opportunities (Xu et al. 2005). Even though the SLCP has some implementation and targeting challenges, it is perceived (by the authors) as a good strategy to overcome the economic hurdles of restoring natural capital.

Such programs indicate that a strategy to accelerate restoration improves the financial benefits received by local landowners and/or users. This can be achieved by remunerating land users, resource users, or poor communities with active interventions, such as helping to propagate restoration material, establish vegetation, and erect exclusion infrastructure.

The last but not least important opportunity lies in accessing the social-responsibility budgets of large corporations. The Rio Bravo Carbon Sequestration program in Belize, Central America, involves twenty-seven organizations in a program that has sequestered an estimated 4.4 million metric tons of carbon since 1995 (Katoomba Group 2005). In a study by

TABLE 29.1

Market framework for ecosystem services in Costa Rica

Service	Individual clients	National clients	International clients
Carbon		National government corporations and private companies (Public Private Partnerships in developed countries) through fuel taxes and energy supply rates.	World Bank; multinational corporations; developed country governments
Water		Industrial, domestic, and agricultural water users and hydroelectric schemes	
Sustainable, consumptive use	Timber companies— sustainable use through forest stewardship certification		
Nonconsumptive use through tourism in natural landscapes	Tourists: on-farm, nature-based tourism opportunities	National and provincial/ regional travel agencies	
Biodiversity		National corporations and national governments in developed countries	World Bank; multinational corporations; developed country governments

Source: Authors' analysis.

the International Institute for Environment and Development (IIED), the biggest buyers of biodiversity services were found to be private corporations, while the sellers group was dominated by communities, public agencies, and private individuals. Indeed, the buyers tend to focus on the most diverse habitats in terms of species richness or those under the greatest threat of extinction, promote eco-labeling schemes for crops and timber, and seek to be perceived as biodiversity-friendly. However, this runs the risk of becoming mere window dressing, with only a narrow focus on popular species and habitats. Other major buyers are from the horticulture industry, concerned with ecosystem services, as well as bioprospecting pharmaceutical companies (Jenkins et al. 2004).

Legal, Institutional, and Governance Issues Impeding the Restoration of Natural Capital

Discussions of legal, institutional, and governance issues are now being focused on critical success factors for the restoration of natural capital.

Land Tenure

The most commonly perceived legal obstacle to restoring natural capital is a lack of an integrated statutory tool for natural capital management and restoration. Examples of this are frequently manifested because land is typically divided into communally owned, protected or private lands, and most natural capital processes by nature is perceived not to be in competition with direct benefits from the land, which is not always the case. So legislation generally does not allow for weighing up the benefits derived from natural capital against that of direct use of the land. Communally owned, protected lands are often held by governments on behalf of communities and are governed at a distance. Income is minimal, and little attempt is made to maintain natural values other than through benign neglect. Where invasive, alien species are present, natural values typically decline, and local communities are disenfranchised because their decision-making powers are limited. In the case of private land, landowners are sometimes encouraged to maintain past exploitative and extractive management and discouraged to restore natural capital.

Positive and Negative Incentives for the Restoration of Natural Capital

In New Zealand, and previously in Hawaii, for example, incentives discourage landowners from managing for both biodiversity and production. Both island archipelagos were largely forested, but they have been harvested to make way for "more economically productive" landscapes based on introduced species (see also chapters 12 and 25), resulting in biodiversity loss being a key environmental problem (for example, DoC/MfE 2000). To make the restoration of natural capital more problematic, if a landowner establishes native woodlands, there can be difficulties obtaining harvest rights when they mature. In the worst situation, some local government bodies require landowners to obtain consent prior to planting native species for forestry. In addition, tax laws in New Zealand give incentives to establish alien forests but do not recognize native forestry as a viable commercial enterprise. Because conservation of biodiversity, including restoration, is not seen as a business, planting forestry lots that use native species are seen to be acts of conservation and not commercial. Unlike alien plantation forestry, claimable expenses for native plantations are set at trivial levels and are deemed "not in the business of forestry" (Barton et al. 2005), creating a major disincentive to restore natural landscapes.

Hawaii has recently changed its tax structure, giving landowners who manage indigenous forests sustainably the same tax break as those in other forms of agriculture. This is a positive step to overcome the historical legal challenges to restoring natural capital (Bonk 1997). For most other countries, this is less of an issue, as plantation forestry is often based on native species, and hence all forestry can be seen to make positive contributions to natural capital. As discussed earlier, some countries such as Costa Rica have even advanced further by paying landowners for ecosystem services, although this methodology is still in its infancy (Daily and Ellison 2000).

Legal Versus Ecological Timescales

The reclamation of open mines provides a good example of this conundrum. Since legal timeframes are much shorter than ecological recovery, restoration laws can impede restoration of natural capital or even block mine closures if alternatives are not found (chapter 23). Indeed, the success of coal surface-mine reclamation efforts in the eastern United States is usually evaluated after five years by counting the number of trees and assuring that groundcover is at least 90%, although natural forest recovery in this area takes a number of decades (Holl 2002). Consequently, this legislation encourages mining companies to use aggressive, nonnative herbaceous and tree species. As a result, these alien ecosystems may help to meet short-term goals of maintaining water quality and thereby satisfy legislative requirements, yet they may inhibit the establishment of native late-successional species. One of the legal measures therefore should be to consider the timescales of restoring the original vegetation, instead of seeking a "quick fix" with fast-growing exotics.

Community Participation and Successful Restoration of Natural Capital

A small, yet practical, restoration example in Cape St. Francis (South Africa) shows the importance of community participation in a project integrating clearance of dense, invasive, alien plant stands and the establishment of thatch reeds (*Thamnocortus insignes*). Thatch reeds provide valuable roofing material for poor rural and affluent urban and semiurban communities in Africa, combining excellent insulation properties and aesthetic attractiveness. Indeed, beyond the immediate financial value of the reeds, the project is allowing natural processes and capital to be restored. These include normalized fire regimes, restored groundwater resources, improved species diversity, and increased land productivity, all of which foster livelihood opportunities for the local, disadvantaged community. Initial estimates of the value of the thatch are in the order of \$11,100 ha^{-1} (on a four- to six-year harvesting rotation). In this project, one of the bigger challenges was community participation as a key factor for overall success; community representation on forums and involvement in the project activities have been essential for the processes of coalition and strategy building at local levels (Stewart and Collett 1998).

Encouraging project ownership through participative processes has been an effective tool, but participation in smaller projects does not always happen easily. However, if the restoration of natural capital is to succeed, the benefits and consequences must be understood and appreciated by those involved. Hence, it is important that the restoration of natural capital is not seen as being dominated by short-term benefits (employment and remuneration) but perceived as a long-term process (Nicholls 2004).

Mechanisms with Potential for the Restoration of Natural Capital

Legal mechanisms to facilitate the trading of conservation credits are being developed in a number of countries. The Mexican Forestry Fund has been under design since 2002, while the United States has had a wetlands mitigation program in place since the 1980s. Australia is in the process of developing legislation that will transform *biodiversity credits* into property rights for private landholders who set aside land with biodiversity value. They will then be able to sell these credits into a pool where property and industrial developers can buy them. If applied responsibly, these initial legal developments bode well for the restoration of natural capital in Australia (Jenkins et al. 2004).

Education, Media, and Marketing

Culture, in the broadest sense, gave humans domination over the natural environment through religious beliefs and economic ideologies (see chapters 2 and 26). It is time to change our culture, which in turn will change our behavior. Today, the idea that ever-increasing consumption and economic growth are essential to our well-being is omnipresent globally (see chapter 1). A strategy for restoring natural capital, which prioritizes sustainability through fulfilled relationships rather than through consumption and growth, flies in the face of this dominant culture. Advocates of the restoration of natural capital must therefore overcome cultural obstacles if the message is to gain support. There is little point in achieving a scientific consensus concerning the implications of declining natural capital, if the power elites or the general public are unaware. Hence, one of the main tasks in changing our culture is to increase awareness about natural capital depletion and degradation and how to rectify the situation through restoration. Changing a dominant cultural paradigm is notoriously difficult, but not impossible. It can be achieved through an energetic and coherent communications strategy, with alternative positions promoted effectively through the media, education, and marketing.

The conservation movement has had some striking, but still very limited, success in communicating alternatives to the dominant culture through these channels. Since restoration and conservation are closely linked, this provides a considerable jump-start for the restoration of natural capital. As this chapter is written, climate change is the subject of a special edition of *Time* magazine (3 April 2006). *Sustainable development* is already part of everyday discourse within the context of people, space, time, and distribution. Restoration of natural capital could add significant meaning and substance to this debate, since it has the unique potential to appeal to a very wide public audience, by articulating a new cultural relationship with the environment.

First, it draws on both economic and ecological principles, thus mainstreaming its message in a language that the media, politicians, and the general public can easily grasp. Second, it has the advantage of a very positive orientation: the message of restoring natural capital is not so much "stop raping the earth" as "start healing the earth's wounds." Third, and crucially, as many chapters in this book reflect, restoration is rooted as much in the experience and needs of developing countries as it is in the developed world. The consultation and consent of all stakeholders, including the poorest of the poor, is a core principle of restoring natural capital.

The concept of sustainability, however, is rather variable. Depending on the concept one follows (see, for example, Dobson 1998), the severity of the ecological crisis and the need for cultural change is also variable. This means that those working in the field of restoring natural capital should communicate in a reasoned way to prevent a perception of arrogance, which may even be counterproductive.

A Strategy to Communicate the Need to Restore Natural Capital

The message of restoration should draw on the best practices of conservation organizations, but it also needs to break new ground. Such a communication strategy should lead toward the development of new economic and legal instruments and tools to foster the integration of natural and social capital, the culture of stewardship, and the restoration and maintenance of natural capital. Here, we make some suggestions.

Media

- Move beyond, while not neglecting, the obvious focus on environmental correspondents, feature pages, and supplements. Instead, target senior journalists from business, economics, and opinion pages, and editors, inviting them to conferences and symposia, where the arguments for restoration can be discussed in depth. In other words, mainstream the human and natural drama of depletion and restoration narratives to news pages and news analysis. The restoration of natural capital could be the biggest theme of the twenty-first century.

- Stress that the restoration of natural capital is also a development strategy, which encompasses the needs of both people and the environment, while the well-being and even the survival of humanity is inextricably linked to the health of the natural world.
- Campaign for the recognition of ecosystem services as natural capital assets and for the inclusion of all natural capital assets within economic indicators, such as the gross domestic product. Thus, this will mainstream our relationship with the environment to the heart of political debate and economic indicators.

Education (schools and colleges)

- Campaign for the inclusion of ecology within the curriculum at all levels, with links to economics, so that students can grasp the connection between economic activity and the environment. These courses should feature hands-on field trips to help on projects related to the restoration of natural capital (see, for example, http://ca.audubon.org/LSP/slews.htm).

Education (continuing education)

- The internet is a vital tool for enabling adults to relate to ecological issues. See http://www.climatecare.org/ for a site that simultaneously educates and prompts individuals to take positive action.
- Prepare television programs on the theme of restoration. The television audience's appetite for "nature" programs is insatiable, and advertising revenues are attractive to television companies. In 2006, *Planet Action*, an innovative conservation series aimed at younger viewers, attracted advertising worth $150 million.
- Collective adult education can also occur outside any formal setting, for example, through public works programs, where farmers are paid to restore biodiversity. The experiences of the Working for Water program in South Africa, the PSA in Costa Rica, and the SLCP in China illustrate both the great potential and some of the pitfalls of education through public works programs (Working for Water 2003; Jenkins et al. 2004; Xu et al. 2005).

Marketing

- Develop a branding campaign, along the lines of the fair trade and organic food campaigns, which clearly identifies the products of those public and private companies that are restoring natural capital. This not only enables environmentally aware consumers to send a message to the market, it also contributes to the mainstreaming of restoration policies among consumers.
- Likewise, with public works programs, simple tools can have broad impact. The

yellow-and-green t-shirt of the Working for Water program is a daily reminder for thousands of South Africans of the benefits of restoring natural capital.

Restorationists have long argued that community participation in restoration is a transformative experience, achieving a shift in thinking in a way no transfer of abstract information ever can. But a more general paradigm shift is necessary to persuade people to participate in or just support a strategy of restoring natural capital, in the first place. An effective communications program is essential to achieving this aim and should also contribute to overcoming the economic and legal/institutional obstacles we have outlined.

Contribution

It is not only essential for society to have a cultural paradigm shift to engage actively in the restoration of natural capital, it is also necessary to work actively toward such a shift through an effective communications program.

The task of those engaged in the restoration of natural capital in every culture is to seek out and stress the traditions that encourage restoration to thrive. On illustrating the link between culture and the restoration of natural capital, irrespective of the local context, the key message behind the restoration of natural capital is that such an activity is of mutual benefit to human beings and to the biodiversity and health of the entire planet (see the definition provided in chapter 1). The underlying determinant of democratic politics, however, is public opinion. It is in overcoming the obstacles to restoration in the dominant culture that the battle for sustainability will be lost or won. Society has to start accepting that the true benefits from restoring natural capital accrue only much later. Hence, the economic opinion makers need to integrate the long-term benefits of well-functioning ecosystems into current economic models, using low, neutral, or negative discount rates when valuating ecosystem services.

Both short- and long-term benefits lie in using restoration for poverty relief and the building of rural communities, while securing and enhancing the earth's natural capital for the future. In the long term, the loop should be closed between the drivers of degradation and the rationale for the restoration of natural capital, by rethinking economic models and revising legal instruments and institutions in order to change the culture of global society.

However, for successful restoration of natural capital, project developers are adamant that irrespective of how good the design is, and how noble its objectives, if its implementation is lacking skilled personnel, the program will not work. The key to all successful implementation is management, management, and management.

Chapter 30

Overcoming Obstacles at a Global Scale to Restore Natural Capital

ROBERT J. SCHOLES, REINETTE BIGGS, ERICA J. BROWN GADDIS,
AND KAREN D. HOLL

The ambition of restoring natural capital extends far beyond patching up landscape fragments degraded by human activities. It represents a desire to shift the environmentalist strategy from being essentially defensive—the protection of islands of near naturalness—to expansive, by seeking to restore essential ecological functionality throughout the biosphere (Jordan 2003; chapters 1 and 3). These goals may have seemed quixotic a decade ago, yet now they are considered to be essential by studies such as the Millennium Ecosystem Assessment (MA 2005a; chapters 5 and 6) for the persistence of humankind as a whole at a reasonable level of well-being.

This chapter, in contrast to the previous two that considered the restoration of natural capital at a landscape scale, explores potential trajectories of the global ecosystem and the associated potential for long-term restoration of natural capital. The first section provides some definitions and conceptual background. The second section takes a theoretical approach to mapping out potential outcomes of a linked human–biosphere system. The third section gives examples of existing global-scale efforts aimed at the restoration of natural capital. Based on these examples, the final section suggests some preconditions for successful global-scale restoration of natural capital.

What Do We Mean by Restoring Natural Capital at the Global Scale?

What we *do not* mean is the return to some former, premodern, or prehuman state of the globe. Return to such a condition is now impossible at the global scale, as was indicated on a landscape scale in chapter 7, even if it was desirable. Irreversible changes, such as species extinction, have already occurred or are under way (Sala et al. 2000; Thomas et al. 2004; MA 2005b). Nor do we envisage global restoration to mean simply the global sum of local-scale restoration activities. What we *do* mean is the assisted recovery of ecological processes operating at very large scales, that is, hundreds to tens of thousands of square kilometers that are necessary for the sustainable delivery of ecosystem services. In some cases, global restoration may be achieved through the aggregated effect of uncoordinated local actions. In other instances, which we focus on, it will require coordinated action aimed directly at large-scale processes.

The restoration of natural capital is possible only if we have not crossed a threshold in the coupled biosphere–human system beyond which the supply of essential ecosystem services is *irreversibly* reduced. By irreversible we mean that restoration is not possible within several generations with a realistic expenditure of effort.

A more ambitious target for global restoration would be to aim for global optimization of ecosystem services. Optimization includes considering sufficient margins for risks and uncertainties (due to ignorance and inherent system unpredictability) and for the maintenance of resilience, that is, a reasonable probability of assured continuation of service supply, when facing variations due to internal or external dynamics. However, ecosystem services cannot all be maximized simultaneously, given that the tradeoffs and human preferences for services change over time. Complexity is introduced when local and global benefits need to be traded, and efficient mechanisms for this seldom exist (chapters 32 and 33).

The Uniqueness of Now

There are certain attributes of the current era, the "Anthropocene" (Crutzen 2002), that make it qualitatively different from any previous time. The human footprint is for the first time globally pervasive and plausibly larger than the sustainable limit (Living Planet Report 2005). To date, "restoration" of local ecosystems has frequently been achieved by displacement of impacts to less-regulated locations or less-exploited resources. For instance, much of the forest regrowth in Europe has been made possible by importing agricultural products from elsewhere. This is not an option at the global scale because few underexploited systems remain from which we can subsidize our local overconsumption of services. The oft-mooted idea of moving to another planet is infinitely less probable than achieving a transition to a sustainable future on the single planet known to support life (Horneck and Schellnhuber 2004).

Previous warnings of the earth's finiteness have been based largely on material limitations (Meadows et al. 1972). Thus far, these predictions have proved unfounded, mostly because human inventiveness has found substitutes for diminishing resources. A key strategy to date has been the increasing use of energy, especially for transport, to alleviate primary constraints such as the local availability of soil nutrients or water. The limitations to development are increasingly linked with the side effects of human activities, rather than an insufficient resource supply. The buildup of greenhouse gases, excessive aquatic nutrients, and global biodiversity declines provide examples. Substitution of these sink and regulating functions may prove more difficult than input substitution.

The human population is projected to peak at nine to twelve billion before the end of the twenty-first century (Nelson et al. 2005) and then stabilize or decline slowly. Importantly, the reduced population growth rate is not predicted primarily for "Malthusian" reasons, such as starvation and other forms of density-dependent and scarcity-induced mortality. It is rather because the birth rate in most places is declining to match (or in some cases, falling below) the mortality rate, which began a steep decline about two centuries ago. The best available hypothesis for the declining birth rate is that it is enabled by rising human well-being, promoting increased parental investment in fewer offspring. Lower child mortality allows families to risk smaller recruitment, while contraceptive technologies and the education and empowerment of women have made this achievable.

The implications for restoration at a global scale, within the context of a human population no longer increasing exponentially, are profound. Will the pressure on global ecosystems ease as a result? Consumption per capita continues to rise even when populations stabilize, and it has a long way to go in most parts of the developing world before reaching developed-world levels. Is the paradigm of ever-expanding consumption, central to most economic theories (chapter 2), valid in the presence of a globally stable or declining population? Many economists argue that economic and population growths are disconnected because the economy is increasingly dematerialized and thus decoupled from the limitations of individual consumption (chapter 3). A consequence of this argument (if valid) would be ultimately a decreasing need to harvest resources and transform ecosystems, opening up the possibility for the extensive restoration of natural capital. In reality, however, there are very few cases of a decline in per capita resource use or waste generation with an increasing gross domestic product.

Life at *K*

Ecologists use the symbol *K* to denote the maximum size that a population reaches if left to its own devices in a given environment. It is often referred to as *carrying capacity*. Given a stabilization of world population, we need to adapt to life around the level of K_{humans} (it is possible for K_{humans} to change over time). This requires a different set of strategies from those deployed by a rapidly growing population far below *K*: an enhanced emphasis on efficiency and risk avoidance, rather than maximum resource capture. There are important unresolved issues regarding the tradeoffs between efficiency (i.e., maximum, short-term productivity) and resilience, which involve the maintenance of a secure supply in the long term. Increasing efficiency usually leads to reduced resilience (Scheffer et al. 2000).

A near-stable human population will require the restoration of natural capital at a global scale for two reasons. First, some (perhaps many) ecosystem services are already significantly degraded, but not yet irreversibly, it is hoped. Second, the quest for optimality is likely to lead to a reallocation of ecosystem uses around the globe. This will mean intensification in some places (for instance, crop agriculture has a comparative advantage in warm, moist, fertile areas) and a deintensification elsewhere, creating restoration opportunities. While speculating on these issues, it is not a foregone conclusion that maintaining the population at nine to twelve billion is possible in the long term. Many of the required technologies have yet to be developed or have been deployed for only a few decades. Their long-term consequences and sustainability are untested.

Which Values Win?

The people–environment relationship is ultimately based on values (Jordan 2003; chapters 2 and 26). Here we focus on three values that are self-evident.

- Comfort: that we strive for this as individuals, families, and communities results in consumption at a level above minimal subsistence. This can be considered a *duty to ourselves*.
- Equity: that all people should have a fair chance of a good life. This is a *duty to others*.
- Sustainability: that the rights of comfort and equity extend to future generations. This is a *duty to the future*.

These three values constitute an approximately finite-sum game, if the population is close to or above K_{humans}, since all interactions between them are neutral or negative. There remains some scope for "efficiency" improvements (many to date are simply displacements of the problem). We can probably achieve equity and sustainability, but this will require substantial compromises on comfort by the more affluent. Currently, our political and economic systems maximize the comfort of privileged groups with lip service to equity and sustainability.

Possible Evolution of a Future Human-Dominated Biosphere

Most recent century-term environmental forecasts, such as by the *Intergovernmental Panel on Climate Change* (IPCC) (Nakicenovic and Swart 2001), the MA (2005c), and the *Global Environmental Outlook* (UNEP 2002), have concluded that the uncertainties are so great that all scenarios must be considered equally probable. Yet when future scenarios are built (as these were) by plausible projections from current states and trends, they tend to eliminate the unthinkable. In this chapter, trying a different approach, we consider the theoretical dynamical modes that a coupled biosphere–human system could exhibit as it approaches K_{human}, without a priori asking if they are plausible. We then explore the necessary circumstances for those outcomes to occur, or not.

The coupled human–biosphere system can be thought of as a *predator–prey* system (figure 30.1), with humans and the biosphere as the predators and prey, respectively (Brown and Roughgarden 1995). The dynamics of predator–prey systems are well understood and can provide a metaphor while allowing for particular features of human–biosphere interaction, such as the biosphere supplying many resources to humans, not all of which change in unison. The simple system depicted in figure 30.1 has four possible outcomes, which depend on the particular values chosen: (1) the replenishment rate of the biosphere; (2) the extraction rate; (3) the conversion efficiency factor; and (4) the decline rate of humans, if resources are restricted.

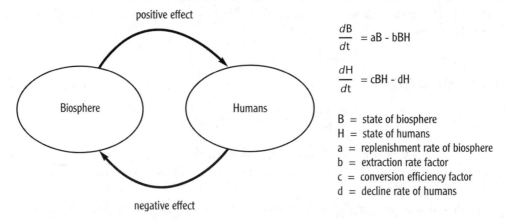

$$\frac{dB}{dt} = aB - bBH$$

$$\frac{dH}{dt} = cBH - dH$$

B = state of biosphere
H = state of humans
a = replenishment rate of biosphere
b = extraction rate factor
c = conversion efficiency factor
d = decline rate of humans

FIGURE 30.1. Representation of the coupled human–biosphere system as a formal predator–prey system.

Terminal Overshoot

This is predicted when the consumption rate of natural capital and its conversion into further demand are high. However, the feedback mechanisms from the state of the biosphere to the rate of human consumption of natural capital are insufficiently strong to avoid the irreversible loss of one or more nonsubstitutable and essential ecosystem services. Consumption of ecosystem services (bBH) peaks at a level substantially higher than their replenishment rate (aB), allowing the human population to temporarily exceed K_{human}. Subsequently, the human population declines to zero, taking many other species with it (Lenton et al. 2004).

Extinctions are an example of an element of the biosphere being driven irreversibly to zero. While some commercial fish species may not be extinct, they persist at low stock levels, perhaps irreversibly (Hutchings 2000). As yet these crashes have not caused corresponding human population declines because other food sources remain exploitable.

An example of a terminal overshoot at a global scale would be runaway, human-induced warming. There are many demonstrated positive feedback loops between the biosphere and global warming, but for out-of-control warming to occur they have to be strongly positive. Fortunately, as yet, none of these positive feedback loops have been shown to be strong enough.

It is tempting to perceive the apparent rise and fall of empires in terms of a "boom-and-bust" cycle, where societies grow until they overreach their resource base. This may be the mechanism in some cases, but it is clear that in most civilization collapses, many processes are at work simultaneously (Tainter 1990; Diamond 2005). It is undeniable that some civilizations grew spectacularly and then appear to have failed rather suddenly—usually not to a zero base, but to a substantially reduced form. In the past such collapses had local to regional consequences. Is it inconceivable that an effectively global society could fail globally?

Human societies are currently on the path to overshoot but have a unique capacity to learn and react, and presumably we would not knowingly succumb to this outcome. However, in the presence of a strong positive feedback, the human response would have to be swift and effective. The restoration of natural capital is implicitly based on the assumption that overshoot is the current trajectory of many (perhaps most) human–ecosystem interactions. It suggests that we have exceeded the supply of ecosystem services in many instances and have caused sufficient damage to prevent system recovery without some form of active human intervention. If we wish to avoid terminal decline, and to meet human well-being objectives in terms of comfort, equity, and sustainability, the restoration of natural capital is required.

Oscillation: Convergent, Divergent, or Constrained

If consumption overshoots ecosystem service supply but then declines fast enough to allow the sufficient recovery of natural capital for renewed extraction, the system itself will oscillate. The oscillations result from a harmonic interplay between the strength of the feedback loops and time delays. The overshoots can become progressively smaller (convergence) or larger (divergence), depending on the parameter values, with the former likely if societies learn from their mistakes. Learning is hard when large amounts of social capital are lost in the process. Diverging oscillations are likely if the fundamental basis of ecosystem service production is progressively eroded by each overshoot. In the worst case, the ultimate outcome is the same as *terminal overshoot*, simply delayed by several boom-and-bust cycles.

An intermediate situation exists, called a *stable limit cycle*, where oscillations remain indefinitely within nonzero bounds. Stable limit cycles are often portrayed as the archetypal behavior of theoretical predator–prey systems, but they may not be common in the real world (Berryman and Millstein 1989; Turchin and Taylor 1992), and especially not in self-adaptive (learning) systems. At the global scale there are many simultaneous out-of-phase oscillations occurring, which combine to have a much steadier effect on the human–biosphere interaction.

If the human–biosphere system is indeed cyclic, it would be useful to understand its periodicity. For the simple model described (figure 30.1), this could be easily determined if the values of a, b, c, and d were known; but reality is significantly more complex. Time lags and inertias can create or destroy harmonic behavior, while technology and learning characteristic of the human–biosphere system may induce changes in the periodicity. In fact, the typical time-constants of human demography, economic and political systems, and ecosystem service recovery suggest that if cycles do exist, they would last centuries to millennia.

Cyclic systems imply restoration, be it either deliberate or autonomous. Preemptive restoration of global ecosystem services may also conceivably turn an incipiently oscillating system into a stable state, by cushioning the decline rate after an overshoot.

Suboptimal Stabilization

This outcome is quite widely observed in the real world where inertia in the human response system means that consumption of ecosystem services sufficiently overshoots their regeneration capacity. The consequence is that irreversible damage is done to the productive capacity of the biosphere before the human population begins to decline. If no essential, nonsubstitutable ecosystem service declines to zero, human populations and ecosystem service-consumption levels stabilize at a new, lower level where human well-being is significantly impaired. Other species would also be negatively affected. These patterns are clearly shown during the nineteenth century by the inappropriate European livestock and grazing practices introduced into large areas of North and South America, southern Africa, and Australia (for example, see Dean and McDonald 1994; chapters 7, 9, 13, 15, 21, 22, and 29).

Monotonic Stabilization

This is the most desired outcome, when human consumption of ecosystem services stabilizes at a level where the generation of those services is reasonably unimpaired. This does not necessarily have to be at the maximum theoretical K_{human}. As a result, many other species can persist in viable populations, albeit at lower than prehuman abundances, while some would also be driven to extinction. The necessary conditions to reach this outcome are strong and prompt feedback from the biosphere to humans, causing timely reduction in human impact. The latter condition necessitates restoration of some aspects of natural capital at a global scale, since two-thirds of the world's ecosystem services are assessed as being consumed at levels at or above their capacity (MA 2005a).

The observation that human population stabilization seems not to be directly driven by biophysical resources, but by social feedback mechanisms, provides hopeful opportunities.

However, ecosystem service consumption continues to rise (both on an absolute and per capita basis), so it may already be too late to avoid some degree of overshoot.

Servicing the needs of the anticipated nine to twelve billion people at peak global population remains a great challenge and will inevitably cause further natural capital loss. The natural capital used could be minimized by striking an appropriate balance between intensification of service production (i.e., increasing yields per unit area) and agricultural extensification (increasing the area cropped). In current locations so intensively used that off-site, ecosystem-service impacts are greater than on-site impacts, deintensification (i.e., restoration) is required. In much of the developing world, careful intensification will arguably be less damaging to natural capital than rampant extensification.

It is unlikely that the rate of future intensification can keep up with the simultaneous growth of the human population and per capita consumption. Most of the increase in per capita food supply between 1960 and 2000 came from intensification rather than extensification, though there was nevertheless a substantial increase in the area under cultivation. Further expansion will generally be into areas of increasingly marginal agricultural potential. Thus, whereas this scenario may provide opportunities for restoration of natural capital in regions such as Europe and North America, the prospects in the rest of the world seem slim over the next century.

Examples of the Restoration of Natural Capital at the Global Scale

Several examples exist of global-scale restoration attempts, which have had mixed success.

Fixing the Ozone Hole

The depletion of stratospheric ozone, particularly over the South Pole, was detected somewhat by chance in the 1970s. Advances in the understanding of global atmospheric circulation and halocarbon photochemistry (WMO 1985) allowed sufficient confidence in the causes and consequences of stratospheric ozone depletion to permit the signing of the Montreal Protocol in 1987, limiting the emission of ozone-depleting substances. Several factors contributed to its successful implementation, which is credited with an observed reduction in the rate of loss of stratospheric ozone. First, substitutes for the ozone-depleting chemicals were already available and comparable in cost-effectiveness. Second, a very small number of producers were involved, and the cause of the problem was relatively uncontested (Parson 2003). Third, there was a clear economic benefit to all parties for maintaining the ozone layer.

Mitigation of Climate Change

The threshold capacity of biospheric absorption of greenhouse gases (GHG) has clearly been crossed, as evidenced by their continual increase in atmospheric concentrations (IPCC 2001). Stabilization of radiant forcing, that is, a decline in GHG emissions to below the level of long-term sustainable biospheric sinks (about 5% of current emission levels) requires restoration at a global scale. Hence, the United Nations Framework Convention of Climate

Change has the objective of "avoiding dangerous climate change," with its Kyoto Protocol as the first step to achieve this. However, these treaties are struggling to meet their objectives, due to (1) the complexity of climate science, allowing disputation of the causes and consequences; and (2) global warming being a "commons" problem with potential for "free-riders" and no obvious enforcement capacity.

Marine Fisheries

Over three-quarters of marine fisheries are exploited at or beyond their limit (Pauly et al. 2006). No significant fishery-resource species are known to be biologically extinct, but many are economically extinct. There are examples of regional-scale recoveries of marine fish stocks when restorative actions have been taken. The complete protection of up to 30% of the species habitat appears to be an effective mechanism (e.g., NRC 2001). This has been achieved where the fishery falls within the Economic Exclusion Zone of one or a few nations with an enforcement capability. However, some regional stocks such as the Atlantic cod (Hutchings 2000) show no signs of recovery even after the cessation of fishing.

Reversing Desertification

Desertification can be defined as the persistent reduction in the supply of ecosystem services in dry lands (MA 2005d). It is the environmental problem that materially affects the most people globally, that is, nearly one in five. The United Nations Convention on Combating Desertification is the formal institutional response to this problem, but it is almost universally regarded as a failure. Why? Desertification is widely perceived among the economically powerful nations as a developing world problem with local causes and local solutions. As a result, it has attracted much less scientific attention and restorative resources than issues that currently affect far fewer people. The reality is that desertification has both local and global causes and consequences. However, complete restoration of degraded drylands is an extremely slow and expensive process that occurs over several decades once the causes of degradation have been alleviated. Hence, the avoidance of further degradation is often a better economic option than restoration, yet for people with no alternative livelihoods, there is no choice: restoration is an imperative.

Multinational River Systems

Most of the world's large rivers flow through several countries. In general, the nations involved have been able to work out collaborative management schemes. In several cases, such as the Rhine and Danube in Europe, nations have cooperated to achieve improved water quality (Wieriks and Schulte-Wülwer-Leidi 1997; Schmidt 2001). However, examples where restoration of multinational basins has not been achieved include the Aral Sea between Russia and Kazakstan, the Colorado River between the United States and Mexico, and the Olifants River between South Africa and Mozambique. All these cases show a strong asymmetry between the relative power of the upstream and downstream countries, with the more powerful getting first use.

Eutrophication

Many aquatic habitats adjacent to population centers and agriculture have experienced increased nutrient loading, ecosystem changes, and even the development of anoxic "dead zones" (Turner and Rabalais 1994). Restoration is possible but will be slow. Nevertheless, catchment-scale interventions, such as riparian buffer strips, can slow down the leaching of nutrients into the aquatic environment, while wetland restoration helps to remove them.

Necessary Conditions for Global Restoration

Consideration of these cases suggests the following minimum conditions are required:

1. Effective and responsive mechanisms for decision making and enforcement need to exist at a scale that matches the degradation and restoration processes. For regional- and global-scale processes, this requires supranational governance institutions, which link to institutions at smaller scales. The Montreal Protocol is an example.
2. Effective control of common resources that allows them to function as regulated-access rather than open-access systems. The current precarious state of marine fisheries is largely attributable to a failure in establishing an effective regulation regime in the oceans. Privatization or nationalization of resources is not the only possible solution. Many examples of communal resources that are well managed exist, provided that the user group is well defined, the rules are clear and accepted, information is reliable, and a mechanism operates for enforcing sanctions.
3. Shared and accepted knowledge with respect to ecosystem service issues. This in turn requires international, free- and open-access monitoring, assessment, and research systems. The combination of the major earth-system research programs and the Intergovernmental Panel on Climate Change appear to provide a major impetus for making climate change a tangible issue. In contrast, the prevention of biodiversity loss and combating desertification languish.

Contribution

Restoration of key aspects of natural capital at a global scale is a prerequisite if humans are to avoid some potentially unpleasant future dynamics. There is a body of economic theory, the *inverse Kunznets curve* (Panayatou 1995), that suggests that natural capital is initially depleted while human-made capital accumulates. At some point, the well-being of society is sufficient that its focus shifts from satisfying basic needs to improving its quality of life, and the society begins to invest in the restoration of natural capital. The examples given relate to local- or regional-scale phenomena, such as urban water and air quality. Is this theory applicable to global-scale phenomena? If regional restoration is fundamentally achieved by shifting pollution and resource consumption to poorer communities elsewhere, it probably does not apply at the global scale. This is because those communities have nowhere left to unburden themselves, and as the ultimate recipients, they cannot raise their own well-being sufficiently to allow reinvesting in nature.

If this is the case, global-scale, cost-redistributive mechanisms are needed to maximize global welfare. There are many nascent examples of such transfer-payment schemes for ecosystem services, such as debt-for-nature swaps, carbon trading, and catchment-protection payments (see chapter 31). These are probably inadequate if they continue to be considered a variety of philanthropy rather than a necessary condition for restoration and collective welfare. A global mechanism may be needed to stabilize the global human–biosphere system, for example, one that funds restoration and welfare gains in poorer communities out of progressive taxation on excess consumption.

True cost accounting of human activities would assist in making rational development choices at all scales, including the global. The declines in natural capital must at least be balanced by increases in manufactured capital (Arrow et al. 2003), minimum stocks of natural capital need to be respected, external costs need to be internalized, and perverse subsidies and incentives that lead to overconsumption of ecosystem services need to be eliminated.

Chapter 31

Managing Our Global Footprint Through Restoration of Natural Capital at a Global Scale

JOSHUA FARLEY, ERICA J. BROWN GADDIS, WILLIAM E. REES, AND KATRINA VAN DIS

Human activities have come to rival geological forces in their impacts on natural capital at a global scale (Vernadsky 1998) but are generally guided by decisions that consider only short-term, local outcomes. We fail to account for the impacts our activities have on global ecosystems and fail to understand the role of ecosystems in maintaining the life-support functions essential for our well-being and survival. We also fail to distinguish between the interest on natural capital, that is, the renewable flow of raw materials and essential services provided by healthy ecosystems, and the capital stock itself. Important long-term changes in global ecosystems may pass unnoticed (Ehrlich 2000). As a result, our sustaining ecosystems are being destroyed and degraded at a rate unprecedented in human history, and our ecological footprint has exceeded the productive area of global ecosystems (Wackernagel and Rees 1996).

Addressing this mismanagement requires not only a reduction in our global footprint but also active investments in restoring natural capital on a global scale. Indeed, with human population growing at the rate of eighty million per year, we must increase the total physical stock of productive natural capital just to maintain per capita natural capital and income at a constant level.

However, adequate restoration of natural capital on a global level presents unique challenges (see chapter 30). System uncertainty increases along with system boundaries, making it more difficult to decide what to restore. At the same time, institutional capacity for implementing policies to promote restoration diminishes once we move beyond the scale of sovereign nations. Success will require an economically efficient and socially just mechanism that creates incentives for restoring adequate amounts of globally valuable natural capital and ecosystem services without threatening national sovereignty. Whereas the previous chapter considered various high-level scenarios and plausible outcomes related to the restoration of natural capital at a global scale, this chapter considers the challenges of the restoration of natural capital at such a large scale. This chapter also outlines some policy principles with which effective institutions and financing mechanisms should comply and assesses existing institutions and restoration strategies in the light of these principles. We then sketch out a potentially suitable instrument for implementing global natural capital restoration and present our conclusions.

The Unique Challenges of Restoring Natural Capital at the Global Scale

While restoring natural capital is challenging at any scale, these challenges are compounded at the global level. Global restoration should be focused on sustaining global ecosystem services, but we are highly uncertain how these services are provided, and how much restoration will be required to sustain them. As an investment in nonmarketed public goods, restoring natural capital requires nonmarket institutions, but we lack global institutions capable of providing global public goods, and sovereignty constrains our policy options. These points deserve elaboration.

Issues of Uncertainty at the Global Scale

Perhaps the greatest challenge to determining the *optimal* amount of conservation and restoration of natural capital is the uncertainty that surrounds global ecosystem functions. Some uncertainty results from a lack of adequate data, while other uncertainty is the result of natural processes and is therefore irreducible. In either case, it is very difficult to determine when human activities will lead us to cross irreversible and catastrophic thresholds beyond which these systems can no longer sustain us, and the presence of time lags and positive feedback loops in complex systems means that we might be slow to realize it when we do cross them. Growing evidence suggests that we are dangerously near irreversible thresholds for atmospheric gases such as carbon dioxide and ozone, oceanic fisheries (Kura et al. 2004), and even oceanic pH (Royal Society 2005). In addition, the loss of seemingly unimportant species or "local" ecosystems could cause chain reactions with tragic and unpredictable global consequences (Holling 1986). We are still learning how to restore critical natural capital, and what little we do know may change as we cross unknown ecological thresholds.

Economic Globalization and the Global Ecosystem

Economic globalization can be described as the increasing irrelevance of political boundaries to market commerce, in which financial capital and raw materials move to wherever private production costs are cheapest, and commodities to where prices are highest. Conventional wisdom praises globalization as the new driver of economic growth and argues that only more growth will provide us with the resources necessary to restore our degraded natural capital to health (Beckerman 1992; IBRD 1992; for an alternative view, see Stern et al. 1996).

The truth is, however, that during the last century real physical output of the global economy has increased ninefold per capita. The law of diminishing marginal utility suggests that each incremental increase provides less benefit than the one before, and empirical evidence shows that in the wealthier countries the benefits of continued increases in consumption are negligible (Lane 2000; Max-Neef 1995). However small its marginal benefits, consumption comes at a real cost. A ninefold increase in consumption accompanied by a quadrupling of the global population translates into a thirty-six-fold increase in throughput, that is, the rate at which raw materials are extracted from global ecosystems, transformed into commodities, and ultimately returned to the ecosystem as waste (Daly and Farley 2004). Both extraction and waste production increase our ecological footprint (Wackernagel and Rees 1996) and threaten the integrity of the global ecosystems that sustain us (Georgescu-Roegen 1971).

Since these ecosystems and functions have grown scarcer per capita, their marginal value has increased, as has the marginal value of restoring them. As ecosystems reach critical ecological thresholds beyond which they can no longer sustain themselves, the marginal benefits of restoring natural capital will become incalculably high (chapter 3). Far from providing resources to solve this problem, economic globalization is providing increasingly unimportant market goods at the expense of increasingly important ecosystem goods and services.

Political boundaries have always been irrelevant to the flux of ecosystem services, many of which are inherently global and demand global institutions to protect and restore them (Kaul et al. 1999; Sandler 2004). Though the creation of institutions necessary to manage a global economic system receives far more attention, the need for institutions and mechanisms to manage an inherently global ecosystem, to protect and restore global natural capital, is far more critical, with greater consequences if ignored.

Nonexcludability of Public Goods: The Free-Rider Problem

One of the most serious challenges to the restoration of natural capital at the global scale is that most decisions regarding global natural capital are based on market economics. In theory, free and competitive markets put resources to their highest and best use. However, in reality, markets function only when resources are excludable. Excludability is a legal construct that allows the owner of a resource to prevent others from using that resource. If one cannot prevent others from using a resource whether or not they pay for it, they (the *free riders*) are unlikely to pay, and market forces are unlikely to provide or protect the resource. As a result, nonexcludable resources are generally underprovided or overexploited in a market economy (Samuelson 1954; Randall 1993).

Clearly, many functions that natural capital delivers are nonexcludable. For example, a coastal mangrove ecosystem can protect against coastal erosion and flooding, while also providing important fish nurseries. Those who benefit from these services do so whether or not they contribute to the preservation and restoration of the wetland, leading to the problem of free riding on the willingness of the property owner to provide the service for free, or on the willingness of others to pay for its provision (Olson 1965). Although the services provided by ecosystems are nonexcludable, ecosystem structure and site are often treated as market goods (Daly and Farley 2004). Because so many ecosystem services are nonmarket goods and often "the precise contribution of a functional element in the ecosystem is not known—indeed is probably unknowable—until it ceases to function" (Vatn and Bromley 1994, 133), market forces favor conversion over restoration, regardless of their relative contributions to social welfare. As a result, many mangrove ecosystems are converted to provide a higher, short-term profit to the owner (i.e., shrimp aquaculture), even if the long-term global value of coastal protection, biodiversity, and seafood production linked to providing fish nurseries is much higher.

Under these circumstances, conventional economists generally consider the loss of ecosystem services as a negative externality of shrimp aquaculture. Demsetz (1967, 350) argues that "property rights develop to internalize externalities when the gains of internalization become larger than the cost of internalization," suggesting that markets can expand to cover such problems. Thus in recent decades we have seen the creation of markets in sinks (e.g., waste absorption capacity for SO_2) and sources (e.g., some fisheries) (Colby 2000).

Ecological economists, however, point out that virtually all economic production has some negative impacts on natural capital (Georgescu-Roegen 1979), and internalizing all of these externalities would require a centralized economic system that is the antithesis of free markets (Daly and Farley 2004). Finally, adequate institutions do not exist for assigning property rights to the global commons, and since the restoration of natural capital generates nonexcludable benefits directly, not as an externality, the free-rider problem arises at the global level.

Additional Problems with the Property Rights Solution: Nonrival Resources

Even if it were possible to develop property rights to all the services natural capital renders, exclusive property rights can actually generate serious inefficiencies. This happens when we use the price mechanism to ration use of nonrival resources, defined as resources for which consumption by one person does not prevent simultaneous consumption by another. Take the example of a watershed restoration project that reduces the likelihood and severity of flooding in a town downstream, beautifies previously degraded hillsides, and provides great hiking opportunities. One household "consuming" flood protection and beautiful views will not reduce the amount of those benefits other households can consume. Except in the case of very heavy use leading to congestion, the same is true for hiking. Since it would be virtually impossible to prevent individuals from enjoying the flood protection or views, it would not be possible to charge for them, but this is not true for hiking. However, if people were charged each time they went hiking, they would presumably hike less than if it were free — prices would ration use to those who valued the hike as much or more than the cost. This would actually reduce the total benefits provided by the forest, without in any way reducing restoration costs, clearly an inefficient outcome for society. In other words, markets in nonrival goods lead to nonoptimal levels of consumption (Daly and Farley 2004). Rivalry is a physical characteristic of resources that cannot be affected by policy (though the Bible tells the parable in which Jesus made a rival good nonrival, feeding the multitudes from a single loaf of bread and a single fish, mere mortals cannot do this!). Other examples of nonrival resources include climate stability, the ozone layer's protective function, and information.

While most nonrival ecosystem services are inherently nonexcludable as well, this is not true for information, which can be made excludable through patents. The restoration of natural capital is in its infancy, and we still have much to learn. Information improves through use, but patents limit use to those who pay royalties. Patents on the restoration of natural capital techniques are therefore likely not only to reduce the use of those techniques, resulting in less restoration, but also to slow the growth rate of knowledge (Heller and Eisenberg 1998; Jaffe and Lerner 2004).

When a resource is both nonexcludable and nonrival, it is a public good. Economists have long recognized that markets lead to the inefficient provision and consumption of public goods, and nonmarket institutions such as governments are required to provide them. Many functions that natural capital provides, however, are global public goods and absolutely essential to our well-being, yet adequate global institutions for providing them do not exist (Kaul et al. 1999; Sandler 2004). We see this as the primary challenge to achieving the restoration of natural capital at a global scale.

Provision of Global Public Goods by Sovereign Nations

While economists generally recognize that governments must provide or protect public goods, virtually all terrestrially based natural capital falls under the control of sovereign nations or corporate owners. These in turn generate a variety of ecosystem services with different spatial distributions that do not respect political boundaries or property regimes. In economists' jargon, ecosystems provide *joint products*. Thus, an ecosystem may produce services that are "consumed" at the local, national, and global scales simultaneously (Farnsworth et al. 1983; Sandler 1993). For example, mangrove ecosystems can buffer local communities from storm surges and protect offshore habitats and the fisheries they sustain by absorbing wastes flowing from the land. Storm protection is a local service, but coastal wetlands serve as a nursery for up to 80% of commercial seafood species in some regions (Hamilton and Snedacker 1984), benefiting fishing communities beyond the local political jurisdiction. Local and national authorities may be capable of creating property rights that internalize local externalities but are likely to ignore regional, global, and intergenerational externalities. Furthermore, no global government or institutions exist to ensure the adequate restoration of natural capital to support global-level processes. Clearly, balancing the cost of restoring natural capital against only local and national benefits will lead to inadequate levels of restoration from the global perspective (Daly and Farley 2004).

Design Principles for Restoring Natural Capital at a Global Scale: Institutions and Policies

Prior to assessing existing institutions, mechanisms, and financing schemes, and suggesting new ones, it is useful to sketch a few basic design principles for managing the collective provision of natural capital restoration at a global scale.

Principle of Subsidiarity: Institutions Are Required at the Scale of the Problem

Because natural capital provides benefits at local, regional, national, and global levels, institutions and financial commitments are required at each of these levels. Collective global efforts to restore natural capital should not replace or distort local and national efforts, but rather complement them (Cornes and Sandler 1996).

Start from Historical Conditions

Institutions and policies need to respect national sovereignty as well as existing political and economic systems in different countries. Appropriate mechanisms should rely on incentives for the restoration of natural capital rather than penalties, such as fines or taxes, as these cannot be imposed on sovereign countries. In fact, many developing nations view international efforts to slow and reverse ecological degradation as an example of *eco-imperialism*. That is, wealthy nations grew wealthy by exploiting their own natural resources and now want to prevent other countries from doing the same (Bin Mohamed and Lutzenberger 1992; Driessen 2003). Any mechanisms that fail to respect this fact are bound to fail.

Achieve Macrolevel Goals While Allowing a Maximum of Microlevel Flexibility

To a limited extent, central institutions can gather most available knowledge and make the best educated guesses concerning global benefits of the restoration of natural capital, as has been attempted with the International Protocol on Climate Change and the Millennium Ecosystem Assessment. However, it is considerably more difficult for any global institution to centralize knowledge about the local costs and benefits of the restoration of natural capital, which can vary dramatically across short distances within the same ecosystem. An appropriate mechanism should, like markets, be able to use "knowledge which is not given to anyone in its totality" (Hayek 1945). Local knowledge is best qualified to determine the local costs and benefits of restoration (see chapters 15, 16, and 29). In other words, a mechanism should strive for macrolevel goals of adequate restoration, while allowing a maximum of microlevel freedom and variability (Daly and Farley 2004). This requires investment in local capacity building, which can be difficult and expensive yet essential for implementing effective restoration plans and management at the local level.

Precautionary Principle for Managing the Stock of Natural Capital

Due to the uncertainty among scientists regarding ecological thresholds and the optimal approaches for restoring natural capital, policies should leave a margin of error when dealing with the biophysical environment, applying the *precautionary principle* (Myers 1993; O'Riordan and Cameron 1994; Daly and Farley 2004). When deciding how much natural capital should be restored, we should also err on the side of caution, favoring the restoration of too much too early, a reversible decision, over too little too late, which may prove an irreversible one. While a sample size of one for unique ecosystems (e.g., the earth) does not allow us to reduce uncertainty through scientific experiments, we can explicitly incorporate uncertainty into systems models (for example, Burman 2005; Halpern et al. 2006), and use them to help avoid the most potentially catastrophic outcomes. Many ecosystems appear to have crossed or to be nearing irreversible thresholds, and thus they demand restoration. Yet, as mentioned, we are often uncertain precisely how to best restore natural capital, and only learning by doing can help us reduce this uncertainty. To be capable of carrying out urgent interventions when the precautionary principle demands it, we must acquire and disseminate the necessary knowledge as rapidly as possible.

Mechanisms Must Be Adaptive to a Changing Biophysical World

Suitable mechanisms for global restoration must work in a rapidly changing global system. Ecological (global climate change, ozone depletion, pollution, soil loss, invasive species, etc.), economic (globalization, recessions, etc.), and political factors (wars, resource disputes, etc.) have and will continue to have dramatic impacts on the restoration of natural capital. Policies must be flexible enough to work under different institutional frameworks and respond to unexpected changes (Daly and Farley 2004).

Policies Should Be Efficient, or at Least Cost Effective

An efficient policy would exactly balance the increasing marginal costs of implementation with the diminishing marginal benefits: we should restore natural capital as long as the benefits of doing so outweigh the costs. Lacking exact knowledge of costs and benefits, we should strive for cost effectiveness: we should meet a specific restoration goal at the lowest possible cost or provide the greatest possible benefits for a given cost. This demands a transdisciplinary approach. Systems biologists, restoration and conservation planners, landscape ecologists, and others should concentrate on criteria for determining where the ecological benefits of restoration are likely to be the greatest, while social scientists should focus on the socioeconomic benefits, as well as the mechanisms and institutions to minimize costs. This is a principle in line with the definition provided for the restoration of natural capital in chapter 1. Relevant benefits include the knowledge gained through restoration projects, which can be increased through free dissemination. Relevant costs include implementation, transaction, opportunity, monitoring, and enforcement, which can be reduced by working through existing international agreements, such as the Kyoto Protocol and the Convention on Biodiversity. Cost-effective approaches should complement existing local and national restoration incentives, and investments in local capacity are likely to offer significant returns. In theory, restoration should be carried out wherever net marginal benefits are positive. While accurate valuation is probably impossible (see chapters 3 and 26), systems models are useful for roughly estimating the net benefits of different interventions.

Existing Mechanisms for Restoring Natural Capital

Applying these principles, the potential effectiveness of the limited number of existing global mechanisms for restoration will be assessed briefly. The restoration of natural capital can, of course, be either active or passive. Passive restoration simply reduces or eliminates the activities causing ecological degradation and allows for self-recovery of the system, though in some cases it requires removal of infrastructure interfering with natural regeneration processes (see chapter 28). In contrast, active restoration involves activities that help initiate and/or speed up the natural processes.

International Treaties and Agreements

A number of these address global restoration of natural capital either directly or indirectly. The Montreal Protocol on substances that deplete the ozone layer, the Kyoto Protocol to the UN Framework Convention on Climate Change (UNFCCC), the UN Convention to Combat desertification (UNCCD), and the Convention on Biological Diversity are all highly relevant to restoration (see also chapter 30). In principle, they all seek to create institutions at the scale of the problem, while respecting national sovereignty. Of course, these two principles can clash when nations that contribute significantly to the problem do not ratify the convention. For example, the United States has failed to ratify the Kyoto Protocol or the Convention on Biological Diversity.

The most successful agreement is arguably the Montreal Protocol, which, although failing to comply with the precautionary principle initially, eventually achieved target reductions through adaptive management. This, however, provides a poor model for the restoration of natural capital. First, no nation has sovereign control of the ozone layer, which is not the case for terrestrial natural capital. Second, a central authority could reasonably estimate the costs of continued degradation, which were unacceptably high. Third, the goal was the total elimination of ozone depleting compounds (ODCs), making microflexibility irrelevant in the long run (Parson 2003).

Like the Montreal Protocol, the UNFCCC addresses ecosystem services over which no nation has sovereign control. The Joint Implementation–Activities Implemented Jointly (JI-AIJ) and the Clean Development Mechanism allow businesses in wealthier countries to finance carbon sequestration projects in poorer countries, including reforestation projects (principle of microflexibility). However, the Kyoto Protocol still allows increasing concentrations of greenhouse gases, thus failing to satisfy the precautionary principle. Furthermore both the JI-AIJ and Clean Development Mechanism involve detailed agreements between partners in different countries and therefore high transaction costs (Haites and Yamin 2000; Michaelowa et al. 2003, and chapter 29).

The Convention on Biodiversity seeks to conserve biological diversity, promote the sustainable use of its components, and promote the fair and equitable sharing of its benefits. Achievement of its goals (UNEP 2004) will require natural capital restoration. Unfortunately we lack the science to effectively measure biodiversity loss (Balmford et al. 2005), and the convention's role in privatizing biodiversity to create incentives has actually stifled scientific efforts to study conservation and restoration issues (ten Kate 2002).

Most international agreements emphasize specific forms of natural capital functions and fail to recognize that healthy ecosystems provide many services simultaneously. This flaw has led, for example, to promoting (under the Kyoto Protocol) exotic monocrop plantations of eucalyptus in Brazil and elsewhere over restoration of more diverse native forests. A more holistic and efficient approach would favor projects that meet the goals of several agreements simultaneously.

Finally, most nations that are signatories of these treaties and agreements are also members of the World Trade Organization (WTO). The WTO and its predecessor, the General Agreement on Trade and Tariffs, have challenged environmental regulations as being barriers to trade (Wallach and Sforza 1999), and many are concerned that this stance will undermine other accords (Vogel 2004). Equally serious, the WTO enforces intellectual property rights, forcing countries to pay royalties on patents, which could slow dissemination of information related to the restoration of natural capital.

Payments for Ecosystem Services

The Clean Development Mechanism is one of a group of mechanisms known as direct payments for ecosystem services (see chapters 19 and 32). Payments for ecosystem services take a variety of forms, have often proven successful in practice, and show considerable promise for expansion (Pagiola et al. 2002; Landell-Mills and Porras 2002). The problem is that most of these mechanisms are local or national in scope, whereas managing global restoration of natural capital will require global mechanisms.

The Global Environmental Facility

The Global Environmental Facility (GEF) is an "independent financial organization that provides grants to developing countries for projects that benefit the global environment and promote sustainable livelihoods in local communities." Seeking to provide the incremental costs necessary to transform "a project with national benefits into one with global environmental benefits," GEF's goals are a step in the right direction, but between 2002 and 2006 it has dedicated only some three billion dollars to such projects, which is virtually nothing relative to the pressing need for restoration. GEF funds are awarded to individual projects, and each project requires a proposal. The major transaction costs of this mechanism come in writing, reviewing, administering, and reporting on the grants. In the words of colleagues familiar with GEF, the process involves "massive" transaction costs, which may "turn the entire GEF mechanism inefficient." It also "limits flexibility and promotes a bureaucratic approach" in which "it's possible to tick all the boxes without achieving anything."

Proposed Method: Adapting Brazil's ICMS Ecologico to the Global Scale

None of the current mechanisms for funding the restoration of natural capital at the global scale satisfy all of the basic principles for effective mechanisms and institutions outlined earlier. Here, we propose a mechanism that does satisfy these criteria based on an adaptation of a policy (ICMS Ecologico) used in some Brazilian states to increase the supply of ecosystem services.

To briefly summarize, Brazilian states capture most of their revenue from sales taxes (Loureiro and Rolim de Moura 1996), returning 25% to municipalities. Some states have chosen to award a portion of this money to municipalities according to how well they maintain watersheds and conserve land (Perrot-Maitre and Davis 2001; Vogel 1997). The goals of the policy are to compensate municipalities for the loss of revenues from conservation areas, develop new incentive mechanisms for environmental management, stimulate the creation of new protected areas, conserve biodiversity, and recompense municipalities for the environmental services that they provide. The net result is an incentive to maintain and expand land uses that provide critical ecosystem services through the restoration of natural capital (May et al. 2002).

So far, the model appears to be quite successful and satisfies the principles outlined earlier. Surface area under conservation and the quality of the conservation units have increased, as have the number of states participating in the program and the revenues dedicated to it (Loureiro and Rolim de Moura 1996). In particular, transaction costs have been very low: four years after implementing this model, approximately US$30 million were redistributed at an incremental administrative cost of only $30 thousand (Vogel 1997).

Such a model could readily be adapted to provide incentives for the restoration of natural capital at the global scale. The approach would require a global institution with adequate financing that would allocate its annual budget to countries in proportion to the quantity and quality of the restoration of natural capital, active or passive, they undertake. Such a global-scale institution would provide additional incentives for national or local mechanisms already in place, thus complying with the principle of subsidiarity. Participation would be entirely voluntary, imposing no threats on sovereignty. Reward criteria could utilize the best scientific knowledge concerning the global values of specific ecosystems, but the ultimate decisions about restoration would be made at the local level, as local people are best informed about opportunity costs. This would satisfy the criterion of allowing maximum microflexibility to achieve macrolevel goals. To satisfy the precautionary principle, we would need to implement this policy as quickly as possible, with adequate funding. The best way might be simply to modify an existing institution such as the GEF. As we implement this instrument we will learn more about its strengths and weaknesses, and over time we will also learn more about the urgency of restoration. Adaptive management would require that the criteria for evaluating restoration efforts and the amount of money available be adjusted accordingly.

The suggested policy is highly cost effective in a number of ways. First, experience in Brazil shows that transaction costs are very low. The growing quality and falling price of satellite imagery, as well as remote sensing, allows for low-cost monitoring, and enforcement costs are essentially zero, as the money would be awarded only for successful efforts. Countries could compete for a fixed pool of money, which would provide constant incentives for greater efficiencies through innovation. Publicly funded research in global ecosystem restoration, freely available, would further increase efficiency.

Successfully financing such an institution is of course a major obstacle. Efficiency demands that each country's contribution equal the benefits it receives from an additional unit of restoration (see chapters 3 and 32). Unfortunately, not only are decisionmakers largely ignorant of the benefits they receive from the restoration of natural capital, but they also have an incentive to underreport their benefits and thus free-ride on the willingness of other countries to finance such restoration activities. A promising alternative would be to tax resource extraction and/or waste emissions, which would simultaneously penalize the destruction of global public goods. For example, a $.016 per liter tax on gasoline consumption would generate some $30 billion dollars per year, which is forty times the current expenditure by the GEF. Such a tax would target the wealthy, who are doing the most damage to natural capital at a global scale. As the wealthy have the lowest benefits from marginal income, this approach also minimizes welfare loss.

Contribution

Growing evidence suggests that restoring natural capital at a global scale is becoming increasingly urgent, and existing mechanisms and institutions are clearly inadequate. Prior to the eighteenth century, market goods were exceedingly scarce. As a market economy emerged over the course of the eighteenth and nineteenth centuries, society had to develop the institutions necessary to make it function (Polanyi 1944). As a consequence, we continue to develop the institutions necessary to promote a global market economy, as described by the academic field of microeconomics. Following the Great Depression of the 1930s, economists and politicians realized that the state played an important role in regulating economic demand, leading to the field of macroeconomics. Today, the scarcest resources at all scales are, increasingly, natural capital, and without these, all economic systems will collapse (see chapter 30). The challenge for this generation is to develop the institutions and mechanisms necessary to ensure ongoing provision of ecosystem services at a global scale. This has motivated the emergence of the field of ecological economics and all the other transdisciplines interested in the restoration of natural capital.

As the nature of scarce resources has changed over the last centuries (from manufactured capital to natural capital), so has the rate of change (Heilbronner 1995). Not only will it take time to develop and implement the institutions and mechanisms necessary to stimulate the global restoration of natural capital, but also, once in place, it will take time to learn how to restore global ecosystems. This knowledge is itself a global public good. The principles suggested offer guidelines for designing appropriate mechanisms. In summary, we cannot promise our proposed instrument will succeed, but we will never know for sure what will work until we try. Given the urgency of the problem, the less we know about solving it, the more urgent it is that we begin implementing potential solutions, if only to learn from our failures. There is no time left for stalling.

Chapter 32

Making Restoration Work: Financial Mechanisms

RUDOLF DE GROOT, MARTIN DE WIT, ERICA J. BROWN GADDIS,
CAROLYN KOUSKY, WILLIAM McGHEE, AND MIKE D. YOUNG

Much of the earth's natural capital has been modified and transformed into urban areas, cultivated fields, and wasteland. The loss and degradation of natural capital and its services is responsible for a host of environmental and societal problems, including reduced water and air quality, soil erosion, and loss of biodiversity. As recognized by the Millennium Ecosystem Assessment (MA), as well as other studies, this degradation of natural capital diminishes human well-being through loss of ecosystem services and the benefits people receive from them (e.g., Vitousek et al. 1997; MA 2005a). While the poor and those directly dependent on nature's services are often the hardest hit by decreased welfare due to deterioration of natural capital, prosperous areas are by no means immune. Besides the direct impact on well-being (for example, decreased water quality or fewer pristine hiking locations), lost services can also hurt communities indirectly through higher costs, as replacing an ecosystem service with built infrastructure, if possible, is often expensive.

With increasing knowledge about the economic value of natural capital and the services they provide (see chapter 26), interest in financial instruments that encourage both their maintenance and, more recently, their restoration is increasing (e.g., Scherr et al. 2004) and is hence also the focus of this chapter. It is also being realized that these financial instruments need to be combined with a mix of institutional and nonfinancial instruments (see chapters 27 and 33) to adequately preserve and restore natural capital.

Why We Need Financial Mechanisms for the Restoration of Natural Capital

There are many reasons why degradation and destruction of natural capital occurs, including the following:

- Misspecification of property rights: A failure to define and assign property rights over either ecosystems in general (largely a developing-country problem) or particular ecosystem services, which eliminates the private incentive to invest in their restoration and protection.
- Market failures: A general failure of markets to reveal the true costs of resource use and to allocate costs and benefits properly. This includes the issue of externalities (e.g., pollution) and public goods (e.g., clean air). With public goods there is again no private

incentive to provide the good, as it would be costly, though all share the benefits. This is the infamous *free-rider* problem, as discussed in the previous chapter.

- Perverse incentives: The presence of government-pricing practices, subsidies, and taxes that reward adoption of degrading practices and/or discourage investment in natural capital, often driven by demands from special interest groups at the expense of society at large.
- Lack of information: In many cases, people understand neither how they benefit from natural capital nor what is needed to maintain access to them.

There has been growing interest in mechanisms that create positive economic incentives for land managers and the users of natural capital to behave in ways that increase, or at least maintain, certain environmental functions and ecosystem services, including *payments for ecosystem services* (PES), also referred to as *markets for environmental services* (MES), and *rewards for ecological services* (RES) (e.g., Landall-Mills and Porras 2002; Bishop 2003; Bayon 2004; chapter 19). The latter may also include nonfinancial compensation. The basic concept is that the beneficiaries of service provision compensate the providers (see also chapters 26 and 31). According to Bishop (2003), the ecosystem services included most in financial mechanisms thus far are (1) carbon sequestration in biomass or soils; (2) provision of habitat for endangered species; (3) protection and maintenance of landscapes that people find attractive (such as the patchwork of hedgerows, cropland, and woodland typical of southern England); and (4) a catchall category of "watershed protection," which involves various hydrological functions related to the quality, quantity, or timing of freshwater flows from upstream areas to downstream users. Most restoration activities aim to restore one or more of these services.

There are basically four types of financial or reward mechanisms, listed here briefly and then discussed in more detail later, whereby service providers can be compensated/rewarded for services delivered (based on, among others, Johnson et al. 2001; Powell and White 2001):

1. Self-organized private market arrangements (e.g., user fees, certification mechanisms, and private contracts)
2. Voluntary private, nonmarket funding mechanisms (e.g., donations and lotteries)
3. Government-supported market creation (e.g., offset and trading schemes to limit access)
4. Financial mechanisms run by government (e.g., public payments and tax incentives)

Self-Organized Private Market Arrangements to Facilitate the Restoration of Natural Capital

Despite the market failures that often plague efforts to restore natural capital, private actors occasionally develop schemes to encourage restoration with no government involvement. These include an array of mechanisms for private deals through the market, such as user fees, transfer payments, land purchases, cost-sharing arrangements, low-interest credit, purchase of land-development rights, and direct payment schemes for ecosystem services (Scherr et al. 2004). In virtually all of these "deals," the beneficiaries pay for the ecosystem services they receive, and arrangements are made to exclude those not paying. This section will give an overview of some of the mechanisms for a variety of different ecosystem services.

Private Payments for Watershed Protection

Downstream users of water are often willing to pay for restoration of the upstream ecosystem to obtain improved water quality. For example, in Colombia, farmers were organized into water user associations involving 97,000 families to restore and protect upstream watersheds, while members financed the restoration through user charges on water consumption (Landell-Mills and Porras 2002). In Costa Rica, a hydroelectric company, Energía Global, paid watershed landowners to restore or maintain forest cover to ensure more constant flows and reduced sediment loads. In the Panama Canal, deforestation has increased the amount of sediment and nutrients (which spur aquatic weed growth) reaching the canal, necessitating expensive dredging. ForestRe, an insurance entity focused on forests and ecosystems, is working with other insurance companies to underwrite a bond to finance watershed reforestation. It has been proposed that companies heavily dependent on the canal buy the bond and receive a reduction in their insurance (against closure of the canal) in exchange (Economist 2005). Private companies like Perrier Vittel in Switzerland have also found that it pays to maintain the functioning of natural capital to protect the source of their mineral water. In the early 1990s, Vittel negotiated contracts with local farmers to alter their practices, in particular reducing nutrient and pesticide runoff into the water, in exchange for payment and other forms of compensation. The total cost for Perrier Vittel was $24.5 million (for the first seven years) (Perrot-Maitre and Davis 2001; Johnson et al. 2001), which presumably is less than it would cost to locate a new source of pristine water, or the amount the brand name would suffer should their water be found to be of lesser quality.

Private Payments to Restore Natural Habitat on/near Farmland

As knowledge about the value of natural capital improves, landowners begin to invest in them. An increase in pollination services, for example, can justify private spending by farmers on the restoration of small amounts of habitat around their fields (Kremen et al. 2004). Restoring riparian or wooded areas, coupled with other practices such as integrated pest management, can eliminate renting honeybee colonies, because the farmers can rely on native pollinators (Nabhan 2001). Native pollinators can also increase harvests, as shown from a study in Costa Rica, where farms near native tropical forests produced greater yields and higher quality coffee (Ricketts et al. 2004). Unfortunately, private benefits from this type of restoration, namely, increased numbers of particular insects, is dependent on the maintenance of habitat patches across an entire landscape, not just on one farm, and free-riding—a failure to contribute—can emerge and become a serious problem.

Private Payments for Amenity Services

Some homeowners have demonstrated their willingness to pay more for the aesthetic beauty and recreation opportunities that restoration can provide. Wild Meadows, a housing development project in Minnesota (USA), includes restored native prairie, woodland, and wetlands around the homes. In fact homeowners pay an annual fee to fund an ecologist to maintain natural areas and to organize community activities, such as nature walks and prairie burns. Recent evidence suggests that this type of demand is increasing, and developers are responding (Amundsen 2006).

Payments through Ecotourism

Many people are willing to pay to see attractive ecosystems, and when this can be captured, it creates a financial incentive for restoration. A good example is the ecotourism company Conservation Corporation (Conscorp) in South Africa. Their business model is based on the fact that land as part of a reserve for hunting or tourism can generate substantially more revenue than from farming or ranching (Heal 1999). Since the presence of lions increases tourist revenue, and lions are at the top of the food chain, supporting them requires restoring the entire ecosystem (Heal 1999). Conscorp therefore signs contracts with landowners to restore farmlands to their original state and stock them with native wildlife.

Product and Regional Certification Mechanisms

Some consumers are willing to pay more for a good if they know its production has maintained or restored natural capital. In global markets, reaching such people can be difficult, as goods are often produced a long way from where they are sold. To enable people to express their willingness to pay more for products contributing to restoration, certification schemes have been established. They all work by establishing a set of performance criteria and standards that participants follow. Participation is usually voluntary. Some schemes focus on organization-oriented or process standards; others are production oriented and require levels of performance and product quality (Mech and Young 2001). An example of the former is the analysis of an entire firm's environmental impact determined by TruCost, using external cost methodology (TruCost 2006), or carbon market schemes such as the Climate Community and Biodiversity standards (http://www.climate-standards.org). Examples of the latter include the certifications established by the Marine Stewardship Council and the Forest Stewardship Council, which guarantee that any product carrying their label is sourced from a sustainably managed resource. These mechanisms work because some people prefer to purchase and even pay higher prices for sustainably managed resources. Sustainable management can, and often does, require considerable restoration investments.

As the demand for better environmental performance increases, wholesalers are beginning to require all their suppliers to meet environmental standards, which can, in turn, be expected to encourage the restoration of natural capital. One of the best-known examples is EUREP-GAP, a scheme that has been developed for the certification of food sold in European supermarkets (e.g., fair trade products). As yet this scheme is silent on the value of ecosystem services, but one could speculate that future developments will require farmers to take maximum advantage of them and avoid practices that diminish ecosystem services.

Voluntary Private, Nonmarket Funding Mechanisms to Restore Natural Capital

Much of the current global environmental restoration activity is sponsored by private donations, trusts and philanthropic funds, conservation organizations, and other voluntary private payments.

Memberships and Donations

Many of the world's top environmental, nongovernmental organizations (NGOs) receive significant portions of their funding from individual memberships, donations, and legacies. The World Wide Fund for Nature (WWF) International, the national WWF organizations, The Nature Conservancy (TNC), the Sierra Club, the Audubon Society, and the Royal Society for the Protection of Birds (RSPB) are all examples. While largely a conservation organization, TNC uses some of its money for the restoration of natural capital, such as the restoration of Glade Wetland in Ohio, and community-based restoration programs. Similarly, the UK-based RSPB, with over one million members (RSPB Annual Report 2002–2003, http://www.rspb.org.uk/), generates at least £25 million per annum through individual membership dues. This allows it to purchase land and to carry out natural capital restoration projects including, for example, restoration of peat bogs and Caledonian pine forest in the Scottish Highlands. In some cases, these membership-based groups are evolving to become "buyers" of ecosystem services.

In addition, in many developed countries, such as Australia, the United States, and Canada, donations to environmental organizations and initiatives are tax deductible, and relief from capital gains tax is applied to the sale of conservation easements (Shine 2004; Parker 2002).

Volunteer Work

People motivated by altruism or a belief in the importance of restoration or conservation donate not only their money, but also, as Holl and Howarth (2000) have noted, their time. The authors note the huge increase in volunteer-based restoration programs. Literature on the "warm glow" individuals receive from giving (time or money) would suggest that this behavior plays at least a small part in funding for restoration.

Lotteries

Many developed countries (e.g., Canada, The Netherlands, France, the United Kingdom, and some states in the US) have long hosted lottery ticket schemes, and some of these support the restoration of natural capital. The Dutch Lottery Fund (*Postcode Loterij*), for example, gives a fixed percentage of each week's national lottery takings directly to WWF Netherlands, which, in turn, assists WWF International to fund pan-European and global initiatives, such as the Forests Reborn and Wild Rivers habitat restoration projects. In the United Kingdom, lotteries support imaginative restoration projects and also act as a spur for consortia of organizations to bring forward national schemes. One such scheme is the Millennium Forest for Scotland (MFS) Initiative, conceived in 1995 and coordinated by WWF Scotland, with support from the government's environmental agency and its forest managers. Over a five-year period (1995–2000) the MFS Initiative successfully targeted the Millennium Commission, one of the large capital project funds within the UK suite of lottery funds, and secured £10 million that was matched by a further £20 million of other funds. This has resulted in 10,000 hectares of restored woodland habitat.

Government-Supported Market Creation to Facilitate the Restoration of Natural Capital

Whereas private markets work for rival excludable goods and services, many goods and services from natural capital are nonexcludable, which requires that they be made excludable using property-right mechanisms and tradable quotas. Nonrival goods must be protected through regulatory mechanisms or government payments for ecosystem services, such as trading schemes, cap-and-trade, and offset systems (World Bank 1997; Sterner 2003). When property rights are well defined (and in some cases created) and enforced by governments, it is possible to use a variety of mechanisms to create incentives to restore and maintain natural capital. Property-right mechanisms range from bubble-licensing arrangements through offset arrangements to full-blown tradable emission permit systems (Bayon 2004). A major issue with the development of these mechanisms is whether *leakage* can be controlled. Leakage occurs when part or all of the restoration in one area is accompanied by loss somewhere else. For example, an area of restored or conserved forest may just push damaging activities elsewhere so that, on net, there is no reduction in degradation; that is, leakage has occurred. This is unfortunately difficult to control or observe and often depends on establishing a credible baseline from which to measure changes, while strict environmental regulations and enforcement arrangements must be in place.

Bubble Licenses

The simplest arrangement is a single license issued to a firm that encompasses all of its facilities and limits its total emissions, rather than emissions from each of its sources. The firm is then free to decide how to minimize its emissions. In this situation, the incentive for the restoration of natural capital is minimal, although the total reduction in emissions can lead to the restoration of some ecosystem services, if they have not been degraded beyond the point of self-repair. Such arrangements are often known as *bubble licenses*.

Offset Investments and Banking

The next step on from a bubble license is a variety of offset arrangements. The most common offset mechanism is one where a government establishes a *no net loss policy*. This requires any person or company proposing to degrade the environment to offset this impact by restoring natural capital elsewhere, so that in aggregate there is no net loss in the provision of ecosystem services. For example, in the case of wetland mitigation in the United States, the functional equivalence of habitat is often interpreted as a no net loss in acreage. A big question in the United States regarding mitigation banking for wetlands is whether no net loss of acres is really the same as no net loss of ecosystem functioning or ecosystem services and values—it usually isn't (Salzman and Ruhl 2001). There is currently a push in the United States to look more closely at this problem and perhaps to address mitigation at a watershed scale. Under some versions of these schemes, particularly those that involve trade between nonpoint and point sources of pollution, exchange ratios of 2:1 or 3:1 (i.e., restoring 2 or 3 ha for every ha lost) are not uncommon (Young and Evans 1998). In a few of the more sophisticated offset schemes, banking is allowed. That is, a company can restore a wetland, for example, so that at some future time another wetland can be drained. To further this ideal and to minimize the impacts on economic development, *wetland mitigation banks* have been established in the United States, and by 2005, more than five hundred wetland mitigation banks were operating and about eight thousand wetland credits had been traded, with a total market volume of about $290 million. Indeed, it is estimated that the total market for wetland mitigation in the United States (not just private) is about $1 billion (Katoomba Group 2006).

Permit Trading

The most sophisticated government-mediated markets involve what are often described as *cap-and-trade* systems. These schemes limit the total amount of development, resource use, or impacts on resources (i.e., through restricting pollution emissions) by keeping all use within a cap. Most cap-and-trade systems operate by issuing permits in proportion to the total amount of emissions that may occur in a given time period and then making these permits tradable. When a firm reduces its quantity of emissions, it is free to sell its permits to someone else. One of the most attractive dimensions of these mechanisms is that those holding permits have a financial interest in ensuring that such systems work, and that it makes compliance cheaper for the firm.

Examples of cap-and-trade schemes include the EU Emission Trading Scheme (ETS) to limit greenhouse gas emissions under the Kyoto Agreement (carbon credits) (Bayon 2004); the Murray Darling Basin Commission's salinity-trading scheme in Australia (Johnson et al. 2001); the U.S. sulfur dioxide trading scheme; and the Individually Tradable Quotas (ITQs), developed for fisheries in many parts of the world (Aranson 2002). The strength of this approach is that it focuses on outcomes and allows firms to profit from investing in restoration or reducing pressure on natural systems. In general, cap-and-trade schemes are preferred over direct regulation as they are a cost-effective way of reaching a given environmental policy target. However, these systems work only for resources that are rival and can be effectively quantified, and where there exists a clear designation of administrative authority.

Financial Mechanisms Run by Government to Facilitate the Restoration of Natural Capital

Very often market failures limit private incentive to engage in the restoration of natural capital. These can be overcome by a variety of government-run mechanisms that change the cost of restoration through payment, taxation, and/or subsidization mechanisms. These schemes may involve governmental agencies at many scales, from local to international.

Direct Public Payment Schemes

Under the Conservation Reserve Program (CRP) in the United States, landowners are paid to retire agricultural land of high conservation value (primarily next to streams) and restore it to a grassland or forested ecosystem (FSA 2005). The amount paid to the farmer for undertaking this restoration is based on the profit that could have been made had the land remained in production. As of 2004, the USDA had placed 14 million hectares (34.7 million acres) under contracts with annual rental payments of $1.6 billion (Barbarika 2004). This is just one example of governmental payments to farmers for improved land management for protection of ecosystem services. Another well-known example is New York City's decision to restore its watershed rather than construct a water treatment plant (see chapter 24).

In an attempt to deliver similar benefits at less cost in Australia, a variety of tender programs are being tried. These reverse-auction schemes involve landowners' submitting bids to the government for the amount they would need to be compensated to undertake certain restoration activities on their land. The government then ranks these bids by a benefit index and the cost asked by the farmer, awarding those with the highest benefit to cost ratio, until their budget is exhausted.

Tax Incentives and Other Fiscal Instruments

Throughout the world, taxation arrangements are used to influence land use. In many cases, however, they can have perverse effects. In Australia, for example, many of the concessions given to agriculture discourage the maintenance of natural capital (Binning and Young 1999). Favorable taxation arrangements, however, can also be developed such that people who restore natural capital are given subsidies or tax breaks—but these tend to be rare (e.g., Reid 2002; Shine 2004). As a general rule, the transaction costs of such arrangements need to be weighted against their socioeconomic and ecological benefits (Parker 2002).

The most common arrangement is to allow deductions for expenses incurred while working on land earmarked for conservation rather than for production. Although uncommon in Europe (Shine 2004), elsewhere certain land-management input costs can be offset, such as the clearing of invasive alien plants in South Africa (Botha 2001). In the United States, farmers may deduct the costs of soil and water conservation from taxable income, limited annually to 25% of the taxpayer's gross income from farming (Bowles et al. 1998), while in Ireland a reduced value-added tax applies to agricultural services in general (Shine 2004).

In southern Ontario significant wetlands, areas of natural interest, and areas of scientific interest (ANSIs) are 100% exempt from property tax (Reid 2002). Similarly, property tax exemptions in Brazil and Guatemala encourage the creation of private nature reserves (Greig-Gran 2000).

Contribution

A large array of financing instruments can be applied to fund the restoration of natural capital. In choosing the right mechanism, a mistake often made is to ask too much of a single instrument. As a general rule, a separate instrument should be used to achieve different objectives. For example, in the development of efficient, equitable, and dependable ways to manage water resource allocation, it is necessary to use at least three financing instruments for achieving the separate goals of efficiency, equity, and dependability (Young and McColl 2005).

Overall, scientific evidence is mounting that financing conservation and the restoration of natural capital is more profitable than conversion to single-function use (Balmford et al. 2002), and that "every dollar invested in [nature conservation and restoration] saves anywhere between US$7.5 and US$200 in avoided damage and repair costs" (Economist 2005).

Chapter 33

Making Restoration Work: Nonmonetary Mechanisms

WILLIAM McGHEE, JOHN CRAIG, RUDOLF DE GROOT,
JAMES S. MILLER, AND KEITH BOWERS

A growing community of people within the academic, governmental, volunteer, and business sectors have shifted thinking from straightforward approaches to ecological restoration and have embraced the more holistic ideal of restoring natural capital, though the phrase itself has not become common in public discourse yet. Theoretical aspects and practical methods for natural capital restoration have evolved and developed through recent responses to increasingly intensive and unsustainable land use. The restoration of natural capital is a necessity rather than a luxury and essential to balance and repair continued and accelerating damage to ecosystems, degradation of habitats, and changes in global and local environmental processes. A key question is whether the capability and the willingness exists (locally, nationally, and internationally) to restore and sustainably manage habitats, ecosystems, and processes so that they preserve the biological diversity and ecosystem services of the original systems.

This chapter examines the restoration of natural capital that is not directly stimulated by monetary or fiscal mechanisms (see chapter 32); rather, it explores examples of the restoration of natural capital that are a consequence of nonmonetary mechanisms, such as regulation, social pressure, corporate conscience, ethics, or faith. It also touches on unintended restoration where habitats, ecosystems, or processes have been regenerated by default through neglect, by accident, or by misfortune. This review of nonmonetary mechanisms has an underlying premise that restoration can occur in spite of human intent or intervention. One of the many challenges facing those involved in restoration is the enumeration, classification, and evaluation of the scale and quality of natural capital restoration. A comprehensive inventory of restoration projects, including their outputs and outcomes, would be valuable in educating and informing decisionmakers and policymakers about the desirability of injecting greater urgency and more resources toward the restoration effort. The question of the extent to which governments, societies, communities, and private companies are motivated, inspired, pressured, or obligated to carry out restoration projects or programs in the absence of direct financial gain is important to those concerned about natural resource depletion.

Finance may not be the main driver of restoration in the examples dealt with in this chapter, but money is almost always applied in active restorative efforts to secure resources or pay for services (chapter 32). The philanthropic gesture, the campaigning cause, the local government planning regulation all have a value or a cost, and therefore a clear distinction has been drawn between mechanisms that have no direct monetary drivers and the cost or price of delivering the restored natural capital.

There is an overlap between nonmonetary and monetary mechanisms. Market protectionism is an example of this overlap and is common practice in all countries and trading blocks. It is used to greatest effect by the European Union, the United States, Japan, and other large economies. Direct government subsidies and restrictions on imported agricultural goods that act as tariffs have detrimental consequences for land use in both exporting and importing economies and societies.

This chapter recognizes five main areas where restitution or restoration of land is the result of a nonfinancial mechanism: (1) governmental intervention by regulation and legislation; (2) intervention through nongovernmental organizations (NGOs); (3) grassroots restoration resulting from local action inspired by faith-based concerns, threats to livelihoods, or loss of culture; (4) action due to changes in private sector behavior, and (5) unintentional restoration, through unpremeditated or accidental activities. We discuss each of these nonmonetary mechanisms, and the chapter concludes with summary recommendations for actions likely to result in an increase in restoration.

Governmental Regulation and Legislation to Promote the Restoration of Natural Capital

Laws and regulations that provide mechanisms and incentives for restoration may be regarded positively by those concerned about resource depletion, such as environmental NGOs, while they may be viewed with greater skepticism by those who feel that their business or livelihoods may be impacted by regulation: extractive industries, industrial agriculture, and commercial forestry and fishing interests.

The degree to which government regulation can promote the restoration of natural capital is unclear. However, it is not unreasonable to assume that the bulk of such restoration to date has been carried out in developed countries such as those within the European Union (EU), in Australia, and in North America. In the United States and Canada "technocratic" motivations are clearly among the principal drivers of restoration (Clewell and Aronson 2006).

Legislation requiring restoration often aims for mitigation for ecologically or environmentally damaging operations that result from real estate development (e.g., the building of new housing or large construction projects) or mitigating habitat impacts by extractive industries and agriculture. The regulatory framework falls within national or state legislation, such as planning law, and it can require reinstatement or restoration of a portion of the disrupted or damaged habitat. This restorative requirement can be in situ or remote from the development.

Sophisticated government intervention, whereby developers are required to offset or mitigate impacts to systems by restoring an area of land or water on a like-for-like basis, is uncommon. The United States is an exception to this and, for example, has, since the mid-1980s, adopted a land-management goal of no net loss (NNL) of wetlands, supported by a system called wetland mitigation banking (see also chapter 32). This is a permit system operated through the U.S. Army Corps of Engineers that obliges developers whose potential actions may damage or destroy wetlands to demonstrate that damage to a wetland is unavoidable, and subsequently to provide *compensatory mitigation* for unavoidable wetland impacts.

Wetland mitigation banking was institutionalized by guidance released in 1995, the *Federal Guidance on the Establishment, Use and Operation of Mitigation Banks* (Shabman and Scodari 2005). A survey in 2005 by the U.S. Army Corps of Engineers established that there were some 450 mitigation banks in operation, with an additional 198 banks in the proposal stage. In 2001 the Environmental Law Institute (ELI) identified 72,437 hectares (173,848 acres) of wetland restoration, creation, enhancement, and/or preservation as a result of mitigation. Yet, wetland mitigation banks are just one of many tools that can be used to mitigate the loss of wetlands, and it should be noted that the majority of wetland mitigation is performed outside of these systems.

The efficacy of the wetland mitigation banking process has received mixed reviews. Many economists regard wetland banking as a monetary incentive that produces compensatory mitigation credits to offset detrimental environmental impacts. However Shabman and Scodari (2005) argue that wetland mitigation banking as currently practiced is a nonmonetary incentive. Regardless of the actual impact and nature of wetland mitigation banking, it has produced quantitative and qualitative wetland restoration and serves as an interesting and partially effective regulatory instrument.

In addition to wetland banking, other government-based mitigation programs include mitigation of endangered species habitat, forests, nutrients, and streams/rivers, all of which also use mitigation banking as a compensatory tool. However, mitigation is only one component of funding the restoration of natural capital. For example, in the United States there are federal programs administered by the U.S. Environmental Protection Agency, Interior Department, and the Natural Resource Conservation Service, among others, which provide funding for restoration projects. These programs are often combined with local and private funding to encourage the restoration of degraded landscapes and habitats.

The EU has taken a different approach than the United States regarding regulation and restoration. Europeans have opted for pan-regional legislation that sets targets for biodiversity through environmental directives. These directives cascade down to national government levels and are manifested in legislation on land management, which has acted as a relatively minor driver in natural capital restoration. The Birds Directive (79/409/EEC) and the Habitats Directive (92/43/EEC) are the main EU legislations governing species and ecosystems; these directives are delivered through a network of sites (aided by a funded program—LIFE Natura 2000), which provide a geographically discrete focus for the management of fauna and flora.

The EU has, since 2001, attempted to move from a position of enforcing conservation directives through the European Environment Agency (EEA)—such as the protection of Ramsar Wetland sites from the threats posed by agriculture, pollution, and water regulation—to a more proactive engagement with groups pursuing biodiversity enhancement and/or restoration. This move toward the restoration of natural capital was expressed in the EU's Working Paper on Prevention and Restoration of Significant Environmental Damage (2001), which built on the EU Treaty, Article 174(2). This article is drafted in terms of the polluter pays principle, where environmental damage is treated as a criminal offence and the polluter is subject to prosecution and a financial penalty for the crime. It did not, however, outline or propose a legal framework for obliging polluters or developers to reinstate or restore habitat, nor did it bind national governments to restoration targets, as is the case in the United States.

Contribution of Nongovernmental Organizations to the Restoration of Natural Capital

Actions to restore natural capital are commonly associated with the *Third Sector*, represented by nongovernmental organizations (NGOs). An NGO usually is a nonprofit group or association that acts outside of institutionalized political structures and pursues matters of interest to its members, by lobbying, persuasion, or direct action.

European and North American environmental NGOs, such as the World Wide Fund for Nature (WWF), The World Conservation Union (IUCN), Conservation International (CI), The Natural Conservancy (TNC), and the Wildlife Conservation Society (WCS), champion endangered species and fight habitat degradation through campaigns, direct action, and practical activity. During the last ten years the larger NGOs, such as WWF and CI, have to a degree become less adversarial with respect to governments and companies and have on occasion opted not to conduct high profile campaigns against products, projects, or companies. Instead, they engaged them in an effort to change private sector thinking and behavior through partnerships to initiate restoration projects. The work of many environmental NGOs focuses on bringing about a change in attitudes and the cultural norms toward the environment, conservation, and restoration through education, public outreach, and projects and by example.

As developing countries become wealthier, their citizens become better equipped to respond to environmental degradation and social injustice, hence the increasing number of new NGOs operating in this field in India, China, Brazil, and South Africa. Many of these NGOs are in the vanguard of natural capital restoration; they bring pressure to bear on politicians and heads of industry to change laws and practice; they provide a focus for public support and a structure for gathering and channeling funding to projects and causes; and they carry out the work of restoration through their own offices or in partnership with other institutions, organizations, or communities.

Both international and local environmental NGOs are moving away from strictly conserving land and habitat and are now employing restoration as a major component of their conservation programs. CI, IUCN, TNC, WCS, and WWF are all doing this on a global scale, and there are many regional and local conservation NGOs in the United States that are also adopting this strategy. Apart from their role in supporting the restoration of natural capital through direct action, most environmental NGOs play a very important role communicating need and helping educate the general public about environmental problems.

Grassroots Action in Support of the Restoration of Natural Capital

Those least able to afford or least equipped to sustainably manage their local resources are often responsible for the degradation of their immediate environments for no better reasons than that they do not have any choice or do not know of any other options (chapter 29). Henao and Baanante (2001) of the International Centre for Soil Fertility and Agricultural Development suggest that African farmers are "mining" the land to such an extent that nutrients are "hemorrhaging" from approximately 85% of African farmland at a truly alarming rate of 30 kg per hectare per annum, and 40% of farmland is losing nutrients at 60 kg or higher per annum.

Communities that suffer the worst degradation of natural resources, willingly or unwillingly, do not have access to resources or funds to carry out restoration, and often these communities are powerless to influence politicians. It is reasonable to assume that most restoration is a result of external pressure or intervention, rather than a direct consequence of community action, and it is often conducted in spite of community opposition. For example, the FACE carbon financed reforestation project in Uganda initially excluded local graziers and fuelwood collectors from the Mount Elgon National Park.

In developed countries, communities and grassroots organizations have been at the forefront of restoration; in Scotland this involvement is a consequence of land reform pressure and in part a response to insensitive and ecologically damaging industrial afforestation carried out by the private sector with support from the UK Forestry Commission (see chapter 14). It is in the developed world, where people have become remote from the land, that interest in the restoration of natural capital appears to be most prevalent. Indeed, environmental concern and the willingness and ability to restore may be regarded as a luxury attendant on wealth and prosperity.

Strong leadership or the efforts of a small group of dedicated individuals often inspire community action. Concerned at the loss of mangroves, coral, and sea grass beds, Pisit Charnsoh founded the Yadfon Association in 1985 to work in coastal villages of Trang in Thailand. Yadfon's work spread throughout thirty villages in the delta and has resulted in nine community-managed forests and 240 hectares of restored mangrove forest. The Thai Forest Service has declared this the first community mangrove forest in Thailand. The success of Yadfon has come as a result of recognizing the need for education, practical outputs, and a decentralized approach to decision making. Economic benefits to local communities have accrued from this restoration initiative through increased catches of seafood, rising by 40% between 1991 and 1994. This community-based action has spread to influence restoration beyond Thailand by way of the Mangrove Action Project (MAP) and is a case of the grassroots influencing an environmental NGO.

Recognizing that grassroots action is effective and sustainable is an important first step for those involved in restoration of natural capital. It should also be recognized that efforts to impose restoration solutions on unwilling communities could be counterproductive. Education, awareness raising, capacity building, and empowerment are all expressions used by those in the NGO sector to describe what needs to be done to assist communities to restore their natural capital. Making people conscious of the desirability of restoration is generally linked to making them aware of the livelihood benefits it can provide to them and their families. For researchers, decisionmakers, and practitioners, this means identifying the problems and solutions associated with restoration and communicating them clearly and simply to communities and grassroots organizations. There is also value in lobbying governments for supporting funds, as has happened with the Land Care funding in Australia and New Zealand.

The Corporate Sector's Contribution to the Restoration of Natural Capital

The private or corporate sector has traditionally been regarded as the perpetrator of environmental degradation and an agent of natural capital destruction. Over the last twenty years, corporations such as BP, Exxon, and Anglo American have attempted to paint a greener picture of themselves and how they operate. Although much remains to be improved, it is incontrovertible that many businesses have attempted to change their stance on the environment and to at least appear to be more understanding of environmental issues. There is little doubt that governmental and public pressure has resulted in a change of corporate behavior. Corporate social responsibility (CSR) has, in the early part of the twenty-first century, become common currency in business jargon, and it is rare to find publicly listed companies of any size operating without some form of CSR policy. CSR covers a plethora of issues, from child labor to the emissions of noxious chemicals, and it is a mechanism by which companies can become directly involved in addressing the effects of their business on the environment (i.e., the environmental *footprint*), or as a way to demonstrate corporate "goodness" by addressing an environmental problem unconnected with the company.

The drivers for companies to operate in a more sustainable manner and/or to become involved in restoration of natural capital were identified in 2001 by Price Waterhouse and Coopers: (1) stakeholder expectation, which encompasses reputational risk; (2) policy and market incentives, including governmental regulation and consumer awareness; (3) changing contexts, with the use of improved technology and alternative materials; and (4) business imperatives, through increased shareholder value and access to capital. Of these, the principal driver was reported to be shareholder value. Yet, restoration is one of many components of being environmentally sustainable, and just because a corporation claims to be environmentally sustainable does not mean that it supports, practices, or is even aware of restoration. The main focus on sustainability within companies is generally with legislative compliance, also termed the *license to operate*, and to a lesser degree with the local environment and communities affected by company business.

Though the restoration of natural capital, that is, restoration integrated with societal welfare, is not on the radar screen of many firms, evidence exists that this notion is changing. Here we offer two such examples of private concerns that endeavor to be world leaders, in all respects.

First, Alcoa World Alumina Australia operates the largest bauxite-producing mine in the world: the Huntly Mine situated in the Darling Range of southwestern Australia. This mine is located in jarrah (*Eucalyptus marginata*) forest within the Perth city water supply. Jarrah forest has high conservation value, containing at least 784 plant species, and is managed by the state government for multiple uses, including water, timber, and recreation. Since 1966, Alcoa has carried out restoration of mined land, with some 550 hectares mined and restored annually. This, in the first instance, was by establishing plantation forestry on reclaimed land; with time, the effort has become more sophisticated and they are now attempting to restore functioning jarrah forest ecosystems. Alcoa maintains that they are doing this restoration because they believe that as a company they need to exceed public and regulatory expectations and to demonstrate procedures of best practices for restoration. In recognition of this restoration effort Alcoa has been rewarded by being included in the United Nations Environment Program's Global 500 Roll of Honor (1990) and was awarded the Model Project Award by the Society for Ecological Restoration International in 2003 (http://www.ser.org/).

Second, QIT Madagascar Minerals (QMM) is a Malagasy company that is a partnership between the government, thorough the *Office des Mines Nationales et des Industries Strategiques de Madagascar*, and the international mining company, Rio Tinto PLC, United Kingdom. The littoral forests that stretch along the eastern coastal plain of Madagascar have largely been reduced to small patches (Consiglio et al., 2006), but they are biologically diverse and rich in endemics. The majority of littoral forests grow on sand, and in some areas in southern Madagascar these sands have been found to be rich in heavy minerals, including ilmenite, a source of titanium dioxide. In 1986, QMM identified a series of sites along the southern coastal plain as appropriate for mining ilmenite, but some of the significant mineral deposits were beneath some of the last remaining littoral forest patches, some intact and others variously degraded. The environmental plan developed by QMM included the compilation of a comprehensive biological inventory of both plants and animals in three forested areas surrounding the communities of Petriky, Mandena, and Sainte Luce (Vincelette et al. 2003). The approximately 60 hectares of forest at Petriky and 160 hectares at Mandena were both degraded, but the nearly 430 hectares in the area of Sainte Luce were still largely undisturbed. The inventory aimed to identify those species of greatest conservation concern, especially those either endemic to the region or already deemed to be threatened or endangered, and the restoration plan includes the goal of maintaining a full complement of the native plants identified in the biological surveys through reforestation efforts in the region. Reforestation will include both native and nonnative plants that can be used by the populations in each village to supply fuelwood, foods, and medicinal plants. Thus the restoration planned, as mining begins in 2006, will meet both the needs for preservation of biological diversity in the region and provision of useful natural resources that may be sustainably used by local inhabitants of the area.

Whether the exemplary efforts of Alcoa or QMM to rebuild natural capital at mined sites have increased their net worth or shareholder value, or whether it has allowed the companies a more flexible license to operate, is not clear. What is clear though is that some companies have started to go above and beyond regulatory standards on their own accord, with respect to the restoration of natural capital.

Why would these companies restore natural capital? A review of incentives that would encourage companies in extractive industries, such as mining, forestry, and agriculture, to follow the lead of companies such as Alcoa and QMM determined that there were three types of corporate incentives: cost incentive, reputational incentive, and market incentive (Greig-Gran 2002). Of these, cost was regarded as the most powerful, with respect to the mining sector.

Unintended Restoration

Unintended restoration of natural capital can occur through various means; here we name four examples: (1) by the removal or exclusion of people from land in war, such as the forest expansion as a result of land mines in Gorongosa in Mozambique or the "accidental paradise" created in the demilitarized zone between North Korea and South Korea; (2) by anthropogenic catastrophe, such as the forest expansion around Chernobyl as a result of radioactive leakage; (3) by natural catastrophe, such as the destruction of shrimp farms in Tamil Nadu by the 2004 tsunami; or (4) through accident or neglect, such as abandoned land on postindustrial sites and former agricultural land left by the rural poor who have migrated to urban centers for employment. None of these instances can be regarded as having been based on "incentives," yet they are agents of restoration and have resulted in a potentially greater scale of natural capital restoration than planned and incentivized activities.

Native hardwood forests across the Appalachian Mountains in the northeastern United States have regenerated after deforestation in the late nineteenth century by mining, railroad expansion, and timber extraction. The recovery of this secondary forest was unplanned and resulted from declining human pressure, but by the early twentieth century, conservation laws were being enacted to protect the recovered forests. The same process is occurring in the montane regions of the Alps of central Europe where a decline in traditional livestock grazing in alpine pastures is resulting in an expansion of secondary and scrub forests, fostering concerns regarding the diminishing area of species-rich alpine meadows. This concern reflects the choices that often need to be made in the restoration of natural capital—not all "recovery" may be good for biodiversity, for ecosystem services, or for human culture and survival. The policy of the Cinque Terre National Park in Italy, which seeks to restrict the advance of natural maquis vegetation in a largely abandoned "constructed landscape" of agricultural terraces, is a case in point. This landscape is designated as a UNESCO World Heritage Site because of its unique cultural value. But the encroaching maquis was also undermining the stability of the terraces and increasing the threat of landslides on the villages below them.

The Chernobyl nuclear disaster in the Ukraine resulted in a human catastrophe on a massive scale. The thirty-kilometer exclusion zone around the *Red Forest*, from which most humans have been displaced, has experienced an explosion in numbers and species of plants and especially animals. Lynx, Przewalski's horse, wolves, brown bears, and eagle owls (*Bubo bubo*) have increased in number with an expansion of birch forest. Although natural processes appear to have recovered, radiation may be taking a toll on the genetics of plants and animals and may change species in ways we cannot yet appreciate.

Likewise, the thirty years of civil war that devastated the human population of Mozambique, notably around the provinces of Sofala and Gorongosa, resulted in a massive expansion of secondary miombo forest. Unfortunately much of this expansion was achieved through the placing of land mines, and there are still various no go (restricted) areas within the buffer zone of the Gorongosa National Park. During the war, farmers and villages were displaced and the trees and vegetation recovered. However, over 20,000 buffalo and many thousands of elephants and large mammals were killed to feed soldiers and local people. The consequence of this is the recovery of the vegetation without functioning populations of keystone species. Just taking people out of the landscape does not automatically lead to "restored ecosystems."

Contribution

The notion of restoring natural capital has growing support from both the scientific and policy-making communities, and nonmonetary mechanisms (regulations, NGOs, grassroot movements, and corporate action) can provide significant and positive incentives for restoration activities. Yet, the examples given in this chapter also show there is significant overlap with monetary incentives (such as subsidies, taxes, and market-based instruments—chapter 32), and both produce value and financial gains for commercial activities. Care should be taken to distinguish between positive incentives and perverse incentives (such as the positive examples covered in Daily and Ellison (2002) versus the perverse outcomes resulting from agricultural subsidies and war). The examples presented here also demonstrate that unintentional regeneration may add significantly to the changes required to preserve the base of natural capital. Acknowledging both and encouraging the positive while working to eliminate the perverse is urgent.

A comprehensive inventory of restoration projects, including their outputs and outcomes, would be valuable in educating and informing decisionmakers and policymakers about the desirability of injecting a greater urgency and more resources toward restoration efforts.

The scale and frequency of restoration activities attributable to nonmonetary mechanisms is a measure or indicator of the value that society places on rectifying human impacts on the environment; they demonstrate that communities and societies are starting to recognize the effects of resource degradation and acknowledge the need for restorative action. The "market" is starting to vote through its actions. This augurs well for the need to mainstream the restoration of natural capital, the theme of the next chapter.

PART IV

Synthesis

This final section of the book comprises two chapters. The first focuses on the importance of mainstreaming the restoration of natural capital, highlighting the fact that restoration is becoming an essential intervention. If the earth is to continue to sustain the human species, then restoration planning should be part of every development plan, national budget, and global convention. It should be taught in schools, practiced in rural and urban areas and discussed in boardrooms. To communicate widely about the importance of natural capital, it is necessary to quantify its spatial distribution and its value to humans. Cowling et al. stress the importance of planning and coordination among initiatives for restoring natural capital, so that maximum benefits can be obtained from this investment.

In the final chapter Milton et al. synthesize the issues that emerged in the book around four themes. These are (1) the valuation of natural capital loss and the restoration activities needed to regain it; (2) the challenge of setting appropriate targets, developing workable approaches, and measuring progress toward natural capital recovery; (3) the need for prioritization and planning of restoration at various scales; and (4) the importance of collaboration and buy-in for successful restoration.

The synthesis concludes with a call for collaborative action. Natural capital is being depleted worldwide. There are no hidden reserves, but we can sustain what we have through a culture of restoration built on an understanding of the real value of natural capital.

Chapter 34

Mainstreaming the Restoration of Natural Capital: A Conceptual and Operational Framework

RICHARD M. COWLING, SHIRLEY M. PIERCE,
AND AYANDA M. SIGWELA

Protected areas alone will never achieve all of the goals and targets required to ensure the persistence of the world's natural capital (e.g., Rosenzweig 2003) and the delivery of services that intact ecosystems supply (Kremen and Ostfeld 2005). Consequently, the burden of conserving (and restoring) natural capital will have to fall increasingly on sectors such as agriculture, transport, forestry, mining, and urban development (e.g., Hutton and Leader-Williams 2003). The *mainstreaming* of biodiversity concerns is one strategy used by the conservation community to respond to the challenge of ensuring the persistence of natural capital and ecosystem services in utilized landscapes (Pierce et al. 2002; Petersen and Huntley 2005a). In essence, mainstreaming may be defined as the process of creating awareness of the value of natural capital in sectors that currently ignore or discount it, to the extent that they will incorporate conservation actions into their routine activities.

A key conservation action in production landscapes is the restoration of degraded or transformed natural capital, as has been pointed out in many of the chapters in this book. Our aim here is to provide a conceptual and operational framework for mainstreaming the restoration of natural capital in production landscapes.

What Is Mainstreaming?

Although mainstreaming is a relative newcomer to the biodiversity and natural capital lexicon, it is an important one, since mainstreaming is a component of the institutions and strategies of some major global biodiversity initiatives. For example, the concept is embedded in several articles of the Convention on Biological Diversity (ratified 1995). It also underpins the ecosystem service approach of the *Millennium Ecosystem Assessment* (MA) and is the explicit objective of the Strategic Priority 2 of the Global Environmental Facility's GEF-3 (2004) Program of Work: "Mainstreaming biodiversity in production landscapes and sectors—to integrate biodiversity conservation into agriculture, forestry, fisheries, tourism, and other production sectors in order to secure national and global environmental benefits" (Petersen and Huntley 2005b). Undoubtedly mainstreaming initiatives will attract considerable resources from funding agencies over the next decade.

According to Petersen and Huntley (2005b) the objective of mainstreaming is "to internalize the goals of biodiversity conservation and sustainable use of biological resources into economic sectors and development models, policies and programs, and therefore into all human behavior." Cowling et al. (2002) identified the following list of desired outcomes of mainstreaming:

- The incorporation of biodiversity considerations into policies governing sectoral activities
- The simultaneous achievement of gains in biodiversity and in the economic sector (the *win-win* scenario)
- Sectoral activity being recognized as based on, or dependent on, the sustainable use of biodiversity
- Situations where sectoral activities result in overall reversal of biodiversity losses

Viewed as a process, mainstreaming is a means to spread the responsibility and benefits of conserving biodiversity and restoring natural capital across a diverse range of sectors. This requires the identification of scenarios that provide benefits for both the natural capital and the targeted sector, and the implementation of actions (for example, the creation of institutions, including incentives) that enable responsible bodies to accomplish these scenarios.

Mainstreaming interventions may happen at all scales of organization and geography, from encouraging backyard conservation of natural capital in a neighborhood to the impact of a multilateral environmental agreement on the global ocean-transport system. Furthermore, a wide range of actors will bear the costs and enjoy the benefits, material and spiritual, associated with mainstreaming, and these will accrue over short and long timescales (Petersen and Huntley 2005b).

There are very few documented cases of effective mainstreaming. Pierce et al. (2002) provide examples from South Africa, and Peterson and Huntley (2005a) from elsewhere in the world. Others, although not explicitly conceptualized as such, appear in Daily and Ellison (2002), Swingland (2003), and Rosenzweig (2003). Pierce et al. (2005) provide a case illustrating how conservation priorities can be mainstreamed into land-use planning through interpretation of scientific products into user-friendly, user-useful maps and guidelines. In addition, Knight et al. (2006) describe how mainstreaming can be integrated into a framework for implementing actions aimed at securing conservation priorities. The latter two are examples of conservation actions that enable or facilitate the restoration of natural capital by identifying restoration priorities.

Conceptual Framework for Mainstreaming the Restoration of Natural Capital

A conceptual framework for restoration boils down to identifying a model of the desired landscape; in other words, what mix of land uses and economic flows are required to meet the needs of different stakeholders (Salafsky and Wollenberg 2000)? Restoration implemented in an ad hoc manner is likely to fail in achieving desirable outcomes (Hobbs and Norton 1996; see also chapter 3), as has been shown for the ad hoc implementation of other conservation actions, such as the location of protected areas (Pressey 1994). Therefore, prior to restoration intervention, stakeholders need to identify an appropriate landscape model characterized by requirements for sustaining biological patterns and processes, and for supporting human needs. Effective restoration requires explicit goals and targets (e.g., Hobbs and Norton 1996) identified in a way that is consistent with a specific landscape model.

For the mainstreaming of restoration to happen, the landscape model must facilitate the identification of plausible and compelling win-win scenarios. Thus, farmers must be convinced that the direct and opportunity costs of restoring native, natural capital on their farms will be outweighed by the benefits of such restoration, for example, in enhanced production through improved pollination services or reduced soil erosion (e.g., Kremen and Ostfeld 2005). Similarly, restoration interventions aimed at achieving nature conservation goals should be guided by the achievement of explicit and defensible targets for biodiversity features, which are set in the process of systematic conservation planning (Pressey et al. 2003).

While there has been much written on conceptual frameworks, goals, and targets for restoration (e.g., Hobbs and Norton 1996), the obvious link between restoration and the systematic, target-driven, conservation planning of landscapes (Margules and Pressey 2000) has only recently been made (e.g., Pressey et al. 2003; Crossman and Bryan 2006). Systematic conservation assessments identify those areas of transformed or degraded natural capital that are required to achieve targets for the conservation of both biodiversity patterns (e.g., species, land classes) and processes (e.g., migration corridors). These areas then become defensible priorities for restoration, as illustrated by Crossman and Bryan (2006) for agricultural landscapes in Australia.

A similar systematic approach is required for the restoration of natural capital for ecosystem service delivery (e.g., Kremen and Ostfeld 2005; Pierce et al. 2005). A few conservation assessments have targeted and incorporated the spatial components of ecosystem services (e.g., Rouget, Cowling, et al. 2003). However, a great deal more research is needed before we can make significant progress in the restoration of natural capital: (1) the natural capital—both intact and degraded—in a particular planning domain needs to be identified and mapped in consultation with those stakeholders who are direct beneficiaries of the services it delivers; (2) the benefits derived from these services and their flows to specific beneficiaries need to be quantified and displayed in ways that are meaningful to stakeholders; (3) targets need to be set for each component of the region's natural capital in a way that is consistent with a landscape model (for example, a certain number of hectares of healthy watershed are required to ensure a sustainable water supply over a specified period); (4) target shortfalls should be identified as priorities for restoration; and (5) mechanisms should be sought to mainstream the restoration of these areas into those sectors that benefit from the services provided by the natural capital.

The major advantage of systematic restoration to achieve the goals for a specific landscape model is that it is target driven and, therefore, defensible, efficient, and effective (Crossman and Bryan 2006). These attributes are likely to greatly facilitate mainstreaming, especially when the actors are cash-strapped government agencies or profit-motivated corporations.

An Operational Framework for Mainstreaming the Restoration of Natural Capital

Cowling et al. (2002) developed an operational framework for mainstreaming biodiversity, informed by eleven South African case studies presented in Pierce et al. (2002). The framework is sufficiently broad to accommodate restoration interventions; along with the establishment of protected areas and effective soil and water conservation, restoration is another tool for conserving biodiversity. The framework comprises four major components:

1. *Prerequisites* essential for mainstreaming to take place
2. *Stimuli*, external and internal to the sector, that catalyze awareness of the need for mainstreaming
3. *Mechanisms* that initiate, enable, or drive mainstreaming
4. *Outcomes* that are measurable indicators of the effectiveness of mainstreaming

In the framework, the mainstreaming process was described as follows: "Given that certain prerequisites are in place, a set of specific stimuli can catalyse activities which then lead to the identification of appropriate mechanisms, with the net result that effective mainstreaming, as measured by outcomes, will happen" (Cowling et al. 2002).

Prerequisites

The prerequisites most frequently cited by the case studies were concern, awareness, and knowledge of biodiversity issues, adequate capacity, and good governance (see also Cowling 2005). The list of prerequisites is consistent with those identified in the lessons learned for enabling effective, community-conservation interventions (e.g., Wells and Brandon 1992; Infield and Adams 1999).

Stimuli

A major stimulus for mainstreaming biodiversity in South Africa was the establishment of democratic and accountable governance after 1994, which all case studies cited (table 34.1). For the first time, South Africa's global biodiversity contributions and responsibilities were formally acknowledged, and access to support from international agencies was possible. More important, however, the new politicians and civil servants were committed to democratic practices and open to biodiversity as a source of socioeconomic development. As a result, there were opportunities for participation by civil society in developing legislation and policy that were biodiversity-friendly. The general point here is that democratic and accountable governance is essential for achieving success in a process as complex as biodiversity conservation (Stephens et al. 2002). Other frequently cited stimuli were (1) a decline in resources (see also chapter 6) and the concern this caused, and (2) the identification of socioeconomic incentives for biodiversity conservation (e.g., in the burgeoning tourism sector since 1994). Indeed, the linkage to socioeconomic delivery is, and will remain, a key stimulus for mainstreaming biodiversity, especially in the developing world.

TABLE 34.1

Most frequently cited elements associated with a framework for mainstreaming biodiversity

Prerequisites	Stimuli	Mechanisms
Democratic and accountable governance	Improved governance	Effective communication
Concern, awareness, and knowledge	Resource decline	Strengthening institutional capacity
Organizational and institutional capacity	Socioeconomic incentives	Enabling legislation and policy

Source: Developed from the analysis of South African case studies (Cowling et al. 2002).

Mechanisms

Interestingly, the most frequently cited mechanism for mainstreaming biodiversity in the South African case studies was effective communication with key stakeholders (table 34.1), including politicians. While communication was invariably part of a package of mechanisms (the others most often being capacity strengthening and the development of enabling legislation and policy), the case studies showed convincingly that communicating the issues in a language that was comprehensible and compelling to stakeholders was absolutely essential for initiating the implementation of mainstreaming actions (see also chapter 29). Others have also emphasized the importance of appropriate and compelling communication in effecting behavior change regarding biodiversity conservation (e.g., Anderson 2001; Schulz and Zelezny 2003).

Outcomes

The mainstreaming outcomes most frequently cited were incorporation of biodiversity issues into sector policies; simultaneous biodiversity and sector gains (win-win); net biodiversity gains exceeding net biodiversity loss through sector activities; and sector activity being based or dependent on sustainable use/management of biodiversity. In all instances, concerns for biodiversity issues were incorporated into routine activities of the specific sector (Pierce et al. 2002). The measurable indicators of these outcomes need to be monitored in order to assess the effectiveness of the mainstreaming interventions.

A Restoration Example

South Africa's *Working for Water Project* (see also chapter 22), arguably the largest restoration program in the developing world (van Wilgen et al. 1998), illustrates well the operational framework for mainstreaming (van Wilgen et al. 2002). The conceptual framework envisages restored landscapes that more effectively deliver an ecosystem service—principally water—to the many sectors that depend on this service. The overall goal of this project is to eradicate populations of invasive alien plant species that demand water and suppress native biodiversity, to secure sustainable water supplies from watersheds, and in doing so empower (through job creation and training) a section of the large number of unemployed and otherwise marginalized South Africans. The project, therefore, seeks to ensure the provision of an ecosystem service in a sustainable and cost-efficient manner but also achieves goals for equity and biodiversity conservation—a win-win-win (van Wilgen et al. 2002).

The prerequisites for the rapid and effective mainstreaming of the project into the national Department of Water Affairs and Forestry were as follows. First, there was a long history of concern for, and considerable awareness and a significant scientific understanding of, the impacts of invasive alien plant species on water supplies and biodiversity (van Wilgen et al. 2002). Second, there was sufficient capacity to both devise and implement a project that would address the problem. It was thus possible for scientists and managers to present a compelling case to stakeholders.

With regard to stimuli, improved governance post-1994—as discussed earlier—played a key role in mainstreaming this restoration project (van Wilgen et al. 2002). So, too, did the alarm aroused by dwindling water supplies to some of South Africa's largest cities that were experiencing unprecedented population increases as a result of massive migration after the abolition of the apartheid legislation in the early 1990s. Moreover, given the establishment of a democratic state in 1994, the social *and* economic benefits of the project were hugely appealing to newly instated government officials, who were very keen to deliver benefits at an early stage of their appointments.

Effective communication to key stakeholders—in this case, politicians and officials in the Department of Water Affairs and Forestry—was a key mechanism employed to mainstream the project (van Wilgen et al. 2002). Another important mechanism for mainstreaming the project in other sectors (e.g., agriculture, local government) was the promulgation of legislation that spelt out obligations for restricting further plantings, as well as invasive alien plant eradication. However, although capacity was strengthened sufficiently to launch, maintain, and grow the project, available capacity falls far short of the requirement to mainstream invasive alien plant control in other sectors, as required by legislation. Until this is done, invasive alien plants will continue to overwhelm many South African landscapes that fall outside of key watershed areas.

Constraints for Mainstreaming the Restoration of Natural Capital

Implementing mainstreaming is not an easy task in developed and developing countries. Governments place emphasis on policies that foster rapid economic growth and do not consider the full costs of degrading natural capital. Some obvious constraints acting on mainstreaming include poor governance (Smith et al. 2003), weak capacity of organizations and institutions in all spheres of government and civil society (e.g., Wells and Brandon 1992; Infield and Adams 1999), a lack of scientific knowledge about biodiversity issues (Raven and Wilson 1992), and a lack of technology skills needed for efficient and cost-effective restoration. Other constraints—less widely documented—include economic paradigms that ignore or discount the value of natural capital (chapters 2 and 3), and a dwindling awareness and concern among most people of the value of natural capital for their livelihoods (see also chapters 29 and 30).

Governance, Capacity, and Knowledge

First, if mainstreaming restoration is to be effective, then governance structures and capacity have to be adequate to implement the requisite projects. Intervening at this level is usually unattractive to donor agencies, is often politically risky, and can consume a great deal of time and money. Second, more attention needs to be given to understanding the spatial and process ecology of ecosystem services (Kremen 2005), and the complex benefit streams they produce. Furthermore, ecosystem scientists, restoration ecologists, and economists need to consider the implementation context of their research and engage the assistance of stakeholders in framing their research questions (Sayer and Campbell 2004).

To generate knowledge on the importance for human well-being of natural capital and its restoration, researchers, enablers, and implementers will need to work in interdisciplinary teams, making a conscious effort to fuse new knowledge systems by breaking down the artificial borderlines among existing ones. The emergence of ecological economics (Costanza et al. 1991) and conservation psychology (Saunders 2003) are promising steps toward achieving the transdisciplinarity (Max-Neef 2005) that we so desperately need to implement projects that address complex problems such as mainstreaming. However, much more remains to be done about integrating the social sciences and humanities with ecosystem and restoration science.

Inappropriate Economic Paradigms

Another major constraint operating on mainstreaming restoration is the pervasive economic paradigm that places primacy on unrestrained economic growth, in the sense that the rhythm and run of market economies are inherently in conflict with those of ecosystems (Orr 2002). According to Orr, "Markets, driven by the logic of self interest, are intended to maximize profits and minimize costs for the owners of capital in the short term. Ecosystems, in contrast, operate by the laws of thermodynamics and processes of evolution and ecology that are played out over the long term" (107). In short, the national and international goals of economic growth and sustainable natural resource management (including restoration) are frequently in conflict (Czech 2003; see also chapter 2). There are, of course, cases where the sustainable management of a resource is compatible with market forces (Adams and Hulme 2001), or where the introduction of incentives can effectively secure sustainability of natural resource use (Hutton and Leader-Williams 2003). We need to learn from these cases but also acknowledge that market economies that discount ecosystem services are unlikely to provide incentives for the restoration of natural capital (see chapters 26 and 32).

Dwindling Awareness of the Importance of Natural Capital

Another constraint acting on mainstreaming is the dwindling awareness among the citizenry of most countries of the world, especially many developed ones (e.g., Orr 2002), of the importance of natural capital and the ecosystem services it provides. In today's highly urbanized world of short attention spans and vicarious thrills, people have become disconnected from nature (e.g., Miller 2005). The human–biodiversity connection may be stronger in developing countries where many people still have regular contact with, and dependence upon, wild nature. However, with increasing urbanization and the global expansion of the consumer culture, circumstances are changing rapidly in even the most remote parts of developing nations. How can we hope to mainstream the restoration of natural capital when most people have no idea of the importance of natural capital in their daily lives (see also chapter 29)?

Related to this dwindling awareness of the importance of natural capital to human well-being are two other issues. First, we need to understand what motivates people to discount the future by ignoring the hugely negative impacts of unmanaged population growth and massive overconsumption on the ecosystem services that underpin their own and their offspring's survival (e.g., Winter and Koger 2004). Penn (2003) argues cogently that major advances in this understanding can be achieved by considering the evolutionary roots of human behavior; in particular; the instinctive nature to exploit short-term interests needs to be curbed. This can be achieved by harnessing, via social pressure, the equally instinctive desire to punish individuals and organizations who behave unfairly and exploit collective interests by, for example, appropriating or despoiling more than their rightful share of limited natural resources.

Second, instead of emphasizing biodiversity per se in efforts to promote awareness of, and concerns for, the plight of the earth's natural capital, we should also focus on the services that these provide for human well-being (Daily 1997), and we should do this in a context that is local and immediate rather than global and remote. Doing this can create opportunities to communicate messages that yield anecdotes and stories that might appeal to human self-interest. These might therefore wield a greater influence on behavioral change in some individuals than might probabilistic statements on extinction rates of obscure species in distant land- and seascapes (e.g., Penn 2003; Winter and Koger 2004).

Contribution

Overcoming the constraints on mainstreaming represents a major challenge. The restoration of natural capital has various key advantages since this process is linked directly to the delivery of ecosystem services; consequently, it should be easier to identify win-win scenarios and therefore ensure participation of important stakeholders. Nonetheless, it is necessary to invest in overcoming barriers to mainstreaming before initiating any major interventions. This will be a long and difficult process, not suited to the short timeframes and logical frameworks so important to governments, nongovernmental organizations, and donors. However, prepared or not, we must continue to work toward solutions by being innovative, reflective, adaptable, and patient.

Acknowledgements

Richard Cowling acknowledges the financial support of the National Research Foundation (Pretoria) and the Nelson Mandela Metropolitan University.

Chapter 35

Restoring Toward a Better Future

SUZANNE J. MILTON, JAMES ARONSON, AND JAMES N. BLIGNAUT

This book brings together economists and ecologists who have collaborated on the science, business, and practice of restoring natural capital. Such collaboration is essential to raise awareness of the urgency of replenishing the natural capital stocks now limiting development, to recharge or revitalize the *income* flow arising from them, and to reduce global demands by people on ecosystem goods and services (Aronson, Blignaut, et al. 2006). To quote from Bill McKibben's (1989) book *The End of Nature*, "In the course of about a hundred years our various engines and fires have released a substantial amount of the carbon that has been buried over time. It is as if someone had scrimped and saved his entire life, and then spent every cent on one fantastic week's debauch. We are living on our capital. . . . It is like taking that week's fling, and, in the process, contracting a horrid disease."

Aronson and van Andel (2006) identified the failure of economic systems to value natural capital as one of the major challenges for restoration. Blignaut and de Wit (2004) called this same failure "one of the major challenges for ecological economics." The vision throughout this book is that the value of restoring the goods and services of nature must be demonstrated and realistically accounted for, because in this densely populated world, both conservative use of natural capital and its continual repair are needed to sustain life-support systems. For natural capital and its restoration to be incorporated into budgets at all levels of decision making, the concept of restoring natural capital should be widely understood, valuation methods established, and working models made easily accessible to decisionmakers.

In the first section of the book, which deals with the meaning of, and rationale for, restoring natural capital, we make it clear that the concept is broader than that of ecological restoration. In common with ecological restoration, natural capital restoration is intended to repair the health, integrity, and self-sustainability of ecosystems for all living organisms. The critical difference is that natural capital restoration focuses on defining and maximizing the value of such restoration while enhancing the well-being of people. The restoration of natural capital embraces both anthropogenic and natural ecosystems. Moreover, restoration projects are designed to build stocks of natural capital that will produce self-sustaining flows of ecosystem goods and services; they contribute to the physical, cultural, and psychological factors that define human well-being. Natural capital restoration complements but does not replace nature conservation and ecological restoration.

The second section of the book presents case studies that discuss targets, approaches, and economic opportunities embedded within the restoration of natural capital at the local level. The final section considers widely applicable guidelines on valuation, understanding, and overcoming physical and socioeconomic barriers at individual project, national, and global levels, and documents policies and institutions contributing to successful restoration.

There are four themes woven throughout the book: (1) the valuation of natural capital loss and the restoration activities needed to regain it; (2) the challenge of setting appropriate targets, developing workable approaches, and measuring progress toward natural capital recovery; (3) the need for prioritization and planning of restoration at various scales; and (4) the importance of collaboration and buy-in for successful restoration.

Values and Valuation

Manufactured capital is derived from natural capital. For this reason, species, habitats, and ecosystem services are all eroded by economic development based on the transformation of natural capital to manufactured and financial capital. Farley et al. (chapter 31) point out that the twentieth century has seen a thirty-six-fold increase in the rate at which raw materials are extracted, transformed, and returned to the ecosystem as waste. The rate of transformation of natural areas decreases once the more productive areas have been cultivated or otherwise exploited (chapter 6), but given the growth in the human population, and in consumption rates, it is unlikely that transformation will cease within the foreseeable future. Globalization enables nations to subsidize their consumption by importing resources from poorer nations without paying environmental costs. This works against conserving and restoring natural capital (chapter 26).

Blignaut et al. (chapter 2) point out that values are imposed partly by ethics and partly by economic systems. Valuation of services such as water purification, goods such as firewood, and the diversity, beauty, and spirituality of living organisms and landscapes differs among individuals and among nations. However, at the global level there is an urgent need to reduce the demand made by the world's enormous, and rapidly growing, human population on ecosystem goods and services. At the same time, we must invest in supplementing life-support services and natural resources. When critical natural capital approaches the lower threshold needed to sustain humanity, restoration of natural capital becomes imperative and independent, at least partially, of conventional monetary valuation (chapter 3). This fact must be recognized and acted upon, at village, national, and global scales, in order to retain and augment sufficient capital to support present and future human societies.

Where natural capital remains above critical levels, its restoration and replenishment are, at present, considered optional, and decisions depend on marginal costs and benefits. However, as Norgaard et al. (chapter 5) point out, the costs and benefits of exploitation and of restoration differ with scale and are not equally borne among individuals. For example, mining companies benefit directly by mining, not by restoration of vegetation. On the other hand, living near an unrestored mine costs local people their health or livelihoods, although they receive no benefits from the enterprise. Moreover, costs of restoration are often beyond the ability to pay of those who most stand to benefit (chapter 23). In such cases particularly, the costs and benefits of restoration should be assessed from a national perspective. Regardless of scale, a holistic evaluation is necessary, since people are motivated to invest in restoration for various reasons (Clewell and Aronson 2006), just as they generally make decisions based on a range of individual and collective values, including, but not limited to, monetary ones.

Nature will flourish without humans (chapter 26), but humans cannot survive without nature. This fact needs to underpin decisions and planning in all sectors of society, and a grassroots environmental or civic pride movement is clearly insufficient to change the way people and nations value and manage natural resources and flows of ecosystem services. The solution is to mainstream (viz., chapter 34) the appreciation and valuation of natural capital, through education, public policy, and pricing, to ensure that the links between nature's services (sensu Daily 1997) and stocks of natural capital are understood and paid for even by city dwellers remote from natural environments. The increase in manufactured capital needs to be offset against the declines in natural capital, external costs need to be internalized, and perverse subsidies and incentives that lead to overconsumption of ecosystem services need to be eliminated (chapter 30).

Targets, Approaches, and Assessment

Global targets for natural capital restoration, for example, the prevention of further global warming, desertification, and species losses (MA 2005a), are generally fewer, clearer, and simpler than local targets. Local-scale priorities are highly diverse, depending on the needs and expectations of particular communities in different cultures. For some the restoration of aesthetic and spiritual qualities or heritage values of sites or landscapes might be a priority. Others might place more value on return of rare species and habitats, or biodiverse, self-sustaining ecosystems that produce flows of food, forage, fuel, medicine, healthy air, water, or other tangible products

Technical approaches vary with project goals, scope, financial, and manpower resources. A sound scientific basis for restoration is obviously desirable (see, for example, chapters 10, 13, and 21), and interinstitutional collaboration, government, and international support can be helpful in advising and funding restoration. But whatever the project, and however small or large the scale, case studies from around the world insist that winning public support, particularly from those whose daily lives are affected by restoration activities, is crucial for project sustainability and success. Success stories from Hawaii, India, Madagascar, New Zealand, Scotland, South Africa, and the United States show how participatory setting of restoration targets and achieving local buy-in to restore natural capital can be achieved (e.g., chapters 8, 12, 14, 16, 17, 20, and 25).

Monitoring progress is fundamental to good management and for making decisions about investment of public funds (Ferraro and Pattanayak 2006), yet there is still much debate about how the restoration of natural capital can best be assessed. Natural science is increasingly mathematical and quantitative. For example, ecologists count species, weigh natural products harvested from restored ecosystems, and measure air and water quality at landscape and regional scales, to assess their progress toward restoring natural capital. Economic science is also quantitative, and economists can track employment levels, incomes, and expenditures in restored and control areas as indicators of natural capital restoration–project progress. Changes in well-being are more complex to measure yet essential for restoration evaluation (see, for example, chapters 8 and 15). Successful restoration of natural capital will ultimately depend upon changes of attitude, locally and globally, and a reorganization of the ways different values are respected or ranked in importance. A major goal of this book is to mainstream restoration of natural capital; to move toward an economy where nature matters, and ecology where people matter. Tracking progress toward this larger goal is even more challenging!

Prioritization and Planning

Restoration activities that are nonessential for immediate human survival compete with other activities and products for investment of time and resources on the part of governments, non-governmental organizations, businesses, and private individuals. Ironically, as illustrated by the case study presented by Aguiar and Roman (chapter 13), areas most degraded by overexploitation are often those where land users are least likely to be able to finance restoration. It is in these cases that the decision to restore natural capital becomes national or international and subject to competing demands on public funds.

Factors that reduce economic prioritization for restoration include, for example, the small size of a damaged area, remoteness from centers of human settlement, and lack of immediate human health hazards posed by the environmental damage. According to Figueroa (chapter 4) biodiversity issues, such as threatened species and habitat loss, as well as ecosystem services currently not benefiting human populations, generally play little part in influencing allocation of national funds for restoration when there are more pressing calls for funding in populous areas. However, the outcomes of cost-benefit analysis or multicriterion decision-making processes will vary over time as values change, new information becomes available, and political and economic pressures shift. The role of community involvement in decisions influencing natural resource management policy, planning, and enforcement is growing in most countries (chapter 27), and mainstreaming of the natural capital restoration concept could create a powerful lobby to drive the restoration of natural capital locally and globally.

The voice of the people in environmental management decisions is growing stronger. Public awareness of the role that natural systems play in the quality of urban and rural life is therefore essential for informed and wise prioritization of funding and land-use planning. Despite the knowledge gaps revealed, the Millennium Ecosystem Assessment report (MA 2005a) should have a major influence on public understanding of environmental issues. An expected outcome is that natural capital restoration should gain higher prioritization and become a routine part of development planning and budgeting.

Public Support and Collaboration

Science tells us that current consumption rates are unsustainable, are damaging the global ecosystem, and will ultimately undermine human welfare, as well as drive many species to extinction (MA 2005a). However, this knowledge will not necessarily result in restoration and sustainable living, because the belief that ever-increasing consumption and "economic growth" are essential to the well-being of the individual and the nation appears to transcend other beliefs and values globally. As Marais et al. (chapter 29) put it, "a strategy for restoring natural capital, which prioritizes sustainability over consumption and growth, flies in the face of this dominant culture." Changed attitudes and public support are thus essential for driving the restoration of natural capital.

In many ways, the restoration of natural capital is an easy concept to market. The message, *by investing in restoration we can return life-supporting processes to damaged ecosystems and fragmented landscapes*, is not condemnatory, for once, but rather positive and optimistic. It joins and reconciles the languages of economics and ecology and contributes to socioeconomic development and sustainability. It brings hope and reestablishes links between people and nature, while recalling the ancient myth of the phoenix rising from the ashes, of life after death, of ongoing evolution. Often, upon hearing this message, people rally round, roll up their sleeves, and set to work, together.

Although ecological restoration has become a recognized business activity (Cunningham 2002), willingness or ability to pay for this service is often limited. Nonfinancial instruments (chapters 29 and 33), including legislation, social pressure, and business conscience (chapter 19), can be effective. Where these fail, economic instruments, such as tax incentives and product certification, can facilitate restoration of natural capital (chapter 32).

The impacts of many small, disconnected, individual efforts to restore natural capital may have educational value—but their social, economic, and biodiversity outcomes can be greatly increased through coordinated planning, financing, and implementation (chapters 18 and 34). Although often difficult to achieve, the results of such initiatives are generally worth the effort.

Final Challenge

Natural capital largely determines the wealth of families, municipalities, and nations. It motivates wars, exploration, and colonization (Diamond 1997), and the more limiting the resource, the greater the investment of the individual or nation in defending or capturing it. Wars have been fought over land, gold, spices, fish, oil, and water. The outcome of such wars has historically been transfer of tenure or ownership, followed by more rapid depletion of the resource. But now that resources are globally depleted compared to a mere century ago (Aronson, Blignaut, et al. 2006), and ecosystem services are failing to keep pace with human demands for clean air and water, for example, war is no longer an "appropriate" or acceptable response. There is no cleaner, richer part of the world left to conquer. At a global scale, our sample size is equal to one. The costs of transforming natural to manufactured capital (depletion of natural resources) can no longer be transferred from the developed to the developing nations. The time has come for every nation to invest in protecting and restoring its natural capital, and for collaboration across all scales in investment and planning to restore to a better future.

active ingredients: Chemicals in plants, fungi, or animals that have some beneficial (e.g., medicinal) and/or toxic action. Sensitivity to the chemicals may be species-specific, i.e. a sheep may be more or less sensitive than a human being.

adaptive management approach: Management that is informed by research and changes as new information becomes available.

alien species: Fungi, plants, or animals that are not native to the country or region in which they are introduced or naturalized. See also *invasive alien species*.

anadromous fish: Fish that hatch in freshwater, migrate to the ocean to mature, then return to freshwater to spawn.

apartheid era: Period of South African history (1948–1990) when people of different ethnic groups were compelled by law to live apart.

assurance bond: Similar to *development bond*; used when there is uncertainty about the possible nature and size of adverse environmental effects. The bond is refunded if and when it is demonstrated that the known potential environmental damages did not occur.

benefit transfer method: Analytical method whereby the values estimated for a specific good or service elsewhere are directly transferred to the site or operation in question.

benefit-cost analysis: See *cost-benefit analysis*.

benefit-cost ratio: See *cost-benefit ratio*.

bequest values: Refers to an individual's willingness to pay to ensure that an environmental resource will be preserved for the benefit of his or her descendants.

biodiversity: Diversity of life at genetic, species, community, ecosystem, and biome levels.

biogeographic characteristics: Factors that determine the distribution of plants and animals, including history, geography, and the physical environment.

biological control: Control of a pest by the introduction, preservation, or facilitation of natural predators, parasites, or other natural enemies.

biophysical functioning: How biological and physical processes work and interact to maintain ecosystems in a given dynamical trajectory.

breeding system: Breeding behavior characteristic of a population or species (e.g., self-fertilization by a plant versus requirement for pollen from another plant of the same species).

319

browser: Any species of large mammal that feeds preferentially on woody plants (shrubs and trees). Browsers are contrasted with grazers, which are large mammals that feed on herbaceous plants such as grasses and forbs.

brush-pack: Branches laid on base ground to trap water and seeds and reduce erosion by wind and water.

bubble licensing: Licensing scheme under which a firm is issued a license that encompasses all of its operations and limits its total emissions or other adverse environmental effects. The firm can then seek to minimize its emissions, etc., in the most cost-effective manner possible while respecting its social and legal engagements.

cap-and-trade: Method of establishing a market for rival, nonexcludable goods and services. A ceiling or limit is set on total pollution emitted, or other adverse impact on the environment, according to certain norms (e.g., total fish captured or other biological resource extracted). Legal permits are issued for this total limit, and penalties are imposed for exceeding it. Those firms with the highest value use for pollution and/or the lowest cost of achieving pollution reduction can trade their permits, theoretically optimizing resource allocation.

carbon sequestration: Concept that refers to capturing carbon and keeping it from entering the atmosphere for some period, under a greenhouse gas reduction program. See also *Kyoto Protocol*. Carbon is sequestered in carbon sinks, such as forests, soils, or oceans.

cardinal measurement: A variable is cardinally measurable if a given interval between measures has a consistent meaning, i.e., if a measure corresponds to points along a line. For example, height, output, and income are cardinally measurable. See *ordinal measurement*.

cascading effect: Process whereby the loss or change of one component of an ecosystem leads to loss or change of many other components.

conservation easement: Legal contracts that, for conservation purposes and maintenance of environmental services, restrict the use and development of a unit of land usually in perpetuity.

contingent valuation: Use of questionnaires to estimate the willingness of respondents to pay for public projects or programs, or willingness to accept compensation in the case of a loss. Example questions: Would you accept a tax of x to pay for the program? Would you be willing to accept x as due compensation for the loss suffered?

cost-benefit analysis: Economic technique applied to public decision making that attempts to quantify and compare, in monetary terms, the advantages (benefits) and disadvantages (costs) associated with a particular policy.

cost-benefit ratio: Discounted measure of project worth that implies the present worth of the cost stream divided by the present worth of the benefit stream. When the cost-benefit ratio is used, the selection criterion is to accept all independent projects with a ratio of 1 or less when discounted at a suitable discount rate. The discount rate used most often is the opportunity cost of capital (see *opportunity cost*).

decadence management: Using a variety of techniques to promote heart rot in trees, tree death, and tree fall to provide habitat elements that are essential to a variety of microbes, plants, fungi, and animals. The techniques include retaining dead trees in the forest, killing trees to serve as substrate for foraging and cavity excavation by wildlife, and creating tree cavities to be used by wildlife.

degradation: Persistent loss in the capacity of ecosystems to deliver ecosystem goods and services.

designer ecosystem: Ecosystem that contains groups of species that do not generally exist together in nature; conceived and constructed so as to support or deliver some conservation, management, and/or other goals.

development bond (or *performance bond*): Commonly set by the principal to a contract when there is a risk that the agent will not complete agreed activities on time or to the agreed specifications. The agent puts up a bond that is held by a third party. If the activity is completed satisfactorily and on time, the bond monies are returned.

diameter at breast height (DBH): Method for measuring the width of a tree trunk (with a standard height of 1.3 m above ground level).

direct-use value: Direct or extractive and consumptive use of natural biota. Includes wood for construction and timber, as well as for energy purposes; medicinal products; edible fruits, herbs, and vegetables; thatch; and the values of livestock production and game hunting.

discount rate: Interest rate at which an agent discounts future events, preferably in a multi-period model. Often denoted as *r*. A present-oriented (or short-term-oriented) agent discounts the future heavily, yielding a high discount rate.

discounting: Method used to determine the monetary value today of a project's future costs and benefits by weighting monetary values that occur in the future by a value less than 1 (the discount rate).

disturbance: Natural or anthropogenic events or activities that significantly change the structure, content, and/or function of ecosystems. Can lead to *degradation*.

easement: Interest in land owned by another, which entitles the holder limited, specific uses of the land. See *conservation easement*.

ecological restoration: Defined by the Society for Ecological Restoration International (SER) as "the process of assisting the recovery of an ecosystem that has been degraded, damaged, or destroyed."

ecology: Study of factors determining the abundance and distribution of plant, animal, fungal, and microbial species, including the interaction of all such organisms with one another and with their physical environment.

econometric: Mathematical models conventionally used to predict future changes either in monetary or nonmonetary terms. Can also be used to estimate losses and gains of natural capital.

economic incentive: Instrument that affects costs and benefits of alternative actions of agents, including, for example, paying people for conserving and restoring ecosystem goods and services; access to credit for farmers that adopt best practices; and premium-paying markets for certified or green-labeled products.

ecosystem assessment: Evaluation of ecosystem attributes relative to a predetermined restoration goal or other benchmark, such as a natural ecosystem in a specified reference site.

ecosystem goods and services: Conditions and processes through which natural ecosystems sustain and fulfill human and other forms of life. Examples include delivery of fuelwood (goods); providing clean water; climate maintenance (carbon sequestration); crop pollination; and fulfillment of human cultural, spiritual, and intellectual needs (services). Also known as *environmental services*.

ecosystem metamodel: General model that includes multiple, more specific models. Mathematically, the more specific models can be the general model with different variables held constant.

ecosystem well-being index: Broad measure of the state of a local environment covering (1) the quality and diversity of its natural land ecosystems; (2) water quantity and quality; (3) air quality, including the emissions of greenhouse gases; and (4) the conservation of genes.

ecosystem: Complex of living organisms, and their associated nonliving environment, interacting as an ecological unit.

ecotype: Population within a species that has a distinct set of morphological and/or physiological characteristics reflecting adaptation to a specific environmental niche.

elasticity: Measure of responsiveness. The responsiveness of behavior measured by variable Z to a change in variable Y is the change in Z observed in response to a change in Y. Specifically, this approximation is common: elasticity = (percentage change in Z) / (percentage change in Y).

emerging ecosystem: Develops after social, economic, and cultural conditions so change the environment that new biotic assemblages colonize and persist for decades with positive or negative social, economic, and biodiversity consequences.

eminent domain: Lawful power of government to expropriate private property without the owner's consent.

endemism: Restriction of the natural distribution of a plant, animal, or other kind of organism to a narrow geographical range (e.g., an island or a single valley).

environmental services: See *ecosystem goods and services*.

excludable goods: Legal construct that allows the owner of a resource to prevent others from using that resource.

existence value: Value that individuals may attach to the mere knowledge of the existence of something, as opposed to having direct use thereof; synonymous with nonuse value.

exit strategy: Plans for mine closure that include measures to address environmental and social concerns.

exogenous natural capital: Sources of natural capital that are not spatially or structurally restricted to an ecosystem.

exotic species: See *alien species*.

externality: Coincidental, beneficial, or adverse side effect of production or consumption, usually associated with economic transactions not taken into account by those directly involved in making the decision. Can be either a benefit or a cost.

forest certification: Process of labeling and marketing wood or other forest products (gum, seeds, ferns, honey) that have been harvested from a sustainably managed forest.

forest grain analysis: Describes the scale of the ecological process of regeneration behavior of a forest community in relation to typical disturbance events.

fynbos: Species-rich, fire-adapted, and shrublands occurring on nutrient-poor soils in the winter-rainfall Cape region of South Africa; characterized by the presence of reeds (Restionaceae) and wintergreen-, wind-, or ant-dispersed shrubs (Ericaceae and Proteaceae).

genetic fitness: Reproductive success of a genotype, usually measured as the number of offspring that survive to reproduce produced by an individual.

genetic integrity: Complete range of genes found within a healthy natural population of a species.

geophyte: Plant with an underground storage organ (e.g., bulb, tuber, corm, or rhizome).

government failure: Situation in which the behavior of optimizing agents in a market would not allocate resources optimally due to the presence and activities of government.

gross domestic product (GDP): Value of the flow of domestic goods and services produced by an economy over a period of time (e.g., one year).

gross national product (GNP): Equal to *gross national income (GNI)*; the value of all final goods and services produced within a nation in a given year, plus income earned by its citizens abroad, minus income earned by foreigners from domestic production.

growth coefficient r: Known as the intrinsic rate of growth, because it measures whether a living population tends to grow ($r > 0$) or shrink ($r < 0$).

habitat fragmentation: Splitting and isolating of patches of an environment in which a population or an individual lives.

hammer mill: Belt-driven cutter for chipping branches or other plant material.

heat scar: Patch of soil subjected to such high temperature during a fire that organic matter is destroyed and physical properties of the soil, such as water-holding capacity, are changed.

herbivore: Animal that feeds on plants.

hierarchy theory: Considers ecosystems as consisting of assemblages of species going through cycles at different speeds.

hogget: Weaned but prereproductive sheep having no more than two permanent front teeth.

home garden: Tree-dominated, multispecies garden providing a variety of food and other useful products.

human capital: Attributes of a person that are productive in some economic context. Often refers to formal educational attainment, with the implication that education is an investment (e.g., education's returns take the form of wages, annual salary, or other formal compensation).

Human Development Index (HDI): Used by the United Nations Development Program to measure social aspects of an economy, incorporating especially life expectancy, income, literacy, and school enrollment.

human well-being: Condition in which all members of society are able to determine and meet their needs and also have a large range of choices to meet their human potential.

Hydropam™ soil binder: Colloid that holds soil particles together, reducing wind erosion.

hydroseeding: Application of a mixture of seeds, water, fertilizer, and binder to a rehabilitation site.

imperfect competition: Market situation where one or more buyers, or sellers, are important enough to have an influence on price.

inbreeding depression: Detrimental result of inbreeding (progeny produced as a result of mating taking place among closely related individuals).

Index of Sustainable Economic Welfare (ISEW): Combines social factors, income inequalities, and environmental deterioration (all components that have a significant impact on quality of life).

indicator: Easily measured surrogate for difficult-to-measure, whole ecosystem attributes or trajectories.

internal rate of return (IRR): Discount rate at which the net present value of an investment is equal to zero; alternatively stated, the rate of discount that would cause the discounted value of all costs to be equal to all future returns.

interpatch: Area between defined landscape patches where natural capital is mobilized, transported, and gained or lost.

interspecific: Interactions or differences between or among species.

intraspecific: Interactions or differences within a single species.

invasive alien plant: Nonindigenous (introduced) naturalized plant species that produce reproductive offspring in very large numbers and thus have the potential to spread over a large area and to disrupt processes of native ecosystems.

keystone species: Species that has a greater influence on other species than would be predicted by its abundance or size. The loss of keystone species can result in a cascade of local extinctions and other irreversible changes in ecosystems.

Kyoto Protocol: Agreement among 159 nations signed in December 1997, specifying the deadlines and specific levels of greenhouse gas reductions that signatory countries are to achieve. Overall, developed countries are to reduce greenhouse gas emissions by 5.2% between 2008 and 2012, as measured against 1990 emission levels.

land-use oversight: Authority over the land-use decisions made by local governments and landowners.

landscape: In ecological terms, an assemblage of ecosystems that interact among themselves in such a way as to produce recognizable patterns. Various other definitions exist in the social sciences (e.g., geography) and in art history. Cultural landscapes are those "sculpted" and maintained by people over generations.

Landscape Function Analysis (LFA): Methodology and procedure for defining landscape structures and measuring landscape surface indicators.

landscape functioning: How landscape processes work and interact to retain, utilize, and cycle natural capital.

landscape surface indicators: Attributes of ecosystems, or assemblages thereof, that can be readily measured by observing the ground.

large stock unit (LSU): Metabolic equivalent of a head of cattle, weighing 450 kg. Used to compare the food consumption of large herbivorous animals of differing sizes through a conversion based on the mass-to-surface-area ratio, which approximates energy consumption.

local skill base: Trained people or people with specific expertise who are native to an area or who reside long enough in an area to be considered citizens.

macroeconomics: Branch of economic theory concerned with the economy as a whole. Deals with large aggregates such as total output, rather than with the behavior of individual consumers and firms.

marginal analysis: Analytical technique that focuses on incremental changes in total values, such as the last unit of a good consumed, or the increase in total cost.

marginal benefit: Increase in total benefit consequential to a one-unit increase in the production of a good.

marginal cost: Increase in total cost consequential to a one-unit increase in the production of a good.

market failure: Situation in which the behavior of optimizing agents in a market would not produce optimal allocation due to market inadequacies. Sources of market failures are, among others, monopolies (a single producer of an item) or oligopolies (a small number of large producers of a single item), that have incentives to underproduce and to price above marginal cost, which then provides consumers incentives to buy less than the optimal allocation and externalities.

mobile resources: Natural capital that can be transported into or out of an ecosystem (e.g., soil nutrients, water, seeds, etc.).

monitoring: Repeated assessments of indicators or ecosystems that are designed to detect change over time.

multiple criteria analysis: Analytical method to assist in decision making, based on a variety of components and elements, some of which could be monetary.

natural capital: Stock of physical and biological natural resources that consist of renewable natural capital (living species and ecosystems), nonrenewable natural capital (subsoil assets, e.g., petroleum, coal, diamonds, etc.), replenishable natural capital (e.g., the atmosphere, potable water, fertile soils), and cultivated natural capital (e.g., crops and forest plantations).

net present value: Present value of the incremental net benefit, or incremental cash flow stream, of a project. Present worth of the benefits less the present worth of the costs of a project.

networked metapopulation: Group of populations or subpopulations of plants or animals that are interconnected so that losses from one population are replaced by migration from another population. Typically such populations would occur in patchy habitat and are managed as a single unit.

nonconsumptive values: Nonconsumptive values comprise those direct-use values that are nonextractive in physical terms (e.g., tourism).

nonmarket recreational value: Value associated with activities taking place during leisure time that are not traded formally in a market. The nonmarket recreational value of a site, for example, can be estimated, using the travel cost method, from the travel and other market costs incurred by the visitors to the site.

offset system: Usually established by a government that requires impacts on one ecosystem site to be mitigated through creation or protection of another site such that there is no net loss in ecosystem services.

open-access resource: Material resource with no property right held by any individual or entity.

open-cast mining: Also known as strip or surface mining, this process removes vegetation and soil over large areas to obtain minerals in or beneath the soil.

opportunity cost: Cost of sacrificing the next best alternative, or the income forfeited as a result of a decision in favor of one option rather than another.

option value: Expression of preference and a willingness to pay for the preservation of an environment against the probability that the individual will make use of it later.

ordinal measurement: Used when one can rank things in relative levels or degrees but cannot measure them precisely. Unlike cardinal measurements, ordinal measurements are not equally spaced but are ranked in order (e.g., A, B, C, etc.).

outcrossing: Mating between two different individual plants; the antithesis of selfing (self-pollination) where the ovary of a plant is fertilized by pollen produced from the same plant.

permaculture: Use of ecology as the basis for designing integrated systems of food production that are ecologically sustainable.

phosphorus- and fecal coliform-restricted basins: Drainage basin of a reservoir in which the fecal coliform (bacteria) and phosphorus (nutrients that lead to eutrophication) levels exceed government standards.

photosynthetic pathway: Photosynthesis is the process by which green plants convert water and carbon dioxide to starch and sugar in the presence of sunlight. There are three major photosynthetic chemical pathways: C_3 (cool moist climates), C_4 (hot moist climates), and CAM (crassulacean acid metabolism), found mainly in warm dry climates.

plant smoke–derivative: Chemical in smoke released by burning plants that stimulates germination in some plant species.

precautionary principle: States that "where there are threats of serious or irreversible environmental damage, lack of scientific certainty should not be used as a reason for postponing measures to prevent environmental degradation." This is the fifteenth principle of sustainability in proceedings of the World Summit on Environment and Development, Rio 1992.

propagation fog bed: Used in propagation of cuttings, this irrigation setup for a nursery releases fine water spray at frequent intervals throughout the day and night, preventing desiccation.

propagule: Seed, egg, or detachable part of a plant or animal that serves as the source of new individuals.

provenance: Geographic place of origin or source of seed or other propagule.

public goods: Goods that are nonrival and nonexcludable, meaning that if one person consumes the good it is not diminished for others or possible to exclude anyone else from consuming the good.

rain-forest biome: Community of living organisms in an evergreen woodland of the tropics distinguished by a continuous leaf canopy and an average rainfall of about 100 inches (2,540 mm) or more per year.

redundancy: Presence of multiple species that play similar roles in processes that maintain ecosystems.

reference site: Natural ecosystem used as the basis for monitoring and evaluating sites targeted for restoration or rehabilition.

rehabilitate: To repair some ecosystem processes at a site (e.g., productivity and other ecological services) without aiming to fully reproduce predisturbance conditions or species composition.

resilience: Capacity of an ecosystem to persist on a given trajectory and with a given set of dynamics in the face of disturbances.

restoration assessment: Evaluation of a restored or rehabilitated ecosystem by comparison of the composition, structure, and/or function with that of a natural ecosystem (see *reference site*).

reverse auction scheme: Auctions in which sellers compete for the right to provide a good. In the case of natural capital restoration, they have been used by governments to have landowners submit bids indicating the amount of compensation they would need to undertake certain land-use practices that increase provision of a given ecosystem service.

ring-barking (or girdling): Removing bark from a continuous circle around a tree trunk. Prevents the movement of sugars from the leaves to the roots.

Safe Harbor Agreement: Voluntary agreement in the United States that promotes management for listed endangered species on nonfederal property while giving assurances to participating landowners that no additional future regulatory restrictions will be imposed.

safe-minimum standard: Restriction that limits the use of resources to levels that are thought to be safe (e.g., conservation of a sufficient area of habitat to ensure the continued provision of ecological functions and services).

savoka: Term used in Madagascar for land left to fallow following slash-and-burn cultivation of hill rice (or other) crops.

scarification: Scratching or disturbance of the soil surface to permit water penetration. Also refers to the nursery technique of cutting or scratching a seed to facilitate germination.

self-incompatibility: Self-sterility.

serotinous: Retention of seeds in fruits or capsules until the death of the mother plant, which in fynbos is usually by fire.

shifting agriculture: Rotational deforestation (by felling and burning trees) to obtain fertile soil for growing food crops. In high-rainfall forested areas with poor, leached soils, fertility typically lasts no more than two to three years.

slash and burn: See *shifting agriculture*.

slimes dam: Impoundment used for the disposal of slimes (water and fine materials such as clay) that are by-products of the extraction of minerals from soil or rock.

socioecological-system approach: Derived through linking biophysical, ecological, and social processes and drivers.

species evaluation: Empirical testing of the suitability of species for use in restoration or other related activities.

stress: Impairment of ecosystem or species functioning caused by abiotic and/or biotic factors.

tavy: Slash-and-burn, cultivation of hill rice in Madagascar.

tiller: Shoot that sprouts from the base of a grass tuft; tillering is the corresponding form of vegetative reproduction in grasses.

tradable emission permit system: See *cap-and-trade*.

traditional ecological knowledge (TEK): Ecological knowledge derived through societal experiences and perceptions accumulated by members of traditional societies during their interaction with nature and natural resources.

traditional society: People living close to nature and natural resources, usually producing distinctly sculptured cultural landscapes of which they are an integral part.

tragedy of the commons: Metaphor for the problem of management of open resources that tends to favor those who take the most. The phrase captures G. Hardin's (1968) notion that, where there is no clear ownership of rights to a natural resource, the users of the resource are likely to overexploit it.

transaction cost: Totality of cost (not including market prices of the sale/purchase) of exchanging and enforcing property rights or of undertaking market/nonmarket transactions.

transformer species: Invasive alien species that transform habitats or landscapes through changing the character, condition, form, or nature of a natural ecosystem over a substantial area.

travel cost method: Analytical method to estimate the value of (especially) a recreation site by determining the cost incurred by travelers to access or visit the site.

tree seal: Product to seal damaged tree tissues so as to prevent fungal infections and drying out of the wound.

trigger-transfer-reserve-pulse(TTRP): A conceptual framework for how landscape systems function in time and space.

vacuum harvesting: Collection of seeds from natural vegetation using a suction system.

vegetation patch structure: Areas within a landscape defined to have a specific type of vegetation with specific attributes.

Wellbeing Index (WI): Combines environmental well-being (see *ecosystem well-being index*) with the quality of human life, which includes indicators pertaining to health, population, household wealth, national wealth, knowledge, culture, freedom and governance, peace and order, household equity, and gender equity.

wild species and genes index: Number of threatened animal and plant species as a percentage of species in that group.

REFERENCES

Abbot, J. 2005. Out of the woodlands, into the fire: Fuelwood and livelihoods within and beyond Lanke Malawi National Park. In *Rural resources and local livelihoods in Africa*, ed. K. Homewood, 15–35. New York: Palgrave Macmillian.

Acocks, J. P. H. 1953. Veld types of South Africa. *Memoirs of the Botanical Survey of South Africa* 28:1–192.

Acocks, J. P. H. 1964. Karoo vegetation in relation to the development of deserts. In *Ecological studies in Southern Africa*, ed. D. H. S. Davis, 100–112. The Hague: W. Junk.

Acocks, J. P. H. 1979. The flora that matched the fauna. *Bothalia* 12:673–709.

Adams, R. G., and D. E. Gallo. 2001. *The impact on Glenn County property tax revenues of public land acquisitions in the Sacramento River Conservation Area.* Report for the U.S. Fish and Wildlife Service. Contract Number 11332-8-G100. Chico: California State University.

Adams, W. M., and D. Hulme. 2001. If community conservation is the answer in Africa, what is the question? *Oryx* 35:193–200.

Agarwal, A., and S. Narain. 1997. *Dying wisdom: Rise, fall and potential of India's water harvesting systems.* New Delhi: Centre for Science and Environment, New Delhi.

Aguiar, M. R., J. M. Paruelo, O. E. Sala, and W. K. Lauenroth. 1996. Ecosystem responses to changes in plant functional type composition: An example from the Patagonian steppe. *Journal of Vegetation Science* 7:381–390.

Aguiar, M. R., and O. E. Sala. 1997. Seed distribution constrains the dynamics of the Patagonian steppe. *Ecology* 78:93–100.

Ainslie, A. 1999. When "community" is not enough: Managing common property natural resources in South Africa. *Development Southern Africa* 16:375–401.

Ainslie, A. 2003. *The South African Millennium Assessment Project Local Level Assessment Scoping Report: The Mid-Great Fish River area.* Unpublished report. Grahamstown, South Africa: Department of Environmental Sciences, Rhodes University.

Allee, W. C. 1949. *Principles of animal ecology.* Philadelphia: Saunders.

Alvarez, G., H. del Valle, and M. Raso. 1992. *Sostenibilidad y pequeña producción. El caso de Colonia Cushamen (Chubut, Argentina).* Buenos Aires: IICA Instituto Interamericano de Cooperación para la Agricultura, INTA.

Amundsen, J. 2006. Subverting the subdivision: Conservation development in the United States. http://ecosystemmarketplace.com/.

Anderson, J. L. 2001. Stone-age minds at work on 21st century science: How cognitive psychology can inform conservation biology. *Conservation Biology in Practice* 2:18–25.

Andriatahina, M. 2003. *Contribution à la promotion de la culture semi-intensive de la vanille pour freiner la dégradation de la forêt classée d'Andravory Vohémar Nord de Madagascar.* Mémoire de fin d'études. Département Eaux and Forêts, Université d'Antananarivo, Antananarivo, Madagascar.

Aranson, R. 2002. A *Review of international experiences with Itqs: An annex to future options for UK fish quota management.* CERMARE Report 58.

Ares, J., A. M. Breeskow, M. Bertiller, M. Rostagno, M. Irisarri, J. Anchorena, G. Defosee, and C. Merino. 1990. Structural and dynamic characteristics of overgrazed lands of Northern Patagonia, Argentina. In *Managed grasslands*, ed. A. Bremeyer, 149–75. Amsterdam: Elsevier Science Publishers.

Aretino, B., P. Holland, A. Matysek, and D. Peterson. 2001. *Cost sharing for biodiversity conservation: A conceptual framework*. Productivity Commission Staff Research Paper. Canberra, ACT: AusInfo.

Armstrong, D. P. 1995. Effects of familiarity on the outcome of translocations. Pt. 2. A test using New Zealand robins. *Biological Conservation* 71: 281–88.

Aronson, J., J. N. Blignaut, S. J. Milton, and A. F. Clewell. 2006. Natural capital: The limiting factor. *Ecological Engineering* 28:1–5.

Aronson, J., A. F. Clewell, J. N. Blignaut, and S. J. Milton. 2006. Ecological restoration: A new frontier for nature conservation and economics. *Journal of Nature Conservation* 14:135–39.

Aronson, J., C. Floret, E. Le Floc'h, C. Ovalle, and R. Pontanier. 1993. Restoration and rehabilitation of degraded ecosystems in arid and semi-arid lands. Pt.1. A view from the South. *Restoration Ecology* 1:8–17.

Aronson, J., and J. Van Andel. 2006. Challenges for ecological theory. In *Restoration ecology: The new frontier*, ed. J. Van Andel and J. Aronson, 223–34. Oxford, UK: Blackwell Science.

Arrow, K. J. P. Dasgupta, and K. G. Mäler. 2003. Evaluating projects and assessing sustainable development in imperfect economies. *Environmental and Resource Economics* 26:647–85.

Ashmole, P. 1994. *Proceedings of the restoring borders woodland conference*. Kidston Mill, Peebles, Scotland: Peeblesshire Environment Concern.

Ashmole, P., and H. Chalmers. 2004. The Carrifran Wildwood Project. *Ecos* 25:11–19.

Aucamp, A. J. 1976. The role of the browser in the bushveld of the Eastern Cape. *Proceedings of the Grassland Society of South Africa* 11:135–38.

Aucamp, A. J., and L. G. Howe. 1979. The production potential of the Eastern Cape Valley Bushveld. *Proceedings of the Grassland Society of South Africa* 14:115–17.

Aucamp, A. J., L. G. Howe, D. W. W. Q. Smith, and J. O. Grunow. 1980. The effect of defoliation on *Portulacaria afra*. *Proceedings of the Grassland Society of South Africa* 15:179–84.

Axelson, E. 1977. A summary of the history of human settlement at Saldanha Bay. *Transactions of the Royal Society of South Africa* 42:215–21.

Ayalew, W., J. M. King, E. Bruns, and B. Rischkowsky. 2003. Economic evaluation of smallholder subsistence production: Lessons from an Ethiopian goat development program. *Ecological Economics* 45:473–85.

Baber, M. J. 1996. Offshore islands and management of Takahe: Can islands support a viable population? MS thesis, University of Auckland, Auckland, NZ.

Badenoch, C. 1994. Woodland origin and loss of native woodland in the Tweed valley. In *Proceedings of the Restoring Borders Woodland conference*, ed. P. Ashmole, 11–26. Kidston Mill, Peebles, Scotland: Peeblesshire Environment Concern.

Bailey, D., J. Gross, E. Laca, L. Rittenhouse, M. Coughenour, D. Swift, and P. Sims. 1996. Mechanisms that result in large herbivore grazing distribution. *Journal of Range Management* 49:386–400.

Bakker, J. P., R. J. Poschlod, B. R. M. Strystra, and K. Thompson. 1996. Seed banks and seed dispersal: Important topics in restoration ecology. *Royal Botanical Society of the Netherlands* 45:461–90.

Baldi, R., S. Albon, and D. Elston. 2001. Guanacos and sheep: Evidence for continuing competition in arid Patagonia. *Oecologia* 129:561–70.

Balmford, A., L. Bennun, B. ten Brink, D. Cooper, I. M. Cote, P. Crane, A. Dobson, et al. 2005. The convention on biological diversity's 2010 target. *Science* 307:212–13.

Balmford, A., and W. Bond. 2005. Trends in the state of nature and implications for human well-being. *Ecology Letters* 8:1218–34.

Balmford, A., A. Bruner, P. Cooper, R. Costanza, S. Farber, R. E. Green, M. Jenkins, et al. 2002. Economic reasons for conserving wild nature. *Science* 297:950–53.

Balmford, A., P. Crane, A. P. Dobson, R. E. Green, and G. M. Mace. 2005. The 2010 challenge: Data availability, information needs, and extraterrestrial insights. *Philosophical Transactions of the Royal Society B.* 360:221–28.

Barbarika, A. 2004. *The Conservation Reserve Program summary and enrollment statistics: Fiscal year 2004*. Report of the Farm Service Agency (FSA). Washington, DC: FSA. http://www.fsa.usda.gov/.

Barnes, J., C. Boyd, and J. Cannon. 2003. Economic incentives for rangeland management in northern Botswana: Implications for biodiversity. In *Proceedings of the 7th International Rangelands Congress*, 26

July–1 August, ed. N. Allsopp, A. R. Palmer, S. J. Milton, K. P. Kirkman, G. I. H. Kerley, C. R. Hurt, and C. J. Brown, 203–12. Durban, South Africa.

Barnes, J. I. 2001. Economic returns and allocation of resources in the wildlife sector of Botswana. *South African Journal of Wildlife Research* 31:141–53.

Barnes, J. I., C. Schier, and G. van Rooy. 1999. Tourists' willingness to pay for wildlife viewing and wildlife conservation in Namibia. *South African Journal of Wildlife Research* 29:101–11.

Bartel, J. R. 2001. New perennial crops: Mallee eucalypts, a model large-scale perennial crop for the wheat-belt. In *Managing agricultural resources proceedings*. Outlook conference 2001. ABARE.

Bartlett, A. 1996. The exponential function XI: The new flat earth society. *Physics Teacher* 34:342–43.

Barton, I., R. MacGibbon, B. Burns, and P. Berg. 2005. *Profiting from diversity: Reducing the impediments to planting native trees*. Pukekohe, New Zealand: Tâne's Tree Trust.

Bayon, R. 2004. *Making environmental markets work: Lessons from early experience with sulfur, carbon, wetlands, and other related markets*. Washington, DC: Forest Trends.

BCD, Town and Regional Planners. 2000. Development framework plan for the Chemfos phosphate mining area, Langebaan Road. Unpublished Report. Cape Town.

Beckerman, W. 1992. Economic growth and the environment: Whose growth? Whose environment? *World Development* 20: 481–96.

Beinart, W. 2003. *The rise of conservation in South Africa: Settlers, livestock, and the environment, 1770–1950*. Oxford, UK: Oxford University Press.

Belnap, J., J. R. Welter, N. B. Grimm, N. Barger, and J. A. Ludwig. 2005. Linkages between microbial and hydrologic processes in arid and semi-arid watersheds. *Ecology* 86:298–307.

Bendini, M., and P. Tsakoumagkos. 1993. *Campesinado y ganadería trashumante en Neuquén*. Grupo de Estudios Sociales Agrarios—Universidad Nacional del Comahue. Buenos Aires: Editorial La Colmena.

Berkes, F. 1999. *Sacred ecology: Traditional ecological knowledge and resource management*. Philadelphia: Taylor and Francis.

Berkes, F., C. Folke, and J. Colding, eds. 1998. *Linking social and ecological systems: Management practices and social mechanisms for building resilience*. Cambridge: Cambridge University Press.

Berryman, A. A., and J. A. Millstein. 1989. Are ecological systems chaotic—and if not, why not? *Trends in Ecology and Evolution* 4:26–28.

Bharvirkar, R., B. Ehlers, K. Nelson, and L. Tonachel. 2003. *Reducing dissatisfaction with the economic impact of habitat acquisition policies in the Sacramento River Conservation Area*. Report. Sacramento, CA: CALFED Bay-Delta Program.

Bickel, U., and J. M. Dros. 2003. The impacts of soybean cultivation on Brazilian ecosystems. http://assets .panda.org/downloads/impactsofsoybean.pdf.

Biggs, R., B. Reyers, and R. J. Scholes. 2006. A biodiversity intactness score for South Africa. *South African Journal of Science* 102:277–283.

Biggs, R., and R. J. Scholes. 2002. Land-cover changes in South Africa 1911–1993. *South African Journal of Science* 98:420–24.

Biggs, R., and R. J. Scholes. 2006. South Africa. In *Our earth's changing land: An encyclopedia of land-use and land-cover change*, ed. H. J. Geist, 542–49. Westport, CT: Greenwood Press.

Biggs, R., E. Bohensky, P. Desanker, C. Fabricius, T. Lynam, A. Misselhorn, C. Musvoto, et al. 2004. *Nature supporting people: The Southern African Millennium Ecosystem Assessment*. Integrated report. Pretoria, South Africa: CSIR.

Bin Mohamed, M., and J. Lutzenberger. 1992. Eco-imperialism and bio-monopoly at the earth summit. *New Perspectives Quarterly* 9:56–58.

Binning, C., and M. Young. 1999. *Talking to the taxman about nature conservation: Proposals for the introduction of tax incentives for the protection of high conservation value native vegetation*. National R and D program on Rehabilitation, Management and Conservation of Remnant Vegetation. Research Report 4/99. CSIRO Wildlife and Ecology Environment, Australia.

Bishop, J. 2003. Payments for ecosystem services: A case of sustainable globalisation? Presentation to a congress on globalisation, localisation and tropical forest management in the 21st century, 22–23 October 2003, Roeterseiland, Amsterdam, The Netherlands.

Bishop, R. C. 1978. Endangered species and uncertainty: The economics of a safe minimum standard. *American Journal of Agricultural Economics* 60:10–18.

Bleek, W. H. I., and I. C. Lloyd. 1911. *Specimens of bushman folklore*. London: George Allen.

Bleibtrau, J. 1970. *The parable of the beast*. London: Paladin.

Blignaut, J. N. 2004a. Towards an economic development ethic. In *Sustainable options: Development lessons from applied environmental economics*, ed. J. N. Blignaut and M. P. de Wit, 408–28. Cape Town: UCT Press.

Blignaut, J. N. 2004b. Reflecting on economic growth. In *Sustainable options: Development lessons from applied environmental economics*, ed. J. N. Blignaut and M. P. de Wit, 32–52. Cape Town: UCT Press.

Blignaut, J. N., and M. P. de Wit. 2004. *Sustainable options: Development lessons from applied environmental economics*. Cape Town: UCT Press.

Bond, W. J., W. D. Stock, and M. T. Hoffman. 1994. Has the Karoo spread? A test for desertification using carbon isotopes from soils. *South African Journal of Science* 90:391–97.

Bonk, K. 1997. Real property tax assessments of native forest in Hawaii county. In *Koa: A decade of growth. Proceedings of the Hawaii Forest Industry Association's 1996 annual symposium*, ed. L. Ferentinos and D. O. Evans, 98–99. Honolulu: Hawaii Forest Industry Association.

Boshoff, A. F, G. I. H. Kerley, and R. M. Cowling. 2001. A pragmatic approach to estimating the distributions and spatial requirements of the medium- to large-sized mammals in the Cape Floristic Region, South Africa. *Diversity and Distributions* 7:29–43.

Boshoff, A. F., G. I. K. Kerley, R. M. Cowling, and S. L. Wilson. 2002. The potential distributions, and estimated spatial requirements, and population size, of the medium- to large-sized mammals in the planning domain of the Greater Addo National Park Project. *Koedoe* 45:85–116.

Boshoff, A. F., C. J. Vernon, and R. K. Brooke. 1983. Historical atlas of the diurnal raptors of the Cape Province (*Aves: Falconiformes*). *Annals of the Cape Provincial Museums* (Natural History) 14:173–297.

Botha, J., E. T. F. Witkowski, and C. Shackleton. 2001. An inventory of medicinal plants traded on the western boundary of the Kruger National Park, South Africa. *Koedoe* 44:7–22.

Botha, M. 2001. *Incentives for conservation on private land: Options and opportunities*. Summary Report 02/2001. Cape Town: Botanical Society, Cape Conservation Unit.

Boucher, C., and M. L. Jarman. 1977. The vegetation of the Langebaan area, South Africa. *Transactions of the Royal Society of South Africa* 42:241–72.

Bowen, D., M. DuPonte, C. Evensen, J. B. Friday, J. Uchida, and G. Uehara. 2000. *CTAHR's role in enhancing/sustaining Hawaii's natural resources and environment*. Honolulu: University of Hawaii, College of Tropical Agriculture and Human Resources.

Bowles, I., D. Downes, D. Clark, and M. Guerin-McManus. 1998. Economic incentives and legal tools for private sector conservation. *Duke Environment Law and Policy Forum* 8:209–43.

Boyer, K. E., and J. B. Zedler. 1996. Damage to cordgrass by scale insects in a constructed salt marsh: Effects of nitrogen additions. *Estuaries* 19:1–12.

Bradshaw, A. D. 1984. Land restoration: Now and in the future. *Proceedings of the Royal Society of London B* 223:1–23.

Bradshaw, A. D. 1987. Restoration: An acid test for ecology. In *Restoration ecology*, ed. W. R. I. Jordan, M. Gilpin, and J. D. Aber, 23–29. Cambridge: Cambridge University Press.

Bromley, D. W. 2000. The two realms of reason: Calculation and sentiment. Paper presented at the 7th Ulvön Conference on Environmental Economics, Sweden, 18–21 June.

Brown, G., and J. Roughgarden. 1995. An ecological economy: Notes on harvest and growth. In *Biodiversity loss: Ecological and economic issues*, ed. C. Perrings, K. G. Maler, C. Folke, C. S. Holling, and B. O. Jansson, 150–89. Cambridge, UK: Cambridge University Press.

Brown, J. H. 1995. *Macroecology*. Chicago: University of Chicago Press.

Brundtland Report. 1987. *Our common future*. World Commission of Environment and Development. Oxford, UK: Oxford University Press.

Bucher, E. 1987. Herbivory in arid and semi-arid regions of Argentina. *Revista Chilena de Historia Natural* 60:265–73.

Buckley, M. 2004. Strategic restoration: Game theory applied to the Sacramento River Conservation Area. In *Proceedings of the Sixteenth International Conference, Society for Ecological Restoration*. Victoria, Canada: Society for Ecological Restoration.

Buckley, M., and B. Haddad. 2006. Socially strategic ecological restoration: A game theoretic analysis. *Environmental Management* 38:48–61.

Buer, K., D. Forwalter, M. Kissel, and B. Stohler. 1989. The middle Sacramento River: Human impacts on physical and ecological processes along a meandering river. In *Proceedings of the California Riparian*

Systems Conference: Protection, management and restoration for the 1990s, ed. D. L. Abell, 22–32. USDA General Technical Report PSW-110.

Burke, A. 2001. Determining landscape function and ecosystem dynamics to contribute to ecological restoration in the southern Namib Desert. *Ambio* 30:29–36.

Burman, J., and S. Levin. 1974. *The Saldanha Bay story*. Cape Town: Human and Rousseau.

Burman, M. 2005. *Risks and decisions for conservation and environmental management*. Cambridge: Cambridge University Press.

Buza, L., A. Young, and P. Thrall. 2000. Genetic erosion, inbreeding and reduced fitness in fragments populations of the endangered tetraploid pea *Swainsona recta*. *Biological Conservation* 93:177–87.

Byrnes, R. M., ed. 1996. *South Africa: A country study*. Federal Research Division, United States Library of Congress. http://countrystudies.us/south-africa.

Cairns Jr., J. 1993. Ecological restoration: Replenishing our national and global ecological capital. In *Nature conservation 3: Reconstruction of fragmented ecosystems*, ed. D. Saunders, R. Hobbs, and P. Ehrlich, 193–208. Chipping Norton, New South Wales: Surrey Beatty and Sons.

Campbell, A., and G. Siepen. 1994. *Landcare: Communities shaping the land and their future*. St. Leonards, Australia: Allen and Unwin.

Campbell, B. M., S. Jeffery, W. Kozanayi, M. Lucket, M. Mutamba, and C. Zindi. 2002. *Household livelihoods in semi-arid regions: Options and constraints*. Jakarta, Indonesia: Center for International Forestry Research.

Campbell, B. M., and M. Luckert. 2001. Towards understanding the role of forests in rural livelihoods. In *Uncovering the hidden harvest: Valuation methods for woodland and forest resources*, ed. B. M. Campbell and M. Luckert, 1–17. London: Earthscan Publications.

Capistrano, D., C. K. Samper, M. J. Lee, and C. Raudsep-Hearne. 2005. *Ecosystems and human well-being: Multiscale assessments*. Vol. 4. *Findings of the sub-global assessments working group of the Millennium Ecosystem Assessment*. Washington, DC: Island Press.

Carey, A. B. 2003. Biocomplexity and restoration of biodiversity in temperate coniferous forest: Inducing spatial heterogeneity with variable-density thinning. *Forestry* 76:127–36.

Carey, A. B., J. Kershner, B. Biswell, and L. D. de Toledo. 1999. Ecological scale and forest development: Squirrels, dietary fungi, and vascular plants in managed and unmanaged forests. *Wildlife Monographs* 142:1–71.

Carey, A. B., B. R. Lippke, and J. Sessions. 1999. Intentional systems management: Managing forests for biodiversity. *Journal of Sustainable Forestry* 9:83–125.

Carlsson, L., and F. Berkes. 2005. Co-management: Concepts and methodological implications. *Journal of Environmental Management* 75:65–76.

Carret, J. C., and D. Loyer. 2003. *Madagascar Protected Area network sustainable financing: Economic analysis perspective*. World Parks Congress/Durban Workshop, Building Comprehensive Protected Areas Systems. World Bank, Washington, D.C., and Agence Française de Développement, Paris.

Carruthers, J. 1995. *The Kruger National Park: A social and political history*. Pietermaritzburg, South Africa: University of Natal Press.

Cashmore, P. B. 1995. Revegetation ecology on Tiritiri Matangi Island and its application to Motutapu Island. MS thesis, University of Auckland, Auckland, NZ.

Caunter, J., A. Cowperthwaite, and E. Gould. 2005. Polluter pays? *Resource Management Journal* 13:1–9.

Cavendish, W. 2000. Empirical regularities in the poverty-environment relationship of rural households: Evidence from Zimbabwe. *World Development* 28:1979–2003.

CCBA (Climate, Community and Biodiversity Alliance). 2004. *Climate, community and biodiversity project design standards* (Draft 1.0). Washington, DC: CCBA. http://www.fao.org/waicent/faoinfo/sustdev/EPdirect/EPan0005.htm.

CES (Coastal and Environmental Services). 1996a. Chemfos mine: Conceptual rehabilitation plan. Unpublished Report. Grahamstown, South Africa: Coastal and Environmental Services.

CES (Coastal and Environmental Services). 1996b. Experimental design for revegetation trials at Chemfos mine. Unpublished Report No. 96/18. Grahamstown, South Africa: Coastal and Environmental Services.

CES (Coastal and Environmental Services). 1997. Detailed work programme for Chemfos revegetation, winter 1997. Unpublished Report No. 97/17. Grahamstown, South Africa: Coastal and Environmental Services.

CES (Coastal and Environmental Services). 1999. Chemfos revegetation programme. Annual report for the 1998–1999 period. Unpublished Report No. 99/12. Grahamstown, South Africa: Coastal and Environmental Services.

CES (Coastal and Environmental Services). 2001. Chemfos closure report. Unpublished Report. Grahamstown, South Africa: Coastal and Environmental Services.

Cessford, G. R. 1995. *Conservation benefits of public visits to protected islands.* Wellington: New Zealand Department of Conservation.

Chiesura, A., and R. de Groot. 2003. Critical natural capital: A socio-cultural perspective. *Ecological Economics* 44:219–31.

Chokkalingam, U., H. Dotzauer, and H. Savenije. 2000. *Towards sustainable management and development of tropical secondary forests in Asia: The Samarinda proposal.* Bogor, Indonesia: CIFOR.

Chopra, K., R. Leemans, P. Kumar, and H. Simons. 2005. *Ecosystems and human well-being: Policy responses.* Vol. 3. *Findings of the Responses Working Group of the Millennium Ecosystem Assessment.* Washington, DC: Island Press.

Christopher, A. J. 1982. Towards a definition of the nineteenth century South African frontier. *South African Geographical Journal* 64:97–113.

CI (Conservation International). 2004. Ecosystem profile: Madagascar and Indian Ocean islands. Critical Ecosystems Partnership Fund, http://www.cepf.net/xp/cepf/where_we_work/madagascar/madagascar_info.xml.

Clewell, A. F. 2000. Editorial: Restoration of natural capital. *Restoration Ecology* 8:1.

Clewell, A. F., and J. Aronson. 2006. Motivations for restoration of ecosystems. *Conservation Biology* 20:420–28.

Clewell, A., and J. P. Rieger. 1997. What practitioners need from restoration ecologists. *Restoration Ecology* 5:350–54.

Coase, R. 1960. The problem of social cost. *The Journal of Law and Economics* 3:1–44.

Coates, D. J., and S. J. van Leeuwen. 1997. Delineating seed provenance areas for revegetation from patterns of genetic variation. In *Proceedings of the second Australian workshop on native seed biology for revegetation,* ed. S. M. Bellairs and J. M. Osborne, 3–14. Newcastle: Australian Centre for Minesite Rehabilitation Research.

Cocks, M. L., A. P. Dold, and I. M. Grundy. 2004. The trade in medicinal plants from forests in the Eastern Cape Province. In *Indigenous forests and woodlands in South Africa: People, policy and practice,* ed. M. Lawes, H. Eeley, C. Shackleton, and B. Geach, 337–66. Scottsville, South Africa: University of KwaZulu Natal Press.

Cocks, M. L., and K. F. Wiersum. 2003. The significance of plant diversity to rural households in eastern Cape Province of South Africa. *Forests, Trees and Livelihoods* 13:39–58.

Coddington, A. 1970. The economics of ecology. *New Society* 15:595–97.

Coetzee, J. H., and M. C. Middelmann. 1997. SWOT analysis of the fynbos industry in South Africa with special reference to research. *Acta Horticulturae* 453:145–52.

Coetzee, K. 2005. Caring for natural rangelands. Pietermaritzburg, South Africa: University of Natal Press.

Coetzee, M., J.-A. Chauvet, H. Goss, C. Meiklejohn, and L. Odendaal. 2002. *Integrated Development Planning (IDP): Local pathway to sustainable development in South Africa.* Pretoria, South Africa: CSIR.

Colby, B. G. 2000. Cap-and-trade policy challenges: A tale of three markets. *Land Economics* 76:638–58.

Conard, N. J. 2003. Hand axes on the landscape and the reconstruction of Paleolithic settlement patterns. In *Erkenntnisjäger - Kultur und Umwelt des frühen Menschen. Festschrift für Dietrich Mania, Band 1,* ed. J. M. Burdukiewicz, L. Fiedler, W.-D. Heinrich, A. Justus, and E. Brühl, 123–44. Halle (Saale), Germany: Landesamt für Archäologie Sachsen-Anhalt: Landesmuseum für Vorgeschichte.

Consiglio, T., G. E. Schatz, G. McPherson, P. P. Lowry II, J. Rabenantoandro, and Z. S. Rodgers. 2006. Deforestation and plant diversity of Madagascar's Littoral Forests. *Conservation Biology* 20:1799–1803.

Conti, G., and L. Fagarazzi. 2005. Forest expansion in mountain ecosystems: "Environmentalist's dream" or societal nightmare? Driving forces, aspects and impacts of one of the main 20th century's environmental, territorial and landscape transformations in Italy. *European Journal of Planning* 11:1–20.

Cornes, R., and T. Sandler. 1996. *The theory of externalities, public goods, and club goods.* 2nd ed. Cambridge: Cambridge University Press.

Costanza, R., and H. E. Daly. 1992. Natural capital and sustainable development. *Conservation Biology* 6:37–46.

Costanza, R., H. Daly, and J. Bartholomew. 1991. Goals, agenda and policy recommendations for ecological economics. In *Ecological economics: The science and management of sustainability*, ed. R. Costanza, 1–21. New York: Columbia University Press.

Costanza, R., R. d'Arge, R. De Groot, S. Farber, M. Grasso, B. Hannon, K. Limburg, et al. 1997. The value of the world's ecosystem services and natural capital. *Nature* 387:253–60.

Costanza, R., and C. H. Perrings. 1990. A flexible assurance bonding system for improved environmental management. *Ecological Economics* 2:57–76.

Cousins, B. 1999. Invisible capital: The contribution of communal rangelands to rural livelihoods in South Africa. *Development Southern Africa* 16:301–18.

Cowling, R. M. 1993. Ecotourism. What is it and what can it mean for conservation? *Veld and Flora* 79:3–5.

Cowling, R. M. 2005. The process of mainstreaming: Conditions, constraints and prospects. In *Mainstreaming biodiversity in production landscapes*, ed. C. Petersen and B. J. Huntley, 18–25. Washington, DC: Global Environment Facility.

Cowling, R. M., and C. Hilton-Taylor. 1994. Patterns of plant diversity and endemism in southern Africa: An overview. *Botanical Diversity in Southern Africa. Strelitzia* 1:31–52.

Cowling, R. M., and P. M. Holmes. 1992. Flora and vegetation. In *The ecology of fynbos. Nutrients, fire and diversity*, ed. R. M. Cowling, 23–61. Cape Town, South Africa: Oxford University Press.

Cowling, R. M., D. Kirkwood, J. J. Midgley, and S. M. Pierce. 1997. Invasion and persistence of bird-dispersed, subtropical thicket and forest species in fire-prone fynbos. *Journal of Vegetation Science* 8:475–88.

Cowling, R. M., S. M. Pierce, and T. Sandwith. 2002. Conclusions: The fundamentals of mainstreaming biodiversity. In *Mainstreaming biodiversity in development: Case studies from South Africa*, ed. S. M. Pierce, R. M. Cowling, T. Sandwith, and K. MacKinnon, 143–53. Washington, DC: World Bank.

Cox, P. A. 1983. Extinction of the Hawaiian avifauna resulted in a change of pollination for the ieie (*Freycinetia arborea*). *Oikos* 41:195–99.

Cox, P. A., and T. Elmquist. 2000. Pollinator extinction in the Pacific Islands. *Conservation Biology* 14:1237–39.

CPWild. 2003. *Sustainable utilization, commercialization and domestication of products from indigenous forest and woodland ecosystems*. Final report, CPWild Project. Stellenbosch, South Africa: Department of Forest Science.

CRA (California Resources Agency). 2000. *Sacramento River conservation area handbook*. Sacramento: Department of Water Resources.

Craig, J. L. 2006. Meta-populations with people. In *Proceedings of greening the city: Bringing biodiversity back into the urban environment conference*, ed. M. I. Dawson, 23–28. Christchurch, NZ: Royal New Zealand Institute of Horticulture and Lincoln University.

Craig, J., S. Anderson, M. Clout, R. Creese, N. Mitchell, J. Ogden, M. Roberts, and G. Ussher. 2000. Conservation issues in New Zealand. *Annual Review of Ecology and Systematics* 33:61–78.

Craig, J. L., and M. E. Douglas. 1986. Resource distribution, aggressive asymmetries and variable access to resources in the nectar feeding bellbird. *Behavioural Ecology and Sociobiology* 18:231–40.

Craig, J. L., N. D. Mitchell, B. Walter, R. Walter, M. Galbreith, and G. Chalmers. 1995. Communities restoring nature's networks: Tiritiri Matangi Island. In *Nature conservation 4: The role of networks*, ed. D. A. Saunders, J. L. Craig, and E. M. Mittisky, 534–41. Chipping Norton, New South Wales: Surrey Beatty and Sons.

Cramer, V. A., and R. J. Hobbs. 2002. Ecological consequences of altered hydrological regimes in fragmented ecosystems in southern Australia: Impacts and possible management responses. *Austral Ecology* 27:546–64.

Craven, A. 1994. 'n Evaluasie van Sommige Plantkundige Faktore wat Kleinwild-digthede in die Weskus Nationale Park Beinvloed. MS thesis, Stellenbosch University, Stellenbosch, South Africa.

CRC (Co-operative Research Centre) for plant-based management of dryland salinity. 2003. *Research portfolio 2002/3, Program 3: New and improved plant species*. University of Western Australia, Co-operative Research Centre Plant-based Management of Dryland Salinity.

Crossman, N. D., and B. A. Bryan. 2006. Systematic landscape restoration using integer programming. *Biological Conservation* 128:369–83.

Crutzen, P. 2002. Geology of mankind. *Nature* 415:23.

Crutzen, P. J., and E. F. Stoermer. 2000. The "Anthropocene." *Global Change Newsletter* 41:12–13.

CSIR (Council for Industrial Research). 1996. *National land-cover database*. Pretoria, South Africa: CSIR.

Cullen, R., E. Moran, and K. F. D. Hughey. 2005. Measuring the success and cost effectiveness of New Zealand multiple-species projects to the conservation of threatened species. *Ecological Economics* 53:311–23.

Cundill, G. 2005. Institutional change and ecosystem dynamics in the communal areas around Mt. Coke State Forest, Eastern Cape, South Africa. MS thesis, Rhodes University, Grahamstown, South Africa.

Cunningham, S. 2002. *The restoration economy—The greatest new growth frontier*. San Francisco: Berrett-Koehler Publishers.

Czech, B. 2003. Technological progress and biodiversity conservation: A dollar spent, a dollar burned. *Conservation Biology* 17:1455–57.

Czech, B., P. R. Krausman, and P. K. Devers. 2000. Economic associations among causes of species endangerment in the U.S. *Bioscience* 50:593–601.

Daily, G., ed. 1997. *Nature's services: Societal dependence on natural ecosystems*. Washington, DC: Island Press.

Daily, G. C., and K. Ellison. 2002. *The new economy of nature: The quest to make conservation profitable*. Washington, DC: Island Press.

Daly, H. 1990. Toward some operational principles of Sustainable Development. *Ecological Economics* 2:1–6.

Daly, H. 1992. Allocation, distribution and scale: Toward an economics that is efficient, just, and sustainable. *Ecological Economics* 6:185–93.

Daly, H. 1996. *Beyond growth: The economics of sustainable development*. Boston: Beacon Press.

Daly, H. E., and J. B. Cobb Jr. 1994. *For the common good: Redirecting the economy toward community, the environment, and a sustainable future*. 2nd ed. Boston: Beacon Press.

Daly, H. E., and J. Farley. 2004. *Ecological economics: Principles and applications*. Washington, DC: Island Press.

Danckwerts, J. E., and P. G. King. 1984. Conservative stocking or maximum profit: a management dilemma. *Journal of the Grassland Society of Southern Africa* 1:25–28.

Davies, R. A. G., and J. D. Skinner. 1986. Spatial utilisation of an enclosed area of the Karoo by springbok *Antidorcas marsupialis* and Merino sheep *Ovis aries* during drought. *Transactions of the Royal Society of South Africa* 46:115–32.

Davies, R. A. G., P. Botha, and J. D. Skinner. 1986. Diet selected by springbok *Antidorcas marsupialis* and Merino sheep *Ovis aries* during a Karoo drought. *Transactions of the Royal Society of South Africa* 46:165–76.

Dawson, N. 1994. The behavioural ecology and management of the takahe. MS thesis, University of Auckland, Auckland, NZ.

De Groot, R. S., M. Stuip, M. Finlayson, and N. Davidson. 2006. *Valuing wetlands: Guidance for valuing the benefits derived from wetland ecosystem services*. RAMSAR Technical Report No.3, CBD Technical Series No. 27. Gland, Switzerland: Ramsar Convention Secretariat.

De Groot, R., M. Wilson, and R. Boumans. 2002. A typology for the description, classification, and valuation of ecosystem functions, goods and services. *Ecological Economics* 41:393–408.

De Lange, J. H., and C. Boucher. 1990. Autecological studies on *Audouinia capitata* (Bruniaceae). P. 1. Plant-derived smoke as a seed germination cue. *South African Journal of Botany* 56:700–703.

De Wet, C. 1995. *Moving together, drifting apart: Betterment planning and villagisation in a South African homeland*. Johannesburg: Witwatersrand University Press

De Wit, M. P., D. J. Crookes, and B. W. Van Wilgen. 2001. Conflicts of interest in environmental management: Estimating the costs and benefits of a tree invasion. *Biological Invasions* 3:167–78.

Dean, W. R. J., and I. A. W. Macdonald. 1994. Historical changes in stocking rates of domestic livestock as a measure of semi-arid and arid rangeland degradation in the Cape Province, South Africa. *Journal of Arid Environments* 26:281–98.

Dean, W. R. J., and S. J. Milton. 2003. Did the flora match the fauna? Acocks and historical changes in Karoo biota. *South African Journal of Botany* 69:68–78.

DEAT (Department of Environmental Affairs and Tourism). 1997. *Environmental potential atlas*, map 8.3. Pretoria, South Africa: DEAT.

Demsetz, H. 1967. Toward a theory of property rights. *American Economic Review* 57:347–59.

Depommier, D. 2002. The tree behind the forest: Ecological and economic importance of traditional agro-forestry systems and multiple uses of trees in India. In *Traditional ecological knowledge, conservation of biodiversity and sustainable development*, ed. D. Depommier and P. S. Ramakrishnan, 42–58. Proceedings of the Indo-French colloquium, Institut Francais de Pondichery, India.

Diamond, J. 1997. *Guns, germs and steel—A short history of everybody for the last 13,000 years.* London: Chatto and Windus.

Diamond, J. 2004. Twilight at Easter. *New York Review of Books* 51: 25 March. http://forests.org/articles/reader.asp?linkid=30194.

Diamond, J. 2005. *Collapse*. New York: Viking Press.

Dixon, J. A., L. F. Scura, R. A. Carpenter, and P. B. Sherman. 1994. *Economic analysis of environmental impacts*. London: Earthscan Publications.

DME (Department of Mineral and Energy Affairs). 1992. *Aide-memoir for the preparation of environmental management programme reports for prospecting and mining*. Pretoria, South Africa: Department of Mineral and Energy Affairs.

Dobson, A. 1998. *Justice and the environment. Conceptions of environmental sustainability and theories of distributive justice*. Oxford, UK: Oxford University Press.

DoC/MfE (Department of Conservation/Ministry for the Environment). 2000. *The New Zealand biodiversity strategy: Our chance to turn the tide*. Wellington, NZ: Department of Conservation.

Dodds, S. H. 2004. The catchment care principle: A practical approach to achieving equity, ecosystem integrity and sustainable resource use. http://www.water.org.au/pubs/publ1_dodds.htm.

Donaldson, J. S. 2002. Biodiversity and conservation farming in the agricultural sector. In *Mainstreaming biodiversity in development. Case studies from South Africa*, ed. S. M. Pierce, R. M. Cowling, T. Sandwith, and K. MacKinnon, 43–56. Washington, DC: World Bank Environment Department.

Donnelly, D. 1992. The potential host range of three seed-feeding *Melanterius* spp. (Curculionidae), candidates for the biological control of Australian *Acacia* spp. and *Paraserianthes (Albizia) lophantha* in South Africa. *Phytophylactica* 24:163–67.

Donnelly, D. 1997. Proposal for biological control of *Acacia cyclops* at Langebaanweg. Unpublished report. Stellenbosch, South Africa: Plant Protection Research Institute.

Dorfman, R. 1997. On sustainability. In *Science with a human face*, ed. R. Dorfman and P. Rogers, 27–51. Cambridge: Harvard School of Public Health and Harvard University Press.

Dovers, S. 1999. Public policy and institutional R and D for natural resource management: Issues and directions for LWRRDC. In *Social, economic, legal, policy and institutional R and D for natural resources management: Issues and directions for LWRRDC*, ed. C. Mobbs and S. Dovers, 78–107. Canberra, ACT: Land and Water Resources Research and Development Corporation.

Downing, B. H. 1978. Environmental consequences of agricultural expansion in South Africa since 1850. *South African Journal of Science* 74:420–22.

Dresp, B. 2006. Restoring is believing. *Ecological Engineering* 28:11–13.

Drewes, S. E., and M. M. Horn. 2000. Study of the variation in active chemical compounds of *Ocotea bullata* (Stinkwood, Unukani) for medicinal use. Unpublished report, CPWild Project. Stellenbosch, South Africa: Department of Forest Science.

Driessen, P. 2003. *Eco-imperialism: Green power, black death*. Bellevue, WA: Free Enterprise Press.

Du Preez, H. M. J. 1988. *Die Vroë Blanke Geskiedenis van Saldanhabaai met Spesifieke Verwysing na die Twee Kompanjiesposte op die Saldanha-skereiland*. Kaapstad: Raad vir Nationale Gedenkwaardigheid.

Duchin, F., and G.-M. Lange. 1994. Strategies for environmentally sound economic development. In *Investing in natural capital*, ed. A. M. Jansson, M. Hammer, C. Folke, and R. Costanza, 250–65. Washington, DC: Island Press.

Dudash, M. R. 1990. Relative fitness of selfed and outcrossed progeny in a self-compatible, protandrous species, *Sabatia angularis* L. (Gentianaceae): A comparison in three environments. *Evolution* 44:1129–39.

Dudley, N. S. 1997. Development of silvicultural practices to promote growth and the quality of *Acacia koa*. In *Koa: A decade of growth: Proceedings of the Hawaii Forest Industry Association's 1996 annual symposium*, ed. L. Ferentinos and D. O. Evans, 45–46. Honolulu: Hawaii Forest Industry Association.

Dunster, J., and K. Dunster. 1996. *Dictionary of natural resource management*. Vancouver: University of British Columbia Press.

Economist. 2005. Are you being served? 23 April.

Efseaff, D. S., J. G. Silveira, F. T. Griggs, F. L. Thomas, and J. Carlon. 2003. Incorporating native grass planting into riparian restoration on the Sacramento River. In *California riparian systems: Processes and floodplain management, ecology, and restoration,* ed. P. M. Faber, 315–22. Proceedings of the 2001 riparian habitat and floodplains conference, Sacramento, CA: Riparian Habitat Joint Venture.

Ehrlich, P. 2000. *Human natures: Genes, cultures, and the human prospect.* Washington, DC: Island Press.

Ekins, P. 2003. Identifying critical natural capital. Conclusions about critical natural capital. *Ecological Economics* 44:277–92.

Ekins, P., C. Folke, and R. de Groot. 2003. Identifying critical natural capital. *Ecological Economics* 44:159–63.

Ekins, P. S., L. Simon, C. Deutsch, C. Folke, and R. de Groot. 2003. A framework for the practical application of the concepts of critical natural capital and strong sustainability. *Ecological Economics* 44:165–85.

Elevitch, C. R., K. M. Wilkinson, and J. B. Friday. 2006. *Acacia koa* (koa) and *Acacia koaia* (koai'a). Ver. 3 In *Traditional trees of Pacific Islands: Their culture, environment, and use,* ed. C. R. Elevitch. Holualoa, HI: Permanent Agriculture Resources. http://www.traditionaltree.org.

Ellis, F. 2000. *Rural livelihoods and diversity in developing countries.* Oxford, UK: Oxford University Press.

Elzinga, A. 1996. Some reflections on post-normal science. In *Culture, perceptions and environmental problems: Interscientific communication on environmental issues,* ed. M. Rolen, 32–46. Stockholm: Swedish Council for Planning and Coordination of Research.

Euston-Brown, D. I. W. 2000. The influence of vegetation type and fire severity, and their interaction, on catchment stability after fire: A case study from the Cape Peninsula, South Africa. Unpublished report. Department of Water Affairs and Forestry: Working for Water Programme.

Euston-Brown, D. I. W., S. Botha, and W. J. Bond. 2002. Influence of fire severity on post-fire vegetation recovery on the Cape Peninsula. Unpublished report. Department of Water Affairs and Forestry: Working for Water Programme.

Evans, L. T., and R. B. Knox. 1969. Environmental control of reproduction in *Themeda australis. Australian Journal of Botany* 17:375–89.

Ewell, J. J., and F. E. Putz. 2004. A place for alien species in ecosystem restoration. *Frontiers in Ecology and the Environment* 2:354–60.

Experts Workshop. 1991. Experts workshop on Community-based Natural Resources Management (CBNRM): Perspectives, experiences and policy issues. Los Banos, Laguna, 19–20 September.

Fabricius, C. 2003. *A social ecology policy for South African national parks.* Pretoria: South African National Parks.

Fabricius, C., M. Burger, and P. A. R. Hockey. 2003. Comparing biodiversity between protected areas and adjacent rangeland in xeric succulent thicket, South Africa: Arthropods and reptiles. *Journal of Applied Ecology* 40:392–403.

Fabricius, C., E. Koch, and H. Magome. 2001. *Community wildlife management in Southern Africa: Challenging the assumptions of eden.* Evaluating Eden Series, no. 6. London: IIED.

Fabricius, C., E. Koch, H. Magome, and S. D. Turner, eds. 2004. *Rights, resources and rural development: Community-based natural resource management in southern Africa.* London: Earthscan Publications.

Fabricius, C., B. Matsiliza, and J. Buckle. 2003. *Community-based natural resource management in Emalahleni and Mbashe: Eastern Cape planning process.* Pretoria, South Africa: GTZ Transform.

Fabricius, C., and D. McGarry. 2004. Frequently asked questions at Macubeni. Pretoria, South Africa: GTZ Transform.

Fairbanks, D. H. K., M. W. Thompson, D. E. Vink, T. S. Newby, H. M. Van den Berg, and D. A. Everard. 2000. The South African Land-cover Characteristics Database: A synopsis of the landscape. *South African Journal of Science* 96:69–82.

Falk, D. A. 2006. Process-centered restoration in a fire-adapted ponderosa pine forest. *Journal for Nature Conservation* 14:140–51.

FAO (Food and Agriculture Organization) 1993. *Forest resources assessment 1990—Tropical countries.* FAO forestry paper no. 112. Rome: FAO.

FAO (Food and Agriculture Organization). 1995. *Forest resources assessment 1990—Global synthesis.* FAO research paper no. 124. Rome: FAO.

FAO (Food and Agriculture Organization). 2001. *Global forest resources assessment 2000: Main report.* Rome: FAO.

FAO (Food and Agriculture Organization). 2005. *Global forest resources assessment 2005: Main report.* Rome: FAO. http://www.fao.org/forestry/foris/webview/forestry2/index.jsp?siteId=101&sitetreeId=16807 &langId=1&geoId=0.

Farley, J., and H. E. Daly. 2006. Natural capital: The limiting factor: A reply to Aronson, Blignaut, Milton and Clewell. *Ecological Engineering* 28:6–10.

Farley, J., J. Erickson, and H. E. Daly. 2005. *Ecological economics: A workbook for problem-based learning.* Washington, DC: Island Press.

Farnsworth, E., T. H. Tidrick, W. M. Smathers, and C. F. Jorda. 1983. A synthesis of ecological and economic theory toward more complete valuation of tropical moist forests. *International Journal of Environmental Studies* 21:11–28.

Fellizar, F. P. 1994. Achieving sustainable development through community based resource management. *Regional Development Dialogue* 15:201–17.

Fenster, C. D., and M. R. Dudash. 1994. Genetic considerations for plant population restoration and conservation. In *Restoration of endangered species*, ed. M. L. Bowles and C. J. Whelan, 34–62. Cambridge: Cambridge University Press.

Ferraro, P. J., and S. K. Pattanayak. 2006. Money for nothing? A call for empirical evaluation of biodiversity conservation investments. *PLoS Biology* 4:e105.

Figueroa, E., R. Asenjo, S. Valdés, and S. Praus. 2002. Definición de criterios y metodologías de valoración económica del daño ambiental. Study for the Inter-American Development Bank and the Council for the State Defense—Chile. Santiago, Chile.

Folke, C., M. Hammer, R. Costanza, and A.-M. Jansson. 1994. Investing in natural capital—why, what, and how? In *Investing in natural capital*, ed. A.-M. Jansson, M. Hammer, C. Folke, and R. Costanza, 1–20. Washington, DC: Island Press.

Forestry Commission. 2005. *Forestry Statistics.* Edinburgh: Forestry Commission.

Fowler, C. W., and L. Hobbs. 2003. Is humanity sustainable? *Proceedings of the Royal Society of London B: Biological Sciences* 270:2579–83.

Fox, J. B., J. E. Taylor, M. D. Fox, and C. Williams. 1997. Vegetation changes across edges of rainforest remnants. *Biological Conservation* 82:1–13.

Frederick, S., G. Loewenstein, and T. O'Donoghue. 2002. Time discounting and time preference: A critical review. *Journal of Economic Literature* 40:351–401.

Freudenberger, D. O., and J. C. Noble. 1997. Consumption, regulation and offtake: A landscape perspective on pastoralism. In *Landscape ecology function and management: Principles from Australia's rangelands*, ed. J. A. Ludwig, D. J. Tongway, D. O. Freudenberger, J. C. Noble, and K. C. Hodgkinson, 35–47. Melbourne, Australia: CSIRO Publishing.

Friday, J. B., C. Cabal, and J. Yanagida. 2000. *Financial analysis for tree farming in Hawaii.* Hilo, HI: College of Tropical Agriculture and Human Resources, Cooperative Extension Service.

Friedman, M. 1953. *Essays in positive economics.* Chicago: University of Chicago Press.

FSA (Farm Service Agency). 2005. Conservation reserve program. http://www.fsa.usda.gov/dafp/cepd/crp.htm.

Gade, D. W. 1996. Deforestation and its effects in highland Madagascar. *Mountain Research and Development* 16:101–16.

Gadgil, M., F. Berkes, and C. Folke. 1993. Indigenous knowledge for biodiversity conservation. *Ambio* 22:151–56.

Gangwar, A. K., and P. S. Ramakrishnan. 1989. Cultivation and use of lesser-known plants of food value by tribals in north-east India. *Agriculture, Ecosystem and Environment* 25:253–67.

Ganzhorn, J. U., S. M. Goodman, and A. Dehgan. 2003. Effects of forest fragmentation on small mammals and lemurs. In *The natural history of Madagascar*, ed. S. M. Goodman and J. P. Benstead, 1228–34. Chicago: University of Chicago Press.

Ganzhorn, J. U., P. P. Lowry II, G. E. Schatz, and S. Sommer. 2001. The biodiversity of Madagascar: One of the world's hottest hotspots on its way out. *Oryx* 35:346–48.

Ganzhorn, J. U., B. Rakotosamimanana, L. Hanna, J. Hough, L. Iyer, S. Olivieri, S. Rajaobelina, C. Rodstrom, and G. Tilkin. 1997. Priorities for Biodiversity Conservation in Madagascar. *Primate Report* 48:1–81.

Gascon, C., T. E. Lovejoy, R. O. Bierregaard, J. R. Malcolm, P. C. Stouffer, H. L. Vasconcelos, W. F. Laurance, B. Zimmerman, M. Tocher, and S. Borges. 1999. Matrix habitat and species richness in tropical forest remnants. *Biological Conservation* 91:223–29.

Geldenhuys, C. J. 1994. Bergwind fires and the location pattern of forest patches in the southern Cape landscape, South Africa. *Journal of Biogeography* 21:49–62.

Geldenhuys, C. J. 1996. Forest management systems to sustain resource use and biodiversity: Examples from the southern Cape, South Africa. In *The biodiversity of African plants*, ed. L. J. G. Van der Maesen, X. M. Van der Burgt, and J. M. Van Medenbach De Rooy, 317–22. Dordrecht: Kluwer Academic Publishers.

Geldenhuys, C. J. 1999a. Requirements for improved and sustainable use of forest biodiversity: Examples of multiple use of forests in South Africa. In *Proceedings forum biodiversity—Treasures in the world's forests*, ed. J. Poker, I. Stein, and U. Werder, 72–82. Schneverdingen, Germany: Alfred Toepfer Akademie für Naturschutz.

Geldenhuys,.C. J. 1999b. *Plots for long-term monitoring of growth and mortality in Weza and Ngome indigenous forests, KwaZulu-Natal: Report on plot establishment and first measurement, 1998.* Report No. FW-01/99. Pretoria, South Africa: Forestwood.

Geldenhuys, C. J. 2004. Bark harvesting for traditional medicine: From illegal resource degradation to participatory management. *Scandinavian Journal of Forest Research* 19:103–15.

Geldenhuys, C. J., and C. Delvaux. 2002. *Planting alternative resources of natural forest tree species for traditional medicine with seedlings collected from a* Pinus patula *stand, Nzimankulu Forest.* Report FW-04/02. Pretoria, South Africa: Forestwood.

Geldenhuys, C. J., P. J. Le Roux, and K. H. Cooper. 1986. Alien invasions in indigenous evergreen forest. In *The ecology and management of biological invasions in Southern Africa*, ed. I. A. W. Macdonald, F. J. Kruger, and A. A. Ferrar, 119–31. Cape Town: Oxford University Press.

Geldenhuys, C. J., and D. Rau. 2004. Experimental harvesting of bark from selected tree species in South Africa. Unpublished internal report, FRP-DFID R8305 project.

Geldenhuys, C. J., D. Rau, and L. Du Toit. 2002. Experimental bark harvesting from selected tree species in the Southern Cape forests—An interim report. Unpublished report, CPWild Project. Stellenbosch, South Africa: Department of Forestry Science.

George, M. R., J. R. Brown, and W. J. Clawson 1992. Application of non-equilbrium ecology to management of mediterranean grasslands. *Journal of Range Management* 45:436–40.

Georgescu-Roegen, N. 1971. *The entropy law and the economic process.* Cambridge: Harvard University Press.

Georgescu-Roegen, N. 1976. *Energy and economic myths: Institutional and analytical economic essays.* New York: Pergamon.

Georgescu-Roegen, N. 1979. Energy analysis and economic valuation. *Southern Journal of Economics* 45:1023–58.

Ginocchio, R. 2004. Mine closure: The Latin American case. Paper prepared for the Global Centre for Post Mining Regeneration. Review of Good Practice World-Wide.

Giradet, S. 2000. Tools for protecting endangered species. PhD thesis, University of Auckland, Auckland, NZ.

Gittinger, J. P. 1983. *Análisis económico de proyectos agrícolas.* 2nd ed. Madrid: Tecnos.

Gobster, P. H. 2006. Invasive species as ecological threat: Is restoration an alternative to fear-based management? *Ecological Restoration* 23:261–70.

Goldstein, J. H., G. C. Daily, J. B. Friday, P. A. Matson, R. L. Naylor, and P. Vitousek. 2006. Business strategies for conservation on private lands: Koa forestry as a case study. *Proceedings of the National Academy of Sciences* 103:10140–45.

Golet, G.'H., M. D. Roberts, R. A. Luster, G. Werner, E. W. Larsen, G. Unger, and G. G. White. 2006. Restoration of large alluvial rivers: A case study of the Sacramento River Project, California. *Environmental Management* 37:862–79.

Golluscio, R., J. Paruelo, and A. Deregibus. 1998. Sustainability and range management in the Patagonian steppes. *Ecología Austral* 8:265–84.

Golluscio, R., M. Román, A. Betelu, D. Rodano, A. Cesa, and A. Frey. 2000. Ganadería de subsistencia: ¿Preservar a los recursos forrajeros o preservar a los pobladores? Third international colloquium on land transformation. *Proceedings of the Association of Universities of the Montevideo Group*, August 2000, Florianópolis, Brazil.

Goodman, P. S., B. James, and L. Carlisle. 2002. Wildlife utilization: Its role in fostering biodiversity conservation in KwaZulu-Natal. In *Mainstreaming biodiversity in development. Case studies from South Africa*, ed. S. M. Pierce, R. M. Cowling, T. Sandwith, and K. MacKinnon, 21–31. Washington, DC: World Bank Environment Department, World Bank.

Gore, J. A., and F. D. Shields. 1995. Can large rivers be restored? *Bioscience* 45:142–52.

Gough, J. D., and J. C. Ward. 1996. Environmental decision-making and lake management. *Journal of Environmental Management* 48:1–15.

Gould, S. J. 1992. *Kropotkin was no crackpot. Bully for Brontosaurus.* New York: W. W. Norton.

Greig-Gran, M. 2000. *Fiscal incentives for biodiversity conservation: The ICMS Ecológico in Brazil.* Environmental Economics Programme Discussion Paper No. DP 00-01. Londres: Institut International pour l'Environnement et le Développement (IIED).

Greig-Gran, M. 2002. *Financial incentives for improved sustainability performance: The business case and the sustainability dividend.* MMSD Report 47. London: Earthscan Publications.

Grime, J. P. 1997. Biodiversity and ecosystem function: The debate deepens. *Science* 277:1260–61.

Groves, R. H., M. W. Hagon, and P. S. Ramakrishnan. 1982. Dormancy and germination of seed of eight populations of *Themeda australis. Australian Journal of Botany* 30:373–86.

Guelke, L. 1979. The white settlers. In *The shaping of South African society 1652–1820*, ed. R. Elphick and H. Giliomee, 1652–780. Cape Town: Longman.

Gunderson, L. H., and C. S. Holling, eds. 2002. *Panarchy: Understanding transformations in human and natural systems.* Washington, DC: Island Press.

Guralnick, L. J., P. A. Rorabaugh, and Z. Hanscom. 1984a. Influence of photoperiod and leaf age on Crassulacean acid metabolism in *Portulacaria afra* (L.) Jacq. *Plant Physiology* 75:454–57.

Guralnick, L. J., P. A. Rorabaugh, and Z. Hanscom. 1984b. Seasonal shifts of photosynthesis in *Portulacaria afra* (L.) Jacq. *Plant Physiology* 76:643–46.

Gutrich, J. J., and D. Donovan. 2001. *Environmental valuation and decision-making: Science as a tool in the management of Hawaiian watersheds.* Report to the U.S. Forest Service. Honolulu, HI: Institute of Pacific Islands Forestry, East-West Center.

Habermas, J. 1993. *Moral consciousness and communication action.* Cambridge: MIT Press.

Hacking, H. 2006. Colusa on offensive against new restoration projects. *Chico Enterprise-Record*, 1 December 2006.

Haites, E., and F. Yamin. 2000. The clean development mechanism: Proposals for its operation and governance. *Global Environmental Change, Pt. A: Human and Policy Dimensions* 10:27–45.

Hajkowicz, S., M. Young, S. Wheeler, D. Hatton MacDonald, and D. Young. 2000. *Supporting decisions: Understanding natural resource management assessment techniques.* A report to the Land and Water Resources Research and Development Corporation, Policy and Economic Research Unit, CSIRO Land and Water. Canberra: CSIRO Land and Water.

Halpern, B., H. Regan, H. Possingham, and M. McCarthy. 2006. Accounting for uncertainty in marine reserve design. *Ecology Letters* 9:2–11.

Hamilton, L. S., and S. C. Snedaker. 1984. *Handbook for mangrove area management.* Paris: UNESCO.

Hansen, A. J., and B. H. Walker. 1985. The dynamic landscape: Perturbation, biotic response, biotic patterns. *Bulletin of the South African Institute of Ecologists* 4:5–14.

Hardin, G. 1968. The tragedy of the commons. *Science* 162:1243–48.

Harris, J. A., R. J. Hobbs, E. Higgs, and J. Aronson. 2006. Ecological restoration and global climate change. *Restoration Ecology* 14:170–76.

Harris, J. A., and R. van Diggelen. 2006. Ecological restoration as a project for global society. In *Restoration ecology: The new frontier*, ed. J. van Andel and J. Aronson, 3–15. Oxford, UK: Blackwell Publishing.

Hassan, R. M. 2002. *Accounting for stock and flow values of woody land resources.* Pretoria, South Africa: CEEPA, University of Pretoria.

Hayek, F. A. 1945. The use of knowledge in society. *American Economic Review* 35:519–30.

Hayes, G. F., and K. D. Holl. 2003. Cattle grazing impacts on annual forbs and vegetation composition of mesic grasslands in California. *Conservation Biology* 17:1694–1702.

Hayman, D. L. 1960. The distribution and cytology of the chromosome races of *Themeda australis* in southern Australia. *Australian Journal of Botany* 8:58–68.

Heal, G. 1999. *Biosphere, markets and governments.* Working Paper 224. College Park, MD: Center for Institutional Reform and the Informal Sector.

Heilbroner, R. 1995. *Visions of the future: The distant past, yesterday, today, and tomorrow.* New York: Oxford University Press.

Heilbronner, R. L. 1985. *The making of economic society.* Upper Saddle River, NJ: Prentice-Hall.

Heller, M. A., and R. S. Eisenberg. 1998. Can patents deter innovation? The anticommons. *Biomedical Research Science* 280:698–701.

Henao, J., and Baanante, C. 2001 Nutrient depletion in the agricultural soils of Africa. In *The unfinished agenda: Perspectives on overcoming hunger, poverty, and environmental degradation,* ed. P. Pinstrup-Andersen and R. Pandya-Lorch, 159–63. Washington, DC: International Food Policy Research Institute.

Henderson, N., and I. Bateman. 1995. Empirical and public choice evidence for hyperbolic social discount rates and the implications for intergenerational discounting. *Environmental and Resource Economics* 5:413–23.

Hendey, Q. B. 1981a. Geological succession at Langebaanweg, Cape Province, and global events of the late Tertiary. *South African Journal of Science* 77:33–38.

Hendey, Q. B. 1981b. Palaeoecology of the late Tertiary fossil occurrences in "E" quarry, Langebaanweg, South Africa, and a reinterpretation of their geological context. *Annals of the South African Museum* 84:1–104.

Hendey, Q. B. 1982. *Langebaanweg: A record of past life.* Cape Town: South African Museum.

Herrick, J. E., and W. G. Whitford. 1999. Integrating soil processes into management: From micro-aggregates to macro-catchments. In *People and rangelands building the future,* ed. D. Eldridge and D. Freudenberger, 91–95. Proceedings of the 6th International Rangeland Congress, Townsville, Australia.

Higgins, S. I., J. Turpie, R. Costanza, R. M. Cowling, D. C. Le Maitre, C. Marais, and G. F. Midgley. 1997. An ecological economic simulation model of mountain fynbos ecosystems. Dynamics, valuation and management. *Ecological Economics* 22:155–69.

Hobbs, R. J., V. A. Cramer, and L. J. Kristjanson. 2003. What happens if we cannot fix it? Triage, palliative care and setting priorities in salinising landscapes. *Australian Journal of Botany* 51:647–53.

Hobbs, R. J., and J. A. Harris. 2001. Restoration ecology: Repairing the earth's ecosystems in the new millennium. *Restoration Ecology* 9:239–46.

Hobbs, R. J., and D. A. Norton. 1996. Towards a conceptual framework for restoration ecology. *Restoration Ecology* 4:93–110.

Hodgkinson, K. C., and J. A. Quinn. 1976. Adaptive variability in the growth of *Danthonia caespitosa* Gaud populations at different temperatures. *Australian Journal of Botany* 24:381–96.

Hodgkinson, K. C., and J. A. Quinn. 1978. Environmental and genetic control of reproduction in *Danthonia caespitosa* populations. *Australian Journal of Botany* 26:351–64.

Hoffman, M. T., and A. Ashwell. 2001. *Nature divided. Land degradation in South Africa.* Cape Town: University of Cape Town Press.

Hoffman, M. T., and R. M. Cowling. 1990a. Vegetation change in the semi-arid Karoo over the last 200 years: An expanding Karoo—fact or fiction? *South African Journal of Science* 86:286–94.

Hoffman, M. T., and R. M. Cowling. 1990b. Desertification in the lower Sundays River Valley, South Africa. *Journal of Arid Environments* 19:105–17.

Hoffman, T. 1997. Human impacts on vegetation. In *Vegetation of southern Africa,* ed. R. M. Cowling, D. M. Richardson, and S. M. Pierce, 507–34. Cambridge: Cambridge University Press.

Holl, K. D. 2002. Tropical moist forest. In *Handbook of ecological restoration,* ed. M. Perrow and A. J. Davy, 539–58. Cambridge: Cambridge University Press.

Holl, K. D., and E. E. Crone. 2004. Local vs. landscape factors affecting restoration of riparian understory plants. *Journal of Applied Ecology* 41:922–33.

Holl, K. D., E. E. Crone, and C. B. Schultz. 2003. Landscape restoration: Moving from generalities to methodologies. *BioScience* 53:491–502.

Holl, K. D., and R. B. Howarth. 2000. Paying for restoration. *Restoration Ecology* 8:260–67.

Holling, C. S. 1978. *Adaptive environmental assessment and management.* Chichester, UK: John Wiley.

Holling, C. S. 1986. The resilience of terrestrial ecosystems: Local surprise and global change. In *Sustainable development of one biosphere,* ed. W. C. Clark and R. E. Munn, 292–317. Cambridge: Cambridge University Press.

Holling, C. S. 2001. Understanding the complexity of economic, ecological, and social systems. *Ecosystems* 4:390–405.

Holling, C. S., L. H. Gunderson, and G. D. Peterson. 2002. Sustainability and panarchies. In *Panarch: Understanding transformations in human and natural systems*, ed. L. H. Gunderson and C. S. Holling, 63–102. Washington, DC: Island Press.

Holloway, L. 2000. Catalysing rainforest restoration in Madagascar. In *Diversité et endémismeàà Madagascar*, ed. W. R. Lorenço and S. M. Goodman, 115–24. Paris: Mémoires de la Société de Biogéographie, Orstom.

Holloway, L. L. 2003. Ecosystem restoration and rehabilitation in Madagascar. In *The natural history of Madagascar*, ed. S. M. Goodman and J. P. Benstead, 1444–51. Chicago: University of Chicago Press.

Holmes, P. M. 2001. Shrubland restoration following woody alien invasion and mining: Effects of topsoil depth, seed source and fertilizer addition. *Restoration Ecology* 9:71–84.

Holmes, P. M. 2002. Depth distribution and composition of seed-banks in alien-invaded and uninvaded fynbos vegetation. *Austral Ecology* 27:110–20.

Holmes, P. M., and R. M. Cowling. 1993. Effects of shade on seedling growth, morphology and leaf photosynthesis in six subtropical thicket species from the Eastern Cape, South Africa. *Forest Ecology and Management* 61:199–220.

Holmes, P. M., and R. M. Cowling.1997a. The effects of invasion by *Acacia saligna* on the guild structure and regeneration capabilities of South African fynbos shrublands. *Journal of Applied Ecology* 34:317–32.

Holmes, P. M., and R. M. Cowling. 1997b. Diversity, composition and guild structure relationships between soil-stored seed banks and mature vegetation in alien plant-invaded South African fynbos shrublands. *Plant Ecology* 133:107–22.

Holmes, P. M., and W. Foden. 2001. The effectiveness of post-fire soil disturbance in restoring fynbos after alien clearance. *South African Journal of Botany* 67:533–39.

Holmes, P. M., and C. Marais. 2000. Impacts of alien plant clearance on vegetation in the mountain catchments of the Western Cape. *Southern African Forestry Journal* 189:113–17.

Holmes, P. M., and D. M. Richardson. 1999. Protocols for restoration based on knowledge of recruitment dynamics, community structure and ecosystem function: Perspectives from South African fynbos. *Restoration Ecology* 7:215–30.

Holmes, P. M., D. M. Richardson, B. W. Van Wilgen, and C. Gelderblom. 2000. Recovery of South African fynbos vegetation following alien woody plant clearing and fire: Implications for restoration. *Austral Ecology* 25:631–39.

Homewood, K. 2005. Rural resource use and local livelihoods in Sub-Saharan Africa. In *Rural resources and local livelihoods in Africa*, ed. K. Homewood, 1–10. New York: Palgrave Macmillian.

Honnay, O., B. Bossuyt, K. Verheyen, J. Butaye, H. Jacquemyn, and M. Hermy. 2002. Ecological perspectives for the restoration of plant communities in European temperate forests. *Biodiversity and Conservation* 11:213–42.

Horneck, G., and H. J. Schellnhuber. 2004. Destiny of humankind from an astrobiology point of view: Does astrobiology provide an exit option for terrestrial mismanagement? In *Earth system analysis for sustainability*, ed. H. J. Schellnhuber, P. J. Crutzen, W. C. Clark, M. Claussen, and H. Held, 91–110. Cambridge: MIT Press.

Hossner, L. R., H. Shahandeh, and J. A. Birkhead. 1997. The impact of acid forming materials on plant growth on reclaimed minesoil. *Journal of Soil and Water Conservation* 52:118–25.

Hufford, K. M., and S. J. Mazer. 2003. Plant ecotypes: Genetic differentiation in the age of ecological restoration. *Trends in Ecology and Evolution* 18:147–55.

Hulme, D., and M. W. Murphree. 2001. Community conservation as policy: Promise and performance. In *African wildlife and livelihoods: The promise and performance of community conservation*, ed. D. Hulme and M. W. Murphree, 280–97. Oxford, UK: James Currey.

Hunt, J. 2003. What does and does not work in native woodland restoration. In *Restoring borders woodland: The vision and the task*, ed. D. Kohn, 78–89. Occasional paper, Monteviot Nurseries, Ancrum, Jedburgh, Scotland: Wildwood Group of Borders Forest Trust.

Hunter, J. C., K. B. Willett, M. C. McCly, J. F. Quinn, and K. E. Keller. 1999. Prospects for preservation and restoration of riparian forests in the Sacramento River valley, CA, USA. *Environmental Management* 24:65–75.

Hutchings, J. A. 2000. Collapse and recovery of marine fishes. *Nature* 406:882–85.

Hutton, J. M., and N. Leader-Williams. 2003. Sustainable use and incentive-driven conservation: Realigning human and conservation interests. *Oryx* 37:215–26.

IBRD (International Bank for Reconstruction and Development). 1992. *World development report 1992: Development and the environment*. New York: Oxford University Press.

IFPRI (International Food Policy Research Institute). 2000. Global study reveals new warning signals: Degraded agricultural lands threaten world's food production capacity. Press release, 21 May 2000. Washington, DC: International Food Policy Research Institute.

IIED (International Institute for Environment and Development). 1994. *Whose eden? An overview of community wildlife management*. London: IIED.

iKhwezi. 2003. *Land use plan for the Macubeni catchment, Emalahleni district, Eastern Cape*. Pretoria, South Africa: GTZ Transform.

INDEC (Instituto Nacional de Estadísticas y Censos). 1988. Censo nacional agropecuario de 1988, resultados generales, características básicas, N° 9, provincia del Chubut. Buenos Aires: INDEC.

Infield, M., and W. A. Adams. 1999. Institutional sustainability and community conservation: A case study from Uganda. *Journal of International Development* 11:305–15.

IPCC (International Panel on Climate Change). 2001. *Climate change 2001: The scientific basis*, ed. J. T. Houghton, Y. Ding, D. J. Griggs, M. Noguer, P. J. van der Linden, X. Dai, K. Maskell, and C. A. Johnson. Contribution of working group I to the third assessment report of the intergovernmental panel on climate change. Cambridge: Cambridge University Press.

IUCN and UNEP (The World Conservation Union and United Nations Environment Programme). 2003. *World database on protected areas*. The World Conservation Union and United Nations Environment Programme.

Jaffe, A. B., and J. Lerner. 2004. *Innovation and its discontents: How our broken patent system is endangering innovation and progress and what to do about it*. Princeton: Princeton University Press.

Jansson, A. M., M. Hammer, C. Folke, and R. Costanza, eds. 1994. *Investing in natural capital: The ecological economics approach to sustainability*. Washington, DC: Island Press.

Janzen, D. H. 2002. Tropical dry forest: Area de Conservacion Guancaste, northwestern Costa Rica. In *Handbook of ecological restoration*, ed. M. R. Perrow and A. J. Davy, 559–8. Cambridge: Cambridge University Press.

Jeanrenaud, S. 2001. Communities and forest management in Western Europe. In *People, forests and policies*, 88–91. Gland, Switzerland: IUCN.

Jenkins, M., S. J. Sherr, and M. Inbar. 2004. Markets for biodiversity services. *Environment* 46:32–42.

Jobbágy, E., and O. Sala. 2000. Controls of grass and shrubs aboveground production in the Patagonian steppe. *Ecological Applications* 10:541–49.

Johnson, B. E., and J. H. Cushman. 2007. Influence of a large herbivore reintroduction on plant invasions and community composition in a California grassland. *Conservation Biology* 21:515–526.

Johnson, C. F., R. M. Cowling, and P. B. Phillipson. 1999. The flora of the Addo Elephant National Park, South Africa: Are threatened species vulnerable to elephant damage? *Biodiversity and Conservation* 8:1447–56.

Johnson, N., A. White, and D. Perrot-Maitre. 2001. *Developing markets for water services from forests: Issues and lessons for innovators*. Washington, DC: The Katoomba Group and Forest Trends.

Jones, L. B. 2005. *Colusa subreach planning project landowner survey*. Report for the Sacramento River Conservation Area Forum. Sacramento: Institute for Social Research, California State University.

Jones, R. 2000. Behavioural ecology and habitat requirements of kokako (*Callaeas cinerea willsoni*) on Tiritiri Matangi Island. MS thesis, University of Auckland, Auckland, NZ.

Jones, T. A., and D. A. Johnson. 1998. Integrating genetic concepts into planning rangeland seedings. *Journal of Range Management* 51:594–606.

Jordan, W. R. 2003. *The Sunflower Forest: Ecological restoration and the new communion with nature*. Berkeley: University of California Press.

Kahn Jr., P. H. 2002. Children's affiliations with nature: Structure, development, and the problem of environmental generational amnesia. In *Children and nature: Psychological, sociocultural, and evolutionary investigations*, ed. P. H. Kahn Jr. and S. R. Kellert, 93–116. Cambridge: MIT Press.

Kangas, P. C. 2004. *Ecological engineering: Principles and practices*. Boca Raton, FL: Lewis Publishers.

Kant, I. 1956. *Critique of the practical reason*. New York: Bobbs-Merrill.

Karfs, R. 2002. Rangeland monitoring in tropical savanna grasslands, Northern Territory, Australia: Relationships between temporal satellite data and ground data. MS thesis, Research School of Tropical Environment Studies and Geography, James Cook University, Townsville, Australia.

Katibah, E. F. 1984. A brief history of riparian forests in the Central Valley of California. In *California riparian systems: Ecology, conservation, and productive management*, ed. R. E. Warner and K. M. Hendrix, 23–29. Berkeley: University of California Press.

Katoomba Group Ecosystem Marketplace. 2005. http://ecosystemmarketplace.com/pages/marketwatch.

Katoomba Group Ecosystem Marketplace. 2006. http://ecosystemmarketplace.com.

Kaval, P. 2004. *The Maungatautari Ecological Island Trust: An economic analysis*. Report prepared for Environment Waikato, NZ.

Kaufman, R. 1995. The economic multiplier of environmental life support: Can capital substitute for a degraded environment? *Ecological Economics* 12:67–79.

Kaul, I., I. Grunberg, and M. A. Stem, eds. 1999. *Global public goods: International cooperation in the 21st century*. New York: Oxford University Press for United Nations Development Program.

Kelley, R. 1989. *Battling the inland sea: Floods, public policy, and the Sacramento Valley*. Berkeley: University of California Press.

Kerley, G. I. H., A. F. Boshoff, and M. H. Knight. 1999. Ecosystem integrity and sustainable land use in the Thicket Biome, South Africa. *Ecosystem Health* 5:104–9.

Kerley, G. I. H., A. F. Boshoff, and M. H. Knight. 2002. The Greater Addo National Park, South Africa: Biodiversity conservation as the basis for a healthy ecosystem and human development opportunities. In *Managing for healthy ecosystems*, ed. D. J. Rapport, W. L. Lasley, D. E. Rolston, N. O. Nielsen, C. O. Qualset, and A. B. Damania, 359–74. Boca Raton: CRC Press.

Kerley, G. I. H., M. H. Knight, and M. De Kock. 1995. Desertification of subtropical thicket in the Eastern Cape, South Africa: Are there alternatives? *Environmental Monitoring and Assessment* 37:211–30.

Kessler, J. J., and E. Wakker. 2000. *Forest conversion and the edible oils sector*. Research Paper prepared for WWF (World Wide Fund for Nature) –Switzerland. Zurich: WWF.

Khuzwayo, N. 2002. Improving livelihoods for people in parks. *Afra News* 53:4–6.

Kirkpatrick, J. B., K. McDougall, and M. Hyde. 1995. *Australia's most threatened ecosystem: The south-eastern lowland native grasslands*. Chipping Norton, New South Wales: Surrey Beatty and Sons.

Knapp, E. E., and K. J. Rice. 1994. Starting from seed—Genetic issues in using native grasses for restoration. *Restoration and Management Notes* 12:40–45.

Knight, A. T., R. M. Cowling, and B. M. Campbell. 2006. Planning for implementation: An operational model for conservation planning. *Conservation Biology* 20:549–61.

Knowles, T. 2005. Developing the carbon trading economy through the rehabilitation of subtropical thicket, Baviaanskloof Mega-Reserve. Unpublished report for the Working for Water Program, Cape Town: Department of Water Affairs and Forestry.

Kokot, D. F. 1948. *An investigation into the evidence bearing on recent climatic changes over southern Africa*. Irrigation Department Memoir, 1–160. Pretoria, South Africa: Government Printer.

Kothyari, B. P., K. S. Rao, K. G. Saxena, T. Kumar, and P. S. Ramakrishnan. 1991. Institutional approaches in development and transfer of water-harvest technology in the Himalaya. In *Advance in water resources technology*, ed. G. Tsakiris, 673–78. Rotterdam: A Balkerma.

Kremen, C. 2003. The Masoala Peninsula. In *The natural history of Madagascar*, ed. S. M. Goodman and J. P. Benstead, 1459–66. Chicago: University of Chicago Press.

Kremen, C. 2005. Managing ecosystem services: What do we need to know about their ecology? *Ecology Letters* 8:468–79.

Kremen, C., and R. S. Ostfeld. 2005. A call to ecologists: Measuring, analysing and managing ecosystem services. *Frontiers in Ecology and Environment* 3:540–48.

Kremen, C., N. M. Williams, R. L. Bugg, J. P. Fay, and R. W. Thorp. 2004. The area requirements of an ecosystem service: Crop pollination by native bee communities in California. *Ecology Letters* 7:1109–19

Kropotkin, P. 1902. *Mutual aid: A factor of evolution*. New York: McClure, Phillips and Co. Also available on http://dwardmac.pitzer.edu/Anarchist_archives/kropotkin/mutaidch1.html#NOTES and from Black Rose Books (New Ed edition) (December 1988).

Kudish, M. 2000. *The Catskill forest: A history*. Fleischmanns, NY: Purple Mountain Press.

Kura, Y., C. Revenga, E. Hoshino, and G. Mock. 2004. *Fishing for answers: Making sense of the global fish crisis*. Washington, DC: World Resources Institute.

Lal, R., G. F. Wilson, and B. N. Okigbo. 1979. Changes in properties of an Alfísol produced by various cover crops. *Soil Science* 127:377–82.

Lamb, D. 1998. Large-scale ecological restoration of degraded tropical forest lands: The potential role of timber plantations. *Restoration Ecology* 6:271–79.

Lamb, D., and M. Tomlinson. 1994. Forest rehabilitation in the Asia-Pacific region: Past lessons and present uncertainties. *Journal of Tropical Forest Science* 7:157–70.

Landell-Mills, N., and I. T. Porras. 2002. *Silver bullet or fools' gold? A global review of markets for forest environmental services and their impact on the poor*. London: IIED.

Lane, R. 2000. *The loss of happiness in market democracies*. New Haven: Yale University Press.

Langholz, J., J. Lassoie, and J. Schelhas. 2000. Incentives for biological conservation: Costa Rica's private wildlife refuge program. *Conservation Biology* 14:1735–43.

Lanly, J.-P. 1982. *Tropical forest resources*. Forestry paper no. 50. Rome: FAO (Food and Agriculture Organization).

Larsen, E. J., and C. Marais. 2001. The effect of alien vegetation on water resources planning for the town of George. Unpublished report for the Working for Water Programme, Cape Town: Department of Water Affairs and Forestry.

Lavelle, P. 1997. Faunal activities and soil processes: Adaptive strategies that determine ecosystem function. *Advances in Ecological Research* 27:93–132.

Lavelle, P., D. Bignell, M. Austen, P. Giller, G. Brown, V. Behan-Pelletier, V. Garey, et al. 2004. Vulnerability of ecosystem services at different scales: Role of biodiversity and implications for management. In *Sustaining biodiversity and functioning in soils and sediments*, ed. D. H. Wall, 193–224. Washington, DC: Island Press.

Lavelle, P., R. Dugdale, A. Asefaw Behere, E. Carpenter, A. M. Izac, D. Karl, J. Lemoalle, F. Luizao, P. Treguer, and B. Ward. 2005. Nutrient cycling. In *Millennium Ecosystem Assessment*, vol 1., 331–53. Washington, DC: Island Press.

Lavelle, P., and A. V. Spain. 2001. *Soil ecology*. Dordrecht, The Netherlands: Kluwer Academic Publishers.

Lechmere-Oertel, R. G., G. I. H. Kerley, and R. M. Cowling. 2005a. Patterns and implications of transformation in semi-arid succulent thicket, South Africa. *Journal of Arid Environments* 62:459–74.

Lechmere-Oertel, R. G., G. I. H. Kerley, and R. M. Cowling. 2005b. Landscape dysfunction and reduced spatial heterogeneity in soil resources and fertility in semi-arid succulent thicket, South Africa. *Austral Ecology* 30:615–24.

Lélé, S., and R. B. Norgaard. 2005. Practicing interdisciplinarity. *BioScience* 55:967–75.

Lenton, T. M., K. G. Caldiera, S. A. Franck, G. Horneck, A. Jolly, E. Rabbow, H. J. Schellnhuber, et al. 2004. Long-term geosphere-biosphere coevolution and astrobiology. In *Earth system analysis for sustainability*, ed. H. J. Schellnhuber, P. J. Crutzen, W. C. Clark, M. Claussen, and H. Held, 111–40. Cambridge: MIT Press.

León, R. J. C., and M. R. Aguiar. 1985. El deterioro por uso pasturil en estepas herbáceas patagónicas. *Phytocoenologia* 13:181–96.

Leone, D. 2002. Arrests in koa thefts. *Honolulu Star Bulletin, Hawaii News*, 2 February.

Levin, S. A. 1992. The problem of pattern and scale in ecology. *Ecology* 73:1943–67.

Lewis III, R. R. 2005. Ecological engineering for successful management and restoration of mangrove forests. *Ecological Engineering* 24:403–18.

Li, W. H. 2004. Degradation and restoration of forest ecosystems in China. *Forest Ecology and Management* 201:33–41.

Limpopo Government. 2002. *Gross geographic product and employment data on a district level for the Limpopo province: 2002–2007*. Polokwane: Limpopo Government.

Lindblom, C. E. 1959. The science of muddling through. *Public Administration Review* 19:79–88.

Lindsay, K. A. 2004. The sustainability of natural area tourism: A case study of Tiritiri Matangi Island. MS thesis, University of Auckland, Auckland, NZ.

Lindsey, P. A. 2003. Conserving wild dogs (*Lycaon pictus*) outside state protected areas in South Africa: Ecological, sociological and economic determinants of success. PhD thesis, University of Pretoria, Pretoria, South Africa.

Lingle, L. 2006. *Report of findings for the establishment of the Hawaii experimental tropical forest.* Honolulu: Department of Land and Natural Resources.

Living Planet Report. 2005. http://www.panda.org/news_facts/publications/key_publications/living_planet _report/index.cfm.

Lloyd, J. W., E. C. van den Berg, and A. R. Palmer. 2002. *Patterns of transformation and degradation in the thicket biome, South Africa.* Report no. 39. Port Elizabeth: Terrestrial Ecology Research Unit, University of Port Elizabeth.

Lockwood, J. L., and S. L. Pimm. 1999. When does restoration succeed? In *Ecological assembly rules,* ed. E. Weiher and P. Keddy, 363–92. Cambridge: Cambridge University Press.

Loudat, T. A., and R. Kanter. 1997. The economics of commercial koa culture. In *Koa: A decade of growth. Proceedings of the Hawaii Forest Industry Association's 1996 annual symposium,* ed. L. Ferentinos and D. O. Evans, 124–47. Honolulu: Hawaii Forest Industry Association.

Loureiro, W., and R. P. Rolim de Moura. 1996. Ecological icms (tax over circulation of goods and services): A successful experience in Brazil. *Workshop of the 4th global biodiversity forum,* 30 August–1 September. http://www.biodiversityeconomics.org/pdf/960830-08.pdf.

Low, B., R. Costanza, E. Ostrom, J. Wilson, and C. P. Simon. 1999. Human-ecosystem interactions: A dynamic integrated model. *Ecological Economics* 31:227–42.

Lowry, P. P. II, G. E. Schatz, and P. B. Phillipson. 1997. The classification of natural and anthropogenic vegetation in Madagascar. In *Natural change and human impact in Madagascar,* ed. S. M. Goodman and B. D. Patterson, 93–123. Washington, DC: Smithsonian Books.

Lubke, R. A., A. M. Avis, A. Heydenrych, and J. D. van Eeden. 1998. Rehabilitation of and coastal environments following mining on the Cape west coast, South Africa. Paper presented at 2nd international conference on restoration ecology, Groningen, The Netherlands, 25–30 August. Abstract only.

Lubke, R. A., and J. D. van Eeden. 2001. Rehabilitation methodology in the restoration of Strandveld on Chemfos mine (West Coast Fossil Park), near Langebaan. In *Conference on environmentally responsible mining in southern Africa,* vol. 2, ed. Chamber of Mines, 44–53. Johannesburg: Chamber of Mines.

Luckert, M. L., and B. M. Campbell. 2003. Seeking livelihood improvements from rangelands in southern Africa: Can we get there from here? In *Proceedings of the 7th international rangelands congress.* Durban, South Africa, 26 July–1 August, ed. N. Allsopp, A. R. Palmer, S. J. Milton, K. P. Kirkman, G. I. H. Kerley, C. R. Hurt, and C. J. Brown, 1628–34. Irene, South Africa: Document Transformation Technologies.

Ludwig, J. A., R. W. Eager, G. N. Bastin, V. H. Chewings, and A. C. Liedloff. 2002. A leakiness index for assessing landscape function using remote sensing. *Landscape Ecology* 17:157–71.

Ludwig, J. A., R. W. Eager, R. J. Williams, and L. M. Lowe. 1999. Declines in vegetation patches, plant diversity and grasshopper diversity near cattle watering points in the Victoria River District, northern Australia. *Rangeland Journal* 21:135–49.

Ludwig, D., R. Hilborn, and C. Walters. 1993. Uncertainty, resource exploitation, and conservation: Lessons from history. *Science* 260:17.

Ludwig, J. A., and D. J. Tongway. 1995. Spatial organisation of landscapes and its function in semi-arid woodlands, Australia. *Landscape Ecology* 10:51–63.

Ludwig, J. A., and D. J. Tongway. 1997. A landscape approach to rangeland ecology. In *Landscape ecology, function and management: Principles from Australia's rangelands,* ed. J. A. Ludwig, D. J. Tongway, D. O. Freudenberger, J. C. Noble, and K. C. Hodgkinson, 1–12. Melbourne, Australia: CSIRO Publishing.

Ludwig, J. A., and D. J. Tongway. 2000. Viewing rangelands as landscape systems. In *Rangeland desertification,* ed. O. Arnalds and S. Archer, 39–52. Dordrecht, The Netherlands: Kluwer Academic Publishers.

Ludwig, J. A., and D. J. Tongway. 2002. Clearing savannas for use as rangelands in Queensland: Altered landscapes and water erosion processes. *Rangeland Journal* 24:83–95.

Ludwig, J. A., D. J. Tongway, G. N. Bastin, and C. James. 2004. Monitoring ecological indicators of rangeland functional integrity and their relationship to biodiversity at local to regional scales. *Austral Ecology* 29:108–20.

Ludwig, J. A., J. A. Wiens, and D. J. Tongway. 2000. A scaling rule for landscape patches and how it applies to conserving soil resources in savannas. *Ecosystems* 3:84–97.

Ludwig, J. A., B. P. Wilcox, D. D. Breshears, D. J. Tongway, and A. C. Imeson. 2005. Vegetation patches and runoff-erosion as interacting ecohydrological processes in semiarid landscapes. *Ecology* 86:288–97.

Luken, J. O. 1990. *Directing ecological succession.* London: Chapman and Hall.

MA (Millennium Ecosystem Assessment). 2003. *Ecosystem and human well-being: A framework for assessment.* Washington, DC: Island Press

MA (Millennium Ecosystem Assessment). 2005a. *Ecosystems and human well-being: Synthesis.* Millennium Ecosystem Assessment Series. Washington, DC: Island Press and World Resources Institute.

MA (Millennium Ecosystem Assessment). 2005b. *Ecosystems and human well-being: Biodiversity synthesis.* Millennium Ecosystem Assessment Series. Washington, DC: Island Press and World Resources Institute.

MA (Millennium Ecosystem Assessment). 2005c. *Ecosystems and human well-being: Scenarios.* Vol 2. Synthesis Report Series. Washington, DC: Island Press.

MA (Millennium Ecosystem Assessment). 2005d. *Ecosystems and human well-being: Desertification synthesis.* Millennium Ecosystem Assessment Series. Washington, DC: Island Press and World Resources Institute.

MA (Millennium Ecosystem Assessment). 2005e. *Ecosystems and human well-being: Multiscale assessments.* Vol 4. Synthesis Report Series. Washington, DC: Island Press.

MA (Millennium Ecosystem Assessment). 2005f. *Ecosystems and human well-being.* Synthesis Report Series. Washington, DC: Island Press.

Ma, Q. 2004. Appraisal of tree planting options to control desertification: Experiences from Three-North Shelterbelt Programme. *International Forestry Review* 6:327–34.

Macdonald, I. A. W. 1992. Vertebrate populations as indicators of environmental change in southern Africa. *Transactions of the Royal Society of South Africa* 48:87–122.

MacMahon, J. A. 1987. Disturbed land and ecological theory: An essay about a mutualistic association. In *Restoration ecology,* ed. W. R. Jordan, M. E. Gilpin, and J. D. Aber, 221–37. Cambridge: Cambridge University Press.

MacMahon, J. A., and K. D. Holl. 2001. Ecological restoration: A key to conservation biology's future. In *Conservation biology: Research priorities for the next decade,* ed. M. E Soulé and G. H. Orians, 245–69. Washington, DC: Island Press.

Madzwamuse, M., and C. Fabricius. 2004. Local ecological knowledge and the Basarwa in the Okavango Delta: The case of Xaxaba, Ngamiland District. In *Rights, resources and rural development: Community-based natural resource management in southern Africa,* ed. C. Fabricius, E. Koch, H. Magome, and S. D. Turner, 160–73. London: Earthscan Publications.

Magale-Macandog, D. B. 1994. Patterns and processes in population divergence of *Microlaena stipoides* (Labill.) R. Br. PhD thesis, University of New England, Armidale, Australia.

Magome, H., and C. Fabricius. 2004. Reconciling biodiversity conservation with rural development: The Holy Grail of community-based natural resource management? In *Rights, resources and rural development: Community-based natural resource management in southern Africa,* ed. C. Fabricius, E. Koch, H. Magome, and S. D. Turner, 93–114. London: Earthscan Publications.

Maikhuri, R. K., R. L. Semwal, K. S. Rao, and K. G. Saxena. 1997. Rehabilitation of degraded community lands for sustainable development in Himalaya, India. *International Journal of Sustainable Development and World Ecology* 4:192–203.

Mander, M. 1998. *Marketing of indigenous medicinal plants in South Africa.* Rome: FAO.

Mander, M., N. Steytler, and N. Diederichs. 2006. The economics of medicinal plant cultivation. In *Commercialising medicinal plants: A southern African guide,* ed. N. Diederichs, 43–52. Stellenbosch, South Africa: African Sun Media.

Manona, C. 1998. The decline in significance of agriculture in the former Ciskei community: A case study. In *Communal rangelands in Southern Africa: A synthesis of knowledge,* ed. T. D. de Bruyn and F. Scogings, 113–18. Alice, South Africa, University of Fort Hare.

Margules, C. R., and R. L. Pressey. 2000. Systematic conservation planning. *Nature* 405:243–53.

Markgraf, V. 1985. Late Pleistocene faunal extinctions in southern Patagonia. *Science* 228:1110–12.

Marshall, G. 1998. *Economics of cost-sharing for agri-environmental conservation.* Paper prepared for the project LPM2, Investment Programs for Effective Natural Resource Management, funded by the Land and Water Resources Research and Development Corporation. Canberra, ACT: Water Resources Research and Development Corporation.

Martin, G. 2002. Tract backtracking; farmers wary of Sacramento River habitat plan. *San Francisco Chronicle,* 17 April.

Mascia, M. B., J. P. Brosius, T. A. Dobson, B. C. Forbes, L. Horowitz, M. A. McKean, and N. J. Turner. 2003. Conservation and the social sciences. *Conservation Biology* 17:649–50.

Matsumura, M. 1994. Coercive conservation, defensive reaction, and the commons tragedy in northeast Thailand. *Habitat International* 18:105–15.

Mattison, E. H. A., and K. Norris. 2005. Bridging the gaps between agricultural policy, land-use and biodiversity. *Trends Ecology and Evolution* 20:610–16.

Max-Neef, M. 1989. *Human scale development: An option for the future.* Development dialogue reprint from 1989 Santiago de Chile: CEPAUR, Dag Hamarskjöld Foundation.

Max-Neef, M. 1995. Economic growth and quality of life: A threshold hypothesis. *Ecological Economics* 15:115–18.

Max-Neef, M. A. 2005. Foundations of transdisciplinarity. *Ecological Economics* 53:5–16.

May, P., F. V. Neto, V. Denardin, and W. Loureiro. 2002. Using fiscal instruments to encourage conservation: Municipal responses to the ecological value-added tax in Paraná and Minas Gerais, Brazil. In *Selling forest environmental services: Market-based mechanisms for conversation and development,* ed. S. Pagiola, J. Bishop, and N. Landell-Mills, 173–99. Sterling, VA: Earthscan Publications.

May, R. M. 1977. Thresholds and breakpoints in ecosystems with a multiplicity of stable states. *Nature* 269:471–77.

McAllister, D. M. 1980. *Evaluation in environmental planning: Assessing environmental, social, economic and political trade-offs.* Cambridge: MIT Press.

McGarry, D. 2004. Use of indigenous forests at Nqabara on the Eastern Cape's wild coast. BS honours dissertation, Department of Environmental Science, Rhodes University, Grahamstown, South Africa.

McIntyre, S., and D. J. Tongway. 2005. Grassland structure in native pastures: Links to soil surface condition. *Ecological Management and Restoration* 6:45–50.

McIver, J., and L. Starr. 2001. Restoration of degraded lands in the interior Columbia River Basin: Passive vs. active approaches. *Forest Management* 153:15–28.

McKenzie, B., ed. 1988. *Guidelines for the sustained use of forests and forest products.* Workshop Report, Forest Biome Project. Pretoria, South Africa: FRD, CSIR.

McKibben, W. 1989. *The end of nature.* New York: Random House.

McNeeley, J. A., ed. 2001. *The great reshuffling: Human dimensions of invasive alien species.* Gland, Switzerland: IUCN.

McNeely, J. A. 1994. Lessons from the past: Forests and biodiversity. *Biodiversity and Conservation* 3:3–20.

Meadows, D. H., D. L. Meadows, J. Randers, and W. W. Behrens. 1972. *Limits to growth. A report to The Club of Rome.* London: Earth Island.

Mech, T., and M. D. Young. 2001. *Designing voluntary environmental management arrangements to improve natural resource management in agriculture and allied rural industries.* Canberra, ACT: Rural Industries Research and Development Corporation.

Meets, M., and C. Boucher. 2004. The determination of the concentration of aqueous smoke solutions used in seed germination. *South African Journal of Botany* 70:313–18.

Mehaffey, M. H., M. S. Nash, T. G. Wade, C. M. Edmonds, D. W. Ebert, K. B. Jones, and A. Rager. 2001. *A landscape assessment of the Catskill/Delaware watersheds 1975–1998.* Washington, DC: United States Environmental Protection Agency (USEPA).

Menard, T. A. 2001. Activity patterns of the Hawaiian hoary bat (*Lasiurus cinereus semotus*) in relation to reproductive time periods. MS thesis, University of Hawaii, Manoa, Honolulu.

Mentis, M. T. 1999. Diagnosis of the rehabilitation of opencast coal mines and the highveld of South Africa. *South African Journal of Science* 95:210–15.

MfE (Ministry for the Environment). 1997. *Guidelines for assessing and managing contaminated gasworks sites in New Zealand.* Wellington, NZ: Ministry for the Environment.

Michaelowa, A., M. Stronzik, F. Eckermann, and A. Hunt. 2003. Transaction costs of the Kyoto mechanisms. *Climate Policy* 3:261–78.

Midgley, J. J. 1991. Valley bushveld dynamics and tree euphorbias. In *Proceedings of the first valley bushveld/subtropical thicket symposium,* ed. P. J. K. Zacharias, G. C. Stuart-Hill, and J. J. Midgley, 8–9. Special Publication. Pietermaritzburg: Grassland Society of Southern Africa.

Midgley, J. J., A. Seydack, D. Reynell, and D. McKelly. 1990. Fine-grain pattern in southern Cape plateau forests. *Journal of Vegetation Science* 1:539–46.

Millar, C. I., and W. J. Libby. 1989. Disneyland or native ecosystem: Genetics and the restorationist. *Restoration and Management Notes* 7:18–24.

Miller, J. R. 2005. Biodiversity conservation and the extinction of experience. *Trends in Ecology and Evolution* 20:430–34.

Mills, A. J., and R. M. Cowling. 2006. Rate of carbon sequestration at two thicket restoration sites in the Eastern Cape, South Africa. *Restoration Ecology* 14:38–49.

Mills, A. J., R. M. Cowling, M. V. Fey, G. I. H. Kerley, J. S. Donaldson, R. G. Lechmere-Oertel, A. M. Sigwela, A. L. Skowno, and P. Rundel. 2005. Effects of goat pastoralism on ecosystem carbon storage in semi-arid thicket, Eastern Cape, South Africa. *Austral Ecology* 30:797–804.

Mills, A. J., and M. V. Fey. 2004. Transformation of thicket to savanna reduces soil quality in the Eastern Cape, South Africa. *Plant and Soil* 265:153–63.

Milton, S. J. 2001. Rethinking ecological rehabilitation in arid and winter rainfall southern Africa. *South African Journal of Science* 97:47–48.

Milton, S. J. 2003. "Emerging ecosystems"—a washing-stone for ecologists, economists and sociologists? *South African Journal of Science* 99:404–6.

Milton, S. J., W. R. J. Dean, M. A. du Plessis, and W. R. Siegfried. 1994. A conceptual model of arid rangeland degradation: The escalating cost of declining productivity. *BioScience* 44:70–76.

Milton, S. J., W. R. J. Dean, and D. M. Richardson. 2003. Economic incentives for restoring natural capital: Trends in southern African rangelands. *Frontiers in Ecology and the Environment* 1:247–54.

Mitchell, N. D. 1985. The revegetation of Tiritiri Matangi Island: The creation of an open sanctuary. *Journal of the Royal New Zealand Institute of Horticulture* 13:36–41.

Mitsch, W. J., J. W. Day Jr., L. Zhang, and R. Lane. 2005. Nitrate–nitrogen retention in wetlands in the Mississippi River Basin. *Ecological Engineering* 24:267–78.

Mitsch, W. J., and S. E. Jørgensen. 2004. *Ecological engineering and ecosystem restoration*. Hoboken, NJ: John Wiley.

MMSD-AS (Mining, Mineral and Sustainable Development-America del Sur). 2002. *Minería, minerales y desarrollo sustentable en América del Sur*. CIPMA, International Development Research Centre. *Equipo MMSD-América del Sur*, CIPMA, IDRC/IIPM.

MOA (Watershed Memorandum of Agreement). 1997. Rules and regulations for the protection from contamination, degradation and pollution of the New York City water supply and its sources. http://www.nysefc.org/home/index.asp?page=294.

Mollison, B. 1988. *Permaculture: A designer's manual*. Tyalgum, NSW: Tagari Publishing.

Monod, J. 1971. *Change and necessity*. New York: Alfred A. Knopf.

Moody, M. E., and R. N. Mack. 1988. Controlling the spread of plant invasions: The importance of nascent foci. *Journal of Applied Ecology* 25:1009–21.

Moolman, H. J., and R. M. Cowling.1994. The impact of elephant and goat grazing on the endemic flora of South African succulent thicket. *Biological Conservation* 67:53–59.

Morrison, M. L. 2002. *Wildlife restoration*. Washington, DC: Island Press.

Mortimer, R. 1993. An economic analysis of Tiritiri Matangi: An application of contingent valuation. Unpublished report.

Mortlock, W. 1999. *Native seed in Australia: A survey of collection, storage and distribution of native seed for revegetation and conservation purposes*. Canberra, ACT: FloraBank Project, Greening Australia.

Mortlock, W. 2000. Local seed for revegetation—Where will all that seed come from? *Ecological Management and Restoration* 1:93–101.

Moss, A. J., and C. L. Watson. 1991. Rain-impact soil crust, 3. Effects of continuous and flawed crusts on infiltration and the ability of plant covers to maintain crusts. *Australian Journal of Soil Research* 29:311–30.

Mulongoy, K. J., and S. Chape, eds. 2004. *Protected areas and biodiversity: An overview of key issues*. Montreal: CBD Secretariat and Cambridge, UK: UNEP-WCMC.

Myers, N. 1990. The biodiversity challenge: Expanded hot-spots analysis. *Environmentalist* 10:243–56.

Myers, N. 1993. Biodiversity and the precautionary principle. *Ambio* 22:74–79.

Nabhan, G. P. 2001. Nectar trails of migratory pollinators: Restoring corridors on private lands. *Conservation Biology in Practice* 2:21–27.

Nakicenovic, N., and R. Swart. 2001. *Special report on emissions scenarios*. A report of the Intergovernmental Panel on Climate Change. Oxford, UK: Cambridge University Press.

National Primary Drinking Water Regulations. 2002. 40 C.F.R. § 141.71(b)(2). http://ecfr.gpoaccess
.gov/cgi/t/text/text-idx?c=ecfr&sid=5f67b5e59dea4cbc04fbdce335ea65e4&rgn=div8&view=text&node
=40:22.0.1.1.3.8.16.2&idno=40.

Nelson, G. C., E. Bennett, A. A. Berhe, K. G. Cassman, R. DeFries, T. Dietz, A. Dobso, et al. 2005. Drivers
of change in ecosystem condition and service. In *Ecosystems and human well-being: Scenarios*, vol. 2,
ed. S. R. Carpenter, P. L. Pingali, E. M. Bennett, and M. B. Zurek, 173–222. Washington, DC: Island
Press.

NEPED (Nagaland Environment Protection & Economic Development) and IRRR (Institute for Regional
and Rural Research). 1999. *Building upon traditional agriculture in Nagaland*. Nagaland, India: Nagaland Environmental Protection, and Silang, Phillippines: Economic Development and International Institute of Rural Reconstruction.

Netshiluvhi, T. R., and R. Scholes. 2001. *Allometry of South African woodlands trees*. Research report no.
ENV-P-I 2001-007. Pretoria, South Africa: CSIR.

Neumayer, E. 1999. *Weak versus strong sustainability*. Cheltenham, U.K.: Edward Elgar.

New York City Government website. 2006. Wastewater. http://www.nyc.gov/html/dep/watershed/html/
wastewater.html.

Newton, A. C., and P. Ashmole. 1998. How may native woodlands be restored to southern Scotland. *Scottish
Forestry* 52:168–71.

NHT (Natural Heritage Trust). 2002. *Australian government envirofund helpful hints for completing your envirofund round one 2004–2005*. http://www.nht.gov.au/envirofund/index.html.

Nicholls, H. 2004. The conservation business. *PloS Biology* 2:1256–59.

Norgaard, R. B. 1985. Environmental economics: An evolutionary critique and a plea for pluralism. *Journal
of Environmental Economics and Management* 12:382–94.

Norton, B. G., and M. A. Toman. 1997. Sustainability: Ecological and economic perspectives. *Land Economics* 73:553–68.

Norton, B. G., and R. E. Ulanowicz. 1992. Scale and biodiversity policy: A hierarchical approach. *Ambio*
21:244–49.

Novellie, P. A., and H. Bezuidenhout. 1994. The influence of rainfall and grazing on vegetation changes in
the Mountain Zebra National Park. *South African Journal of Wildlife Research* 24:60–71.

NRC (National Research Council). 1992. *Restoration of aquatic ecosystems*. Washington, DC: National
Academy Press.

NRC (National Research Council). 2000. *Watershed management for potable water supply: Assessing the
New York City strategy*. Washington, DC: National Academy Press.

NRC (National Research Council). 2001. *Marine protected areas: Tools for sustaining ocean ecosystem*.
Washington, DC: National Academy Press.

NRC (National Research Council). 2003. *NEON: Addressing the nation's environmental challenges*. Report
of the Committee on the National Ecological Observatory Network, Board of Life Sciences. Washington, DC: National Academy Press.

Nunes, P. A. L. D., J. C. J. M. Van den Berg, and P. Nijkamp. 2003. *The ecological economics of biodiversity*.
Cheltenham, UK: Edward Elgar.

O'Connor, M. 2000. *Natural capital*. EVE Policy Research Brief No. 3, Cambridge Research for the Environment. Cambridge: Department of Land Economy, University of Cambridge.

OECD (Organisation for Economic Co-operation and Development). 1994. *Project and policy appraisal:
Integrating economics and environment*. Brussels: OECD.

Oesterheld, M., and M. Oyarzábal. 2004. Grass-to-grass protection from grazing in a semi-arid steppe. Facilitation, competition, and mass effect. *Oikos* 107:576–82.

Olson, M. 1965. *The logic of collective action*. Cambridge: Harvard University Press.

Olsson, P., C. Folke, and F. Berkes. 2004. Adaptive co-management for building resilience in social-ecological systems. *Environmental Management* 34:75–90.

Olukoye, G. A., W. N. Wamicha, J. I. Kinyamario, J. I. Mwanje, and J. W. Wakhungu. 2003. Community
participation in wildlife management: Experiences, issues and concerns from northern rangelands of
Kenya. In *Proceedings of the 7th international rangelands congress. 26 July–1 August*, ed. N. Allsopp,
A. R. Palmer, S. J. Milton, K. P. Kirkman, G. I. H. Kerley, C. R. Hurt, and C. J. Brown, 1791–97. Durban, South Africa.

O'Riordan, T., and J. Cameron, eds. 1994. *Interpreting the precautionary principle.* London: Earthscan Publications.

Orr, D. W. 2002. *The nature of design: Ecology, culture and human intention.* New York: Oxford University Press.

Ostrom, E., J. Burger, C. B. Field, R. B. Norgaard, and D. Policansky. 1999. Revisiting the commons: Local lessons, global challenges. *Science* 284:278–82.

Page, T. 1977. *Conservation and economic efficiency.* Baltimore: Johns Hopkins University Press for Resources for the Future.

Pagiola, S., J. Bishop, and N. Landell-Mills, eds. 2002. *Selling forest environmental services: Market-based mechanisms for conservation and development.* Sterling, UK: Earthscan Publications.

Pahl-Wostl, C. 2006. The importance of social learning in restoring the multifunctionality of rivers and floodplains. *Ecology and Society* 11:10.

Palmer, M., E. Bernhardt, E. Chornesky, S. Collins, A. Dobson, C. Duke, B. Gold, et al. 2004. Ecology for a crowded planet. *Science* 304:1251–52.

Palmer, R., H. Timmerman, and D. Fay. 2002. *From conflict to negotiation: Nature-based development on South Africa's wild coast.* Pretoria, South Africa: Human Sciences Research Council.

Panayatou, T. 1995. Environmental degradation at different stages of economic development. In *Beyond Rio: The environmental crisis and sustainable livelihoods in the Third World,* ed. I. Ahmed and J. A. Doeleman, 13–36. London: Macmillan.

Pannell, D. J. 2004. Someone has to pay . . . Any volunteers? Voluntary versus regulatory approaches to environmental protection in agricultural landscapes of Australia. Unpublished paper presented to Department of Environment and Heritage. Canberra, ACT: Department of Environment and Heritage. http://www.general.uwa.edu.au/u/dpannell/dp0406.htm.

Parker, D. P. 2002. *Cost-effective strategies for conserving private land. An economic analysis for land trusts and policy makers.* Bozeman, Montana: Property and Environment Research Center. http://www.perc.org/pdf/land_trusts_02.pdf.

Parrotta, J. A. 2001. Restoration forestry for multiple objectives. In *Proceedings of the international symposium on tropical forestry research challenges in the New Millennium,* ed. R. V. Varma, K. V. Bhat, E. M. Muralidharan, and J. K. Sharma. Peechi, India: Kerala Forest Research Institute.

Parson, E. A. 2003. *Protecting the ozone layer: Science and strategy.* New York: Oxford University Press.

Patnaik, S. 2003. Conservation and development potential of non-timber forest products. In *Methodological issues in mountain research: A socio-ecological systems approach,* ed. P. S. Ramakrishnan, K. G. Saxena, S. Patnaik, and S. Singh, 199–217. New Delhi: UNESCO and Oxford: IBH.

Pauly, D., J. Alder, A. Bakun, S. Heileman, K. H. Kock, P. Mace, W. Perrin, et al. 2006. Marine fisheries systems. In *Ecosystems and human well-being: Current state and trends,* vol. 1, ed. R. Hassan, R. J. Scholes, and N. Ash, 477–511. Washington, DC: Island Press.

Pawson, E., and T. Brooking, eds. 2003. *Environmental histories of New Zealand.* Auckland, NZ: Oxford University Press.

PCE (Parliamentary Commissioner for the Environment). 2001. *Weaving resilience into our working lands.* Wellington, NZ: Parliamentary Commissioner for the Environment.

PCE (Parliamentary Commissioner for the Environment). 2002. Creating our future: Sustainable development in New Zealand. Wellington, NZ: Parliamentary Commissioner for the Environment.

PCE (Parliamentary Commissioner for the Environment). 2004. *Growing for good: Intensive farming, sustainability and New Zealand's environment.* Wellington, NZ: Parliamentary Commissioner for the Environment.

Pearce, D. W., and R. K. Turner. 1991. *Economics of natural resources and the environment.* Baltimore: Johns Hopkins University Press.

Pejchar, L., and D. Press. 2006. Achieving conservation objectives through production forestry: The case of *Acacia koa* on Hawaii Island. *Environmental Science and Policy* 9:439–447.

Pejchar, L., K. D. Holl, and J. L. Lockwood. 2005. Hawaiian honeycreeper home range size varies with habitat: Implications for native *Acacia koa* forestry. *Ecological Applications* 15:1053–61.

Penn, D. J. 2003. The evolutionary roots of our environmental problems: Toward a Darwinian ecology. *Quarterly Review of Biology* 78:275–301.

Perelman, S. B., R. J. C. León, and J. P. Bussacca. 1997. Floristic changes related to grazing intensity in a Patagonian shrub steppe. *Ecography* 20:400–406.

Perrin, M. 2003. *Incentives for forest landscape restoration: Maximising benefits for forests and people.* WWF discussion paper. Gland, Switzerland: WWF.

Perrot-Maitre, D., and P. Davis. 2001. *Case studies: Developing markets for water services from forests.* Washington, DC: Forest Trends. http://www.forest-trends.org.

Peterken, G. 1998. Woodland composition and structure. In *Native woodland restoration in southern Scotland: Principles and practice,* ed. A. C. Newton and P. Ashmole, 22–33. Occasional paper no. 2. Monteviot Nurseries, Ancrum, Jedburgh, Scotland: Wildwood Group of Borders Forest Trust.

Peterson, C., and B. J. Huntley, eds. 2005a. *Mainstreaming biodiversity in production landscapes.* Washington, DC: Global Environment Facility.

Peterson, C., and B. J. Huntley. 2005b. What is mainstreaming biodiversity? In *Mainstreaming biodiversity in production landscapes,* ed. C. Petersen and B. J. Huntley, 2–11. Washington, DC: Global Environment Facility.

Pezzey, J., and M. Toman. 2002. *The economics of sustainability: A review of journal articles.* Resources for the Future Discussion Paper 02-03. Washington, DC: Resources for the Future.

Pierce, S. M., R. M. Cowling, A. T. Knight, A. T. Lombard, M. Rouget, and T. Wolf. 2005. Systematic conservation planning products for land-use planning: Interpretation for implementation. *Biological Conservation* 125:441–58.

Pierce, S. M., R. M. Cowling, T. Sandwith, and K. MacKinnon, eds. 2002. *Mainstreaming biodiversity in development: Case studies from South Africa.* Washington, DC: World Bank.

Pimentel, D., L. Lach, R. Zuniga, and D. Morrison. 2000. Environmental and economic costs of nonindigenous species in the United States. *BioScience* 50:53–65.

Polanyi, K. 1944. *The great transformation.* Boston: Beacon Press.

Popper, D. E., and F. J. Popper. 1987. The Great Plains: From dust to dust—A daring proposal for dealing with an inevitable disaster. *Planning* 53:12–18.

Postel, S., and B. D. Richter. 2003. *Rivers for life: Managing water for people and nature.* Washington, DC: Island Press.

Potts, B. M., R. C. Barbour, A. B. Hingston, and R. E. Vaillancourt. 2003. Genetic pollution of native eucalypt gene pools—Identifying the risks. *Australian Journal of Botany* 51:1–25.

Powell, I., and A. White. 2001. *A conceptual framework for developing markets and market-based instruments for environmental services of forests.* Paper prepared for Developing Markets for Environmental Services of Forests, 4 October, Vancouver, BC.

Prescott-Allen, R. 2001. *The wellbeing of nations: A country-by-country index of quality of life and the environment.* Washington, DC: Island Press.

Pressey, R. L. 1994. Ad hoc reservations: Forward or backward steps in developing representative reserve systems? *Conservation Biology* 8:662–68.

Pressey, R. L., R. M. Cowling, and M. Rouget. 2003. Formulating conservation targets for biodiversity pattern and process in the Cape Floristic Region, South Africa. *Biological Conservation* 112:99–127.

Preston-Whyte, R. A., and P. D. Tyson. 1988. *Atmosphere and weather of southern Africa.* Cape Town: Oxford University Press.

Programme Dette Nature. 2003. *Rapport annuel du projet CAF.* Andravory, Madagascar: WWF.

Prosser, I. P., I. D. Rutherford, J. M. Olley, W. J. Young, P. J. Wallbrink, and C. J. Moran. 2001. Large-scale patterns of erosion and sediment transport in river networks, with examples from Australia. *Marine and Freshwater Research* 52:81–99.

Pyle, R. M. 1993. The extinction of experience. *Horticulture* 56:64–67.

Ramakrishnan, P. S. 1991. *Ecology of biological invasion in the tropics.* New Delhi: National Institute of Ecology, International Scientific Publications.

Ramakrishnan, P. S. 1992a. *Shifting agriculture and sustainable development: An interdisciplinary study from North-Eastern India.* Paris: UNESCO-MAB Series and Carnforth, Lancashire, UK: Parthenon Publishers. Republished by New Delhi: Oxford University Press, 1993.

Ramakrishnan, P. S. 1992b. Tropical forests: Exploitation, conservation and management. *Environment and Development Impact (UNESCO)* 42:149–62.

Ramakrishnan, P. S. 2001. *Ecology and sustainable development.* New Delhi: National Book Trust.

Ramakrishnan, P. S. 2003. Jhum/shifting agriculture (slash and burn agriculture). In *Encyclopedia of soil science,* ed. R. Lal, 1–4. New York: Marcel Dekker.

Ramakrishnan, P. S., J. Campbell, L. Demierre, A. Gyi, K. C. Malhotra, S. Mehndiratta, S. N. Rai, and

E. M. Sashidharan. 1994. *Ecosystem rehabilitation of the rural landscape in South and Central Asia: An analysis of issues.* New Delhi: Special Publication, UNESCO (ROSTCA).

Ramakrishnan, P. S., A. N. Purohit, K. G. Saxena, and K. S. Rao. 1994. *Himalayan environment and sustainable development.* New Delhi: Diamond Jubilee Publishers, Indian National Science Academy.

Ramakrishnan, P. S., K. G. Saxena, and U. M. Chandrashekara, eds. 1998. *Conserving the sacred: For biodiversity management.* New Delhi: UNESCO; Oxford, UK: IBH Publishers; and Enfield, NH: Science Publishers.

Ramakrishnan, P. S., R. P. Shukla, and R. Boojh. 1982. Growth strategies of trees and their application to forest management. *Current Science* 51:448–55.

Ramakrishnan, P. S., and P. M. Vitousek. 1989. Ecosystem-level processes and consequences of biological invasions. In *Biological invasions: A global perspective,* ed. J. A. Drake, H. A. Mooney, F. di Castri, R. H. Groves, F. J. Kruger, M. Rejmanek, and M. Williamson, 281–300. New York: John Wiley.

Raman, T. R. S., and D. Mudappa. 2003. Correlates of Hornbill distribution and abundance in rainforest fragments in the Southern Western Ghats, India. *Bird Conservation International* 13:199–212.

Rambler, M. B., L. Margulis, and R. Fester, eds. 1989. *Global ecology: Toward a science of the biosphere.* Boston: Academic Press.

Ramsay, P. 1997. *Revival of the land.* Edinburgh: Creag Meagaidh National Nature Reserve, Scottish Natural Heritage.

Randall, A. 1988. What mainstream economists have to say about the value of biodiversity. In *Biodiversity,* ed. E. O. Wilson, 217–23. Washington, DC: National Academy Press.

Randall, A. 1993. The problem of market failure. In *Economics of the environment,* 3rd ed., ed. R. Dorfman and N. Dorfman, 144–61. New York: W. W. Norton and Co.

Rangan, H. 2001. The muti trade: South Africa's traditional medicines. *Diversity* 2:16–25.

Raven, P. H., and E. O. Wilson. 1992. A fifty-year plan for biodiversity surveys. *Science* 258:1099–1100.

Rawls, J. 1987. The idea of an overlapping consensus. *Oxford Journal of Legal Studies* 7:1–25.

Ray, D., and K. Watts. 2003. Native woodland habitat network development for the Ettrick Forest. In *Restoring borders woodland: The vision and the task,* ed. D. Kohn, 110-16. Occasional paper. Monteviot Nurseries, Ancrum, Jedburgh, Scotland: Wildwood Group of Borders Forest Trust.

Rebelo, A. G., and P. M. Holmes. 1988. Commercial exploitation of *Brunia albiflora* (Bruniaceae) in South Africa. *Biological Conservation* 43:195–207.

Rees, W. E. 1995. Cumulative environmental assessment and global change. *Environmental Impact Assessment Review* 15:295–309.

Rees, W. E. 2004. Why conventional economic logic doesn't protect biodiversity. In *Gaining ground: In pursuit of ecological sustainability,* ed. D. M. Lavigne, 16–19. Guelph, Canada: International Fund for Animal Welfare, and Limerick, Ireland: University of Limerick.

Rees, W. E. 2006. Why conventional economic logic won't protect biodiversity. In *Gaining ground: In pursuit of ecological sustainability,* ed. D. M. Lavigne, 207–26. Guelph, Canada: International Fund for Animal Welfare, and Limerick, Ireland: The University of Limerick.

Reid, R. 2002. Practical options for the greening of Carolinian Canada. http://www.carolinian.org.

Renard, Y. 1991. Institutional challenges for community-based management in the Caribbean. *Nature and Resources* 27:4–9.

Repetto, R. 1993. Government policy, economics and the forest sector. In *Forests for the future: Their use and conservation,* ed. K. Ramakrishna and G. Woodwell, 93–110. New York: Yale University Press.

Rhodes University, Fort Cox College of Agriculture and Forestry, and UNITRA (University of Transkei). 2001. *Participatory monitoring at Machibi, Mt. Coke State Forest, Eastern Cape.* Pretoria, South Africa: Department of Water Affairs and Forestry.

Rice, S. 1997. Koa stewardship: North and south Kona. In *Koa: A decade of growth.* Proceedings of the Hawaii Forest Industry Association's 1996 annual symposium, ed. L. Ferentinos and D. O. Evans, 9. Honolulu: Hawaii Forest Industry Association.

Richardson, D. M., and B. W. Van Wilgen. 1986. The effects of fire in felled *Hakea sericea* and natural fynbos and implications for weed control in mountain catchments. *South African Forestry Journal* 139:4–14.

Richardson, D. M., and B. W. Van Wilgen. 2004. Invasive alien plants in South Africa: How well do we understand the ecological impacts? *South African Journal of Science* 100:45–52.

Ricketts, T. H., G. C. Daily, P. R. Ehrlich, and C. D. Michener. 2004. Economic value of tropical forest to coffee production. *Proceedings of the National Academy of Sciences* 101:12579–82.

Roche, C. J. 2004. "Ornaments of the desert". Springbok treks in the Cape colony, 1774–1908. MA thesis, University of Cape Town, Cape Town, South Africa.

Rodwell, J. S. 1991. *British plant communities. 1. Woodlands and scrub*. Cambridge: Cambridge University Press.

Rodwell, J. S., and G. S. Patterson. 1994. *Creating new native woodlands*. Bulletin 112. London: HMSO.

Román, M. 1993. Diagnóstico de la estructura del sector primario en la producción de lana en la provincia del Chubut. Buenos Aires: Informe para el Consejo Federal de Inversiones (CFI).

Román, M., P. Tsakougmagkos, and L. Araoz. 1992. *Elementos para el análisis de la vinculación entre las políticas públicas y el proceso de desertificación de la Patagonia*. Informe final de consultoría. Proyecto de Lucha contra la Desertificación en Patagonia (LUDEPA). Análisis de la influencia de las políticas públicas sobre el proceso de desertificación en Patagonia. Buenos Aires: INTA/GTZ.

Rookmaker, L. C. 1989. *The zoological exploration of Southern Africa, 1650–1790*. Rotterdam: A. A. Balkema.

Rosenzweig, M. L. 2003. *Win-win ecology: How the earth's species can survive in the midst of human enterprise*. New York: Oxford University Press.

Rotundo, J. L., and M. R. Aguiar. 2004. Vertical seed distribution in the soil constrains regeneration of *Bromus pictus* in a Patagonian steppe. *Journal of Vegetation Science* 15:515–22.

Rouget, M., R. M. Cowling, D. M. Richardson, and R. L. Pressey. 2003. Identifying spatial components of ecological and evolutionary processes for regional conservation planning in the Cape Floristic Region, South Africa. *Diversity and Distributions* 9:191–210.

Rouget, M., D. M. Richardson, R. M. Cowling, J. W. Lloyd, and A. T. Lombard. 2003. Current patterns of habitat transformation and future threats to biodiversity in terrestrial ecosystems of the Cape Floristic Region, South Africa. *Biological Conservation* 112:45–62.

Royal Society. 2005. Ocean acidification due to increasing atmospheric carbon dioxide. Policy document. http://www.royalsoc.ac.uk.

Sagoff, M. 2002. The Catskills parable. A billion dollar misunderstanding. http://www.perc.org/publications/percreports/june2005/catskills.php. January 2006.

SAIRR (South African Institute for Race Relations). 2004. *Fast Facts* 8:1–16.

Sala, O. E., F. S. Chapin III, J. J. Armesto, E. Berlow, J. Bloomfield, R. Dirzo, E. Huber-Sanwald, et al. 2000. Global biodiversity scenarios for the year 2100. *Science* 287:1770–74.

Salafsky, N., and E. Wollenberg. 2000. Linking livelihoods and conservation: A conceptual framework and scale for assessing the integration of human needs and biodiversity. *World Development* 28:1421–38.

Salati, E., and P. B. Vose. 1984. Amazon basin: A system in equilibrium. *Science* 225:129–38.

Salzman, J., and J. B. Ruhl. 2001. Currencies and the commodification of environmental law. *Stanford Law Review* 53:607–94.

Samuelson, P. 1954. The pure theory of public expenditure. *Review of Economics and Statistics* 36:387–89.

Sandler, T. 1993. Tropical deforestation: Markets and market failures. *Land Economics* 69:225–33.

Sandler, T. 2004. *Global collective action*. New York: Cambridge University Press.

SANParks (South African National Parks Board). 2003. *Annual report*. Pretoria, South Africa: SANParks.

SARB (South African Reserve Bank). 2004. *Quarterly bulletin 232, June 2004*. Pretoria: South African Reserve Bank.

SARPN (Southern Africa Regional Poverty Network). 2003. Poverty indicators. http://www.sarpn.org.za/regionalviews/southafrica.php.

Saunders, C. D. 2003. The emerging field of conservation psychology. *Human Ecology Review* 10:137–49.

Sayer, J., and B. M. Campbell. 2004. *The science of sustainable development: Local livelihoods and the global environment*. Cambridge: Cambridge University Press.

Sayer, J., C. Elliott, and S. Maginnis, S. 2003. Protect, manage and restore: Conserving forests in multi-functional landscapes. Paper presented at the World Forestry Congress, Canada.

Scheffer, M., W. Brock, and F. Westley. 2000. Socio-economic mechanisms preventing optimum use of ecosystem services: An interdisciplinary theoretical analysis. *Ecosystems* 3:451–71.

Scheffer, M., and S. R. Carpenter. 2003. Catastrophic regime shifts in ecosystems: Linking theory to observation. *Trends in Ecology and Evolution* 18:648–56.

Schelling, T. 1992. Some economics of global warming. *American Economic Review* 82:1–14.

Scherr, S., A. White, and A. Khare. 2004. *For services rendered: The current status and future potential of mar-*

kets for the ecosystem services provided by tropical forests. International Tropical Timber Organization, Technical Series No. 21. Yokohama, Japan: International Tropical Timber Organization.

Schmidt, K. F. 2001. A True-Blue vision for the Danube. *Science* 294:1445–47.

Scholes, R. J., and C. L. Bailey. 1996. Can savannas help balance the South African greenhouse gas budget? *South African Journal of Science* 92:60–61.

Scholes, R. J., and R. Biggs, eds. 2004. *Ecosystem services in southern Africa: A regional assessment.* Pretoria, South Africa: Council for Scientific and Industrial Research.

Scholes, R. J., and R. Biggs. 2005. A biodiversity intactness index. *Nature* 434:45–49.

Scholes, R., N. Gureja, M. Giannecchinni, D. Dovie, B. Wilson, N. Davidson, K. Piggott, et al. 2001. The environment and vegetation of the flux measurement site near Skukuza, Kruger National Park. *Koedoe* 44:73–83.

Scholes, R. J., and M. Van der Merwe. 1996. Sequestration of carbon in savannas and woodlands. *Environmental Professional* 18:96–103.

Schoonover, J. E., K. W. J. Williard, J. J. Zaczek, J. C. Mangun, and A. D. Carver. 2006. Agricultural sediment reduction by giant cane and forest riparian buffers. *Water, Air and Soil Pollution* 169:303–15.

Schulz, P. W., and L. Zelezny. 2003. Reframing environmental messages to be congruent with American values. *Research in Human Ecology* 10:126–36.

Schumacher, E. F. 1973. *Small is beautiful.* London: Harper and Row.

Scoones, I. 1998. *Sustainable rural livelihoods: A framework for analysis.* Brighton, UK: Institute for Development Studies.

Scott, D. F., F. W. Prinsloo, and G. Moses. 1999. Results of the afforested catchment experiments: Range and variability of effects and the controlling variables. Paper presented at the 9th National Hydrology Symposium, November 1999, University of the Western Cape, South Africa.

Scott, J. M., S. Mountainspring, F. L. Ramsey, and C. B. Kepler. 1986. Forest bird communities of the Hawaiian Islands: Their dynamics, ecology and conservation. *Studies in Avian Biology* 9:1–431.

Scowcroft, P. G., and J. Jeffrey. 1999. Potential significance of frost, topographic relief, and *Acacia koa* stands to restoration of mesic Hawaiian forests on abandoned rangeland. *Forest Ecology and Management* 114: 447–58.

Sellin, R., and D. Beesten. 2004. Conveyance of a managed vegetated two-stage river channel. *Water Management* 157:21–33.

Senapati, B. K., S. Naik, P. Lavelle, and P. S. Ramakrishnan. 2002. Earthworm-based technology application for status assessment and management of traditional agroforestry systems. In *Traditional ecological knowledge for managing biosphere reserves in South and Central Asia*, ed. P. S. Ramakrishnan, R. K. Rai, R. P. S. Katwal, and S. Mehndiratta, 139–60. New Delhi: UNESCO, and Oxford: IBH.

SER (Society for Ecological Restoration International). 2002. *Society for Ecological Restoration International's Primer of Ecological Restoration.* http://www.ser.org/Primer.

SER (Society for Ecological Restoration International). 2004. Natural capital and ecological restoration: An occasional paper of the SER Science and Policy Working Group. http://www.ser.org/content/Naturalcapital.asp.

Shabman, L., and P. Scodari. 2005. The future of wetlands mitigation banking. *Choices* 20:65–70.

Shackleton, C. 1998. Comparisons of plant diversity in protected and communal lands in the Bushbuckridge lowveld savanna, South Africa. *Biological Conservation* 94:273–85.

Shackleton, C., and R. Scholes. 2000. Impact of fire frequency on woody community structure and soil nutrients in the Kruger National Park. *Koedoe* 43:75–81.

Shackleton, C., and S. Shackleton. 1997. *The use and potential for commercialisation of veld products in the Bushbuckridge area.* Research report. Nelspruit and Strandgate: DANCED and DARUDEC.

Shackleton, C., and S. Shackleton. 2000. Direct use values of secondary resources harvested from communal savannas in the Bushbuckridge lowveld, South Africa. *Journal of Tropical Forest Products* 6:28–47.

Shackleton, C., and S. Shackleton. 2002. *Use of marula products for domestic and commercial purposes by households in the Bushbuckridge District, Limpopo Province, South Africa.* Research report. Grahamstown: Environmental Sciences, Rhodes University.

Shackleton, C. M., and S. E. Shackleton. 2004a. The use of woodlands for direct household provisioning. In *Indigenous forests and woodlands in South Africa: People, policy and practice*, ed. M. Lawes, H. Eeley, C. Shackleton, and B. Geach, 337–66. Pietermaritzburg, South Africa: University of KwaZulu-Natal Press.

Shackleton, C. M., and S. E. Shackleton. 2004b. Everyday resources are valuable enough for community-based natural resource management programme support: Evidence from South Africa. In *Rights, resources and rural development: Community-based natural resource management in Southern Africa*, ed. C. Fabricius, E. Koch, H. Magome, and S. D. Turner, 135–46. London: Earthscan Publications.

Shackleton, S., C. Shackleton, and B. Cousins. 2000. *Re-valuing the communal lands of southern Africa: New understandings of rural livelihoods*. London: Overseas Development Institute.

Shine, C. 2004. *Using tax incentives to conserve and enhance biological and landscape diversity in Europe*. Report prepared for committee of experts for the development of the Pan-European Ecological Network (STRA-REP), 8th meeting Krakow, 5–6 October. Council of Europe and UNEP.

Shogren, J. F., G. M. Parkhurst, and C. Settle. 2003. Integrating economics and ecology to protect nature on private lands: Models, methods, and mindsets. *Environmental Science and Policy* 6:233–42.

Short, J., and B. Turner. 1994. A test of the vegetation mosaic hypothesis: A hypothesis to explain the decline and extinction of Australian mammals. *Conservation Biology* 8:439–49.

Shukla, R. P., and P. S. Ramakrishnan. 1986. Architecture and growth strategies of tropical trees in relation to successional status. *Journal of Ecology* 74:33–46.

Siegfried, W. R. 1989. Preservation of species in southern African nature reserves. In *Biotic diversity in southern Africa: Concepts and conservation*, ed. B. J. Huntley, 186–201. Cape Town: Oxford University Press.

Sigwela, A. M. 2004. The impacts of land use on vertebrate diversity and vertebrate mediated processes in the thicket biome, Eastern Cape. PhD thesis, University of Port Elizabeth, Port Elizabeth, South Africa.

Silvertown, J. 2004. Sustainability in a nutshell. *Trends in Ecology and Evolution* 19:276–78.

Simmons, P. 1999. Koa resources in Hawaii: Private lands. In *Harvest-to-market: Adding value to Hawaii's woods*, 54. Proceedings of the Hawaii Forest Industry Association symposium, ed. Renewable Resource Extension Program of the University of Hawaii at Manoa. Hilo: Hawaii Forest Industry Association.

Simon, J. 1996. *The ultimate resource 2*. Rev. ed. Princeton: Princeton University Press.

Singh, A. 2004. Environmental conflict along the Sacramento River: Stakeholders' perspectives on habitat restoration. MS thesis, California State University, Chico.

Skead, C. J. 1980. *Historical mammal incidence in the Cape Province*. Vol. 1. Cape Town: Department of Nature and Environmental Conservation, Provincial Administration of the Cape of Good Hope.

Skead, C. J. 1987. *Historical mammal incidence in the Cape Province*. Vol. 2. *The eastern half of the Cape Province, including the Ciskei, Transkei and East Griqualand*. Cape Town: Chief Directorate of Nature and Environmental Conservation of the Provincial Administration of the Cape of Good Hope.

Skinner, J. D. 1993. Springbok *(Antidorcas marsupialis)* treks. *Transactions of the Royal Society of South Africa* 48:291–305.

Smith, L. G. 1993. *Impact assessment and sustainable resource management*. New York: Longman Scientific and Technical.

Smith, P. E., J. K. Horne, and D. C. Schneider. 2001. Spatial dynamics of anchovy, sardine, and hake pre-recruit stages in the California Current. *ICES Journal of Marine Science* 58:1063–71.

Smith, R. J., R. D. J. Muir, M. J. Walpole, A. Balmford, and N. Leader-Williams. 2003. Governance and the loss of biodiversity. *Nature* 426:67–70.

Smuts, J. C. 1926. *Holism and evolution*. London: Macmillan.

Sodhi, N. S., L. H. Liow, and F. A. Bazzaz. 2004. Avian extinctions from tropical and subtropical forests. *Annual Review of Ecology and Systematics* 35:323–45.

Solow, R. 1991. Sustainability: An economist's perspective. In *The environmental ethics and policy book: Philosophy, ecology and economics*, ed. D. Van De Veer and C. Pierce, 438–43. Belmont CA: Thomson/Wadsworth.

Soriano, A. 1956. Aspectos ecológicos y pastoriles de la vegetación patagónica relacionados con su estado y capacidad de recuperación. *Revista de Investigaciones Agrícolas* 10:349–79.

Soriano, A., C. P. Movia, and R. J. C. León. 1983. Deserts and semi-deserts of Patagonia. Vegetation. In *Temperate deserts and semi-deserts*. Vol. 5. *Ecosystems of the World*, ed. N. E. West, 440–54. Amsterdam: Elsevier.

SRCAF (Sacramento River Conservation Area Forum). 2003. *Sacramento River conservation area forum handbook*. Prepared for the Resources Agency of the State of California.

SRCAF (Sacramento River Conservation Area Forum). 2006. Draft good neighbor policy. http://www.sacramentoriver.ca.gov/publications/GoodNeighborPolicy/2GoodNeighborPolicy.pdf.

SSA (Statistics South Africa). 2004. Census 2001: Key results (Report-03-02-01). Statistics South Africa website. http://www.statssa.gov.za.

State of Hawaii, 2001. *Annual report to the Twenty-first Legislature, regular session of 2002, relating to the forest stewardship program.* Honolulu: Department of Land and Natural Resources, Division of Forestry in Wildlife.

Stephens, T., D. Brown, and N. Thornley. 2002. *Measuring conservation achievement: Concepts and their application over the Twizel Area.* Science for Conservation Monograph Series. Wellington, NZ: Department of Conservation.

Stern, D. I., M. S. Common, and E. B. Barbier. 1996. Economic growth and environmental degradation: The environmental Kuznets curve and sustainable development. *World Development* 24:1151–60.

Sterner, T. 2003. *Policy instruments for environmental and natural resource management.* Resources for the Future, World Bank, Swedish International Development Cooperation Agency (SIDA). Washington, DC: RFF Press.

Stevens, L. E., T. J. Ayers, J. B. Bennett, K. Christensen, M. J. C. Kearsley, V. J. Meretsky, A. M. Phillips, et al. 2001. Planned flooding and Colorado River riparian trade-offs downstream from Glen Canyon Dam, Arizona. *Ecological Applications* 11:701–10.

Stevenson-Hamilton, J. 1957. *Wildlife in South Africa.* London: Hamilton and Co.

Stewart, A. M., and J. L. Craig. 2001. Predicting pro-environmental behaviours: A model and a test. *Journal of Environmental Systems* 28:293–318.

Stewart, A. M., and J. L. Craig. 1985. Movements, status and access to nectar in the tui. *New Zealand Journal of Zoology* 12:649–57.

Stewart, M., and P. Collett. 1998. Accountability in community contributions. In *Community and sustainable development. Participation in the future,* ed. D. Warburton, 52–67. London: Earthscan Publications.

Stiven, R. 2005. Native woodland. Topic paper no 4, Scottish Forestry Strategy Consultation. Edinburgh: Forestry Commission.

Stuart-Hill, G. C. 1989. Succulent valley bushveld. In *Veld management in the Eastern Cape,* ed. J. Danckwerts and W. R. Teague, 165–74. South Africa: Department of Agriculture and Water Supply.

Stuart-Hill, G. C. 1992. Effects of elephants and goats on the Kaffrarian succulent thicket of the Eastern Cape, South Africa. *Journal of Applied Ecology* 29:699–710.

Stuart-Hill, G. C., and A. J. Aucamp. 1993. Carrying capacity of the succulent valley bushveld of the eastern Cape. *African Journal of Range and Forage Science* 10:1–10.

Stuart-Hill, G. C., and J. E. Danckwerts. 1988. Influence of domestic and wild animals on the future of succulent Valley Bushveld. *Pelea* 7:45–56.

Swart, M. L., and F. O. Hobson. 1994. Establishment of spekboom. *Dohne Bulletin* 3:10–13.

Swingland, I. R., ed. 2003. *Capturing carbon and conserving biodiversity.* London: Royal Society and Earthscan Publications.

Symstad, A. J., and D. Tilman. 2001. Diversity loss, recruitment limitation, and ecosystem functioning: Lessons learned from a removal experiment. *Oikos* 92:424–35.

Tainter, J. A. 1990. *The collapse of complex societies.* Cambridge: Cambridge University Press.

Talbot, W. J. 1961. Land utilization in the arid regions of southern Africa. Pt. 1. South Africa. A history of land use in arid regions. *Arid Zones Research* 17:299–338.

Taylor, P. J. 2005. *Unruly complexity: Ecology, interpretation, engagement.* Chicago: University of Chicago Press.

Ten Brink, B. J. E. 2000. *Biodiversity indicators for the OECD environmental outlook and strategy.* RIVM Report 402001014. Bilhoven, The Netherlands: National Institute for Public Health and the Environment (RVIM).

Ten Kate, K. 2002. Science and the Convention on Biological Diversity. *Science* 295:2371–72.

Texeira, M., and J. M. Paruelo. 2006. Demography, population dynamics and sustainability of the Patagonian sheep flocks. *Agricultural Systems* 87:12–46.

Thomas, C. D., A. Cameron, R. E. Green, M. Bakkenes, L. J. Beaumont, Y. C. Collingham, B. F. N. Erasmus, et al. 2004. Extinction risk from climate change. *Nature* 427:145–48.

Thompson, K. 1961. Riparian forests of the Sacramento Valley, CA. *Annals of the Association of American Geographers* 51:294–315.

Tidmarsh, C. E. 1948. Conservation problems of the Karoo. *Farming in South Africa* 23:519–30.

Tilton, J. 1995. Assigning the liability for past pollution: Lessons from the US mining industry. *Journal of Institutional and Theoretical Economics* 151:139–54.

Timmermans, H. G. 2004. Rural livelihoods at Dwesa/Cwebe: Poverty, development and natural resource use on the Wild Coast, South Africa. MS thesis, Rhodes University, Grahamstown, South Africa.

Tipping, R. 1998. The application of palaeoecology to native woodland restoration: Carrifran, a case study. In *Native woodland restoration in southern Scotland: Principles and practice*, ed. A. C. Newton and P. Ashmole, 9–21. Occasional paper no. 2. Monteviot Nurseries, Ancrum, Jedburgh, Scotland: Wildwood Group of Borders Forest Trust.

Todkill, W. B. 2001. Towards the rehabilitation of degraded subtropical thicket in the Addo Elephant national park. MS thesis, University of Port Elizabeth, Port Elizabeth, South Africa.

Tompkins, S. 1989. *Forestry in crisis: The battle for the hills.* London: Christopher Helm.

Tongway, D. J. 1993. Functional analysis of degraded rangelands as a means of defining appropriate restoration techniques, vol. 4. In *Proceedings of the fourth international rangeland congress, 22–26 April 1991*, ed. A. Gaston, M. Kernick, and H. LeHouerou, 166–68. Montpellier, France: Service Central d'Information Scientifique et Technique.

Tongway, D. J. 1995. Monitoring soil productive potential. *Environmental Monitoring and Assessment* 37:303–18.

Tongway, D. J., and N. L. Hindley. 1995. *Manual for soil condition assessment of tropical grasslands.* Canberra, ACT: CSIRO Sustainable Ecosystems.

Tongway, D. J., and N. L. Hindley. 2003. *Indicators of ecosystem rehabilitation success: Stage two—Verification of EFA indicators.* Final Report to the Australian Centre for Mining Environmental Research. Brisbane, Australia: Centre for Mined Land Rehabilitation; University of Queensland and Canberra, ACT: CSIRO Sustainable Ecosystems.

Tongway, D. J., and N. L. Hindley. 2004. Landscape function analysis: Procedures for monitoring and assessing landscapes with special reference to minesites and rangelands. Version 3.1. CD. Canberra, ACT: CSIRO Sustainable Ecosystems.

Tongway, D. J., N. L. Hindley, J. A. Ludwig, A. Kearns, and G. Barnett. 1997. *Indicators of ecosystem rehabilitation success and selection of demonstration sites.* Final Report. Canberra, ACT: CSIRO Division of Wildlife and Ecology.

Tongway, D. J., and J. A. Ludwig. 1990. Vegetation and soil patterning in semi-arid mulga lands of Eastern Australia. *Australian Journal of Ecology* 15:23–34.

Tongway, D. J., and J. A. Ludwig. 2002. Desertification, reversing. In *Encyclopedia of soil science*, ed. R. Lal, 343–45. New York: Marcel Dekker.

Tongway, D. J., J. A. Ludwig, and W. G. Whitford. 1989. Mulga log mounds: Fertile patches in the semi-arid woodlands of eastern Australia. *Australian Journal of Ecology* 14:263–68.

Tongway, D. J., and E. L. Smith. 1989. Soil surface features as indicators of rangeland site productivity. *Australian Rangeland Journal* 11:15–20.

TruCost. 2006. TruCost: Taking the environment into account. http://www.trucost.org/.

Turchin, P., and A. D. Taylor. 1992. Complex dynamics in ecological time series. *Ecology* 73:289–305.

Turner, R. E., and N. N. Rabalais. 1994. Coastal eutrophication near the Mississippi River Delta. *Nature* 368:619–21.

Turpie, J. 2004. The role of resource economics in the control of invasive alien plants in South Africa. *South African Journal of Science* 100:87–93.

Turpie, J. K. 2003. The existence value of biodiversity in South Africa: How interest, experience, knowledge, income and perceived level of threat influence local willingness to pay. *Ecological Economics* 46:199–216.

Turpie, J. K., and B. J. Heydenrych. 2000. Economic consequences of alien infestation of the Cape Floral Kingdom's fynbos vegetation. In *The economics of biological invasions*, ed. C. Perrings, M. Williamson, and S. Dalmazzone, 152–82. Cheltenham, UK: Edward Elgar.

Turpie, J., B. J. Heydenrych, and S. J. Lamberth. 2003. Economic value of terrestrial and marine biodiversity in the Cape Floristic Region: Implications for defining effective and socially optimal conservation strategies. *Biological Conservation* 112:233–51.

Turpie, J. K., and A. Joubert. 2001. Estimating potential impacts of a change in river quality on the tourism value, Kruger National Park: An application of travel cost, contingent and conjoint valuation methods. *Water* SA 27:387–98.

Turpie, J. K., R. G. Lechemer-Oertel, A. Sigwela, G. Antrobus, J. Donaldson, H. Robertson, A. Skowno, et al. 2003. The ecological and economic implications of conversion to game farming in the xeric succulent thicket of the Eastern Cape, South Africa. In *An ecological-economic appraisal of conservation on commercial farmland in four areas of South Africa*, ed. J. K. Turpie, 91–124. Cape Town: Conservation Farming Report, National Botanical Institute.

Turpie, J. K., H. Winkler, and G. Midgley. 2004. Economic impacts of climate change in South Africa: A preliminary assessment of unmitigated damage costs. In *Sustainable options: Development lessons from applied environmental economics*, ed. J. Blignaut and M. de Wit, 130–60. Cape Town: UCT Press.

Tyrrell, M. L., M. H. Hall, and R. N. Sampson. 2004. *Dynamic models of land use change in northeastern USA: Developing tools, techniques, and talents for effective conservation action*. New Haven: Yale School of Forestry and Environmental Studies.

UNEP (United Nations Environment Programme). 2002. *Global environment outlook 3*. United Nations Environment Programme. Earthprint.

UNEP (United Nations Environment Programme). 2004. Report on the sixth meeting of the conference of the parties to the convention on biological diversity (UNEP/CBD/COP/6/20/Part 2) strategic plan decision VI/26 (CBD 2002). http://www.biodiv.org/doc/meetings/cop/cop-06/official/cop-06-20-part2-en.pdf.

UNEP and IISD (United Nations Environment Programme and the International Institute for Sustainable Development). 2004. *Exploring the links*. Nairobi, Kenya, and Winnipeg, Canada: UNEP and IISD.

Union Européenne. 1998. *Manuel de vulgarisation de la culture semi-intensive de la vanille*. Antananarivo: European Union Madagascar.

United States Census Bureau, Population Division. 2005. *Annual estimates of housing units for counties in New York: April 1, 2000 to July 1, 2004 (HU-EST2004-04-36)*. http://www.census.gov/popest/housing/HU-EST2004-4.html; and annual estimates of housing units for the United States and States: April 1, 2000 to July 1, 2004 (HU-EST2004-01). http://www.census.gov/popest/housing/HU-EST2004.html. Washington, DC: United States Census Bureau, Population Division.

USEPA (United States Environmental Protection Agency). 2006. *Document provided at public information sessions to receive comment on the status of the New York City Catskill/Delaware watershed filtration avoidance determination*. 17, 25 May and 6, 14 June.

USEPA (United States Environmental Protection Agency). 2000. Assessing New York City's watershed protection program. http://www.epa.gov/Region2/water/nycshed/fadmidrev.pdf.

Van der Bank, M. 2000. Genetic variation of *Ocotea bullata* (Stinkwood). Unpublished report, CPWild Project. Stellenbosch, South Africa: University of Stellenbosch, Department of Forest Science.

Van Kooten, G. C., and E. H. Bulte. 2000. *The economics of nature. Managing biological assets*. Oxford, UK: Blackwell.

Van Wilgen, B. W. 2004. Scientific challenges in the field of invasive alien plant management. *South African Journal of Science* 100:19–20.

Van Wilgen, B. W., M. P. de Wit, H. J. Anderson, D. C. Le Maitre, I. M. Kotze, S. Ndala, B. Brown, and M. B. Rapholo. 2004. Costs and benefits of biological control of invasive alien plants: Case studies from South Africa. *South African Journal of Science* 100:113–22.

Van Wilgen, B. W., D. C. Le Maitre, and R. M. Cowling. 1996. Valuation of ecosystem services: A case study from South African fynbos ecosystems. *BioScience* 46:184–89.

Van Wilgen, B. W., D. C. Le Maitre, and R. M. Cowling. 1998. Ecosystem services, efficiency, sustainability and equity: South Africa's Working for Water Programme. *Trends in Ecology and Evolution* 13:378.

Van Wilgen, B. W., P. R. Little, R. A. Chapman, A. H. M. Görgens, T. Willems, and C. Marias. 1997. The sustainable development of water resources: History, financial costs and benefits of alien plant control programmes. *South African Journal of Science* 100:113–22.

Van Wilgen, B. W., C. Marais, and D. Magadlela. 2002. Win-win-win: South Africa's Working for Water Programme. In *Mainstreaming biodiversity in development: Case studies from South Africa*, ed. S. M. Pierce, R. M. Cowling, T. Sandwith, and K. MacKinnon, 5–20. Washington, DC: World Bank.

Van Wilgen, B. W., D. M. Richardson, D. C. Le Maitre, C. Marais, and D. Magadlela. 2001. The economic consequences of alien plant invasions: Examples of impacts and approaches to sustainable management in South Africa. *Environment, Development and Sustainability* 3:145–68.

Van Zyl, H. 2003. *Benefits of natural resource rehabilitation: The case study of woodlands in Limpopo Province*. Research report. Pretoria, South Africa: Nathan Associates.

Vatn, A., and D. W. Bromley. 1994. Choices without prices without apologies. *Journal of Environmental Economics and Management* 26:129–48.

Vermeij, G. J. 2004. *Nature: An economic history*. Princeton: Princeton University Press.

Vermeulen, W. J., and C. J. Geldenhuys. 2004. Experimental protocols and lessons learnt from strip harvesting of bark for medicinal use in the Southern Cape forests. Unpublished report, FRP-DFID R8305 project.

Vernadsky, V. I. 1998. *Biosphere*. New York: Copernicus.

Victor, J. E., and A. P. Dold. 2003. Threatened plants of the Albany Centre of floristic endemism, South Africa. *South African Journal of Science* 99:437–46.

Vincelette, M., L. Randrihasipara, J.-B. Ramanamanjato, P. P. Lowry II, and J. U. Ganzhorn. 2003. Mining and environmental conservation: The case of QIT Madagascar Minerals in the southeast. In *The natural history of Madagascar*, ed. S. M. Goodman and J. P. Benstead, 1535–37. Chicago: University of Chicago Press.

Vitousek, P. M., H. A. Mooney, J. Lubchenco, and J. M. Melillo. 1997. Human domination of Earth's ecosystems. *Science* 277:494–99.

Vlok, J. H. J., D. I. W. Euston-Brown, and R. M. Cowling. 2003. Acock's Valley Bushveld 50 years on: New perspectives on the delimitation characterisation and origin of subtropical thicket vegetation. *South African Journal of Botany* 69:27–51.

Vlok, J. H. J., and R. I. Yeaton. 1999. The effect of overstorey proteas on plant species richness in South African mountain fynbos. *Diversity and Distributions* 5:213–22.

Vogel, C. H. 1988a. Climatic change in the Cape colony, 1820–1900. *South African Journal of Science* 84:11.

Vogel, C. H. 1988b. 160 years of rainfall of the Cape—Has there been a change? *South African Journal of Science* 84:724–26.

Vogel, C. H. 1994. (Mis)management for droughts in South Africa: Past, present and future. *South African Journal of Science* 90:4–6.

Vogel, D. 2004. Trade and environment in the global economy; Contrasting European and American perspectives. In *Green giants? Environmental policy of the United States and the European Union*, ed. N. Vig and M. Faure, 231–52. Cambridge: MIT Press.

Vogel, J. H. 1997. The successful use of economic instruments to foster sustainable use of biodiversity: Six case studies from Latin America and the Caribbean. *Biopolicy* 2:1–44.

Von Breitenbach, F., and J. von Breitenbach. 1995. *National list of indigenous trees*. Pretoria, South Africa: Dendrological Foundation.

Von Hayek, F. A. 1993. Social of Distributive Justice. In *Justice*, ed. A. Ryan, 117–58. Oxford, UK: Oxford University Press.

Wackernagel, M., and W. E. Rees. 1996. *Our ecological footprint: Reducing human impact on the Earth*. Gabriola Island, BC, Canada: New Society Publishers.

Wackernagel, M., and W. E. Rees. 1997. Perceptual and structural barriers to investing in natural capital: Economics from an ecological footprint perspective. *Ecological Economics* 20:3–24.

Wali, M. K. 1992. *Ecosystem rehabilitation*. Vol. 2. *Ecosystem analysis and synthesis*. The Hague, The Netherlands: SPB Academic Publishers.

Walker, B. H. 1992. Biodiversity and ecological redundancy. *Conservation Biology* 6:18–23.

Walker, B., S. Carpenter, J. Anderies, N. Abel, G. Cumming, M. Janssen, L. Lebel, J. Norberg, G. D. Peterson, and R. Pritchard. 2002. Resilience management in social-ecological systems: A working hypothesis for a participatory approach. *Conservation Ecology* 6(1):14. http://www.consecol.org/vol6/iss1/art14.

Walker, G. J., and K. J. Kirby. 1987. An historical approach to woodland conservation in Scotland. *Scottish Forestry* 41:87–98.

Wallach, L., and M. Sforza. 1999. *Whose trade organization? Corporate globalization and the erosion of democracy*. Washington, DC: Public Citizen.

Ward, S. C., and J. M. Koch. 1996. Biomass and nutrient distribution in a 15.5-year old forest growing on a rehabilitated bauxite mine. *Australian Journal of Ecology* 21:309–15.

Waters, C. M., D. L. Garden, A. B. Smith, D. A. Friend, P. Sanford, and G. C. Auricht. 2005. Performance of native and introduced grasses for low-input pastures. Pt.1. Survival and recruitment. *Rangeland Journal* 27:23–39.

Waters, C. M., D. S. Loch, and P. W. Johnston. 1997. The role of native grasses and legumes for land revegetation in central and eastern Australia with particular reference to low rainfall areas. *Tropical Grasslands* 31:304–10.

Waters, C. M., G. J. Melville, and A. C. Grice. 2003. Genotypic variation among sites within eleven Australian native grasses. *Rangelands Journal* 25:70–84.

Waters, C. M., J. Virgona, and G. J. Melville. 2004. A genecological study of the Australian native grass *Austrodanthonia caespitosa* (Gaudich) H. P. Linder. In *Proceedings of the fifth Australian workshop on native seed biology*, June 2004, pp. 21–23. Brisbane: Australian Center for Minerals Extension and Research.

WCD (World Commission on Dams). 2000. *Dams and development: A new framework for decision-making.* London: Earthscan Publications.

Weiss, S. B. 1999. Cars, cows, and checkerspot butterflies: Nitrogen deposition and management of nutrient-poor grasslands for a threatened species. *Conservation Biology* 13:1476–86.

Weitzman, M. 1998. Why the far-distant future should be discounted at its lowest possible rate. *Journal of Environmental Economics and Management* 36:201–8

Wells, M., and K. Brandon. 1992. *People and parks: Linking protected area management with local communities.* Washington, DC: World Bank.

Wells, M. J., R. J. Poynton, A. A. Balsinhas, K. J. Musil, H. Joffe, E. van Hoepen, and S. K. Abbott. 1986. The history of introduction of invasive plants to southern Africa. In *The ecology and management of biological invasions in southern Africa*, ed. I. A. W. Macdonald, F. J. Kruger, and A. A. Ferrar, 21–36. Cape Town: Oxford University Press.

West, C. J. 1980. Aspects of regeneration on Tiritiri Matangi Island. MS thesis, University of Auckland, Auckland, NZ.

Westoby, M., B. Walker, and I. Noy-Meir. 1989. Opportunistic management for rangelands not at equilibrium. *Journal of Range Management* 42:266–74.

Whalley, R. D. B., and C. E. Jones. 1995. Microlena (*Microlaena stipoides*). *Plant Varieties Journal* 8:27–28.

Whisenant, S. G. 1999. *Repairing damaged wildlands.* Cambridge: Cambridge University Press.

White, P. S., and S. T. A. Pickett. 1985. Natural disturbance and patch dynamics: An introduction. In *The ecology of natural disturbance and patch dynamics*, ed. S. T. A. Pickett and P. S. White, 3–13. London: Academic Press.

Whitehead, D., J. R. Leathwick, and A. S. Walcroft. 2001. Modelling annual carbon uptake for the indigenous forests of New Zealand. *Forest Science* 47:9–20.

Whitesell, C. D. 1990. *Acacia koa* Gray. In *Silvics of North America*, Vol. 2, *Hardwood*, ed. R. M. Burns and B. H. Honkala, 17–28. Washington, DC: USDA Forest Service Agricultural Handbook 654.

Whitesell, C. D., D. S. Debell, T. H. Schubert, R. F. Strand, and T. B. Crabb. 1992. *Short-rotation management of Eucalyptus: Guidelines for plantations in Hawaii.* Gen. Tech. Report PSW-GTR-137. Albany, CA: Pacific Southwest Research Station, USDA Forest Service.

Wieriks, K., and A. Schulte-Wülwer-Leidig. 1997. Integrated water management for the Rhine river basin: From pollution prevention to ecosystem improvement. *Natural Resources Forum* 21:147–56.

Wilcove, D. S., and J. Lee. 2004. Using economic and regulatory incentives to restore endangered species: Lessons learned from three new programs. *Conservation Biology* 18:639–45.

Wilcox, B. P., D. D. Breshears, and C. D. Allen. 2003. Ecohydrology of a resource-conserving semiarid woodland: Effects of scaling and disturbance. *Ecological Monographs* 73:223–39.

Wilcox, J. R. N. 1977. Desert encroachment in South Africa. MA thesis, University of the Witwatersrand, Johannesburg, South Africa.

Wilkinson, K. M., and C. R. Elevitch. 2003. *Growing koa: A Hawaiian legacy tree.* Holualoa, HI: Permanent Agriculture Resources.

Wilson, S. L., and G. I. H. Kerley. 2003. The effects of plant spinescence on the foraging efficiency of indigenous and introduced browsers of similar body size. *Journal of Arid Environments* 55:150–8.

Winkler, E. 1997. Koa economics and resource values. In *Koa: A decade of growth. Proceedings of the Hawaii Forest Industry Association's 1996 annual symposium*, ed. L. Ferentinos and D. O. Evans, 149–52. Honolulu: Hawaii Forest Industry Association.

Winter, D. D., and S. M. Koger, 2004. *The psychology of environmental problems.* 2nd ed. Mahwah, NJ: Lawrence Erlbaum Associates.

WMO (World Meteorological Organization). 1985. *Atmospheric ozone 1985.* Report No.16 of the Global Ozone Research and Monitoring Project, vols. 1–3. Geneva, Switzerland: World Meteorological Organization.

Wogaman, J. P. 1986. *Economics and ethics.* Philadelphia: Fortress Press.

Wohl, E., P. L. Angermeier, B. Bledsoe, G. M. Kondolf, L. MacDonnell, D. M. Merritt, M. A. Palmer, N. L. Poff, and D. Tarboton. 2005. River restoration. *Water Resources Research* 41:1–12.

Wolf, K. 2002. *Advancing wildlife restoration and compatible farming along the Sacramento River*. Chico, CA: The Nature Conservancy.

Woodland, P. S. 1964. The floral morphology and embryology of *Themeda australis*. *Australian Journal of Botany* 12:157–72.

Working for Water. 2003. Working for Water evaluation report: Institutional, organisational and management component. Unpublished Report for the Department of Water Affairs and Forestry, Cape Town.

World Bank. 1997. *Five years after Rio: Innovations in environmental policy*. Environmentally Sustainable Development Studies and Monograph Series, no. 18. Washington, DC: World Bank.

World Bank. 2004. *Sustaining forests: A development strategy*. Washington, DC: World Bank.

WRI (World Resources Institute). 1985. *Tropical forests: A call for action*. Washington, DC: WRI.

WWF (World Wide Fund for Nature). 2006. *Living planet report 2006*. Gland, Switzerland: World Wide Fund for Nature.

WWF/China. (World Wide Fund for Nature/China) 2005. *Panda conservation in the Minshan landscape* (Minshan Project) http://www.wwfchina.org/english/sub_loca.php?loca=26andsub=87.

WWF/IUCN (World Wide Fund for Nature/The World Conservation Union). 2003. *Forests reborn: A workshop on forest restoration*. Gland, Switzerland: WWF and IUCN.

Wynne, B. 1992. Uncertainty and environmental learning: Reconceiving science and policy in the preventative paradigm. *Global Environmental Change: Human and Policy Dimensions* 2:111–27.

Xu, Z., M. T. Bennet, R. Tao, and J. Xu. 2005. China's sloping land conversion program four years on: Current situation, pending issues. *International Forestry Review* 6:317–27.

Yaron, G. 2001. Forest, plantation crops or small-scale agriculture? An economic analysis of alternative land use options in the Mount Cameroon area. *Journal of Environmental Planning and Management* 44:85–108.

Young, A. G., A. H. D. Brown, B. G. Murray, P. H. Thrall, and C. H. Millar. 2000. Genetic erosion, restricted mating and reduced viability in fragmented populations of the endangered grassland herb *Rutidosis leptorrhynchoides*. In *Genetic demography and viability of fragmented populations*, ed. A. Young and G. Clark, 335–59. Cambridge: Cambridge University Press.

Young, A. G., and B. G. Murray. 2000. Genetic bottlenecks and dysgenic gene flow into re-established populations of the grassland daisy, *Rutidosis leptorrhynchoides*. *Australian Journal of Botany* 48:409–16.

Young, M. D. 1992. *Sustainable investment and resource use: Equity, environmental integrity and economic efficiency*. Carnforth, UK: Parthenon Press, and Paris: Unesco.

Young, M. D., and R. Evans. 1998. Groundwater pollution: Can right markets help? *Water* 25:44–47.

Young, M. D., and D. Hatton MacDonald. 2006. How should we discount the future? An environmental perspective. In *Economics and the future: Time and discounting in private and public decision making*, ed. D. J. Pannell and S. G. M. Schilizzi, 121–36. Cheltenham, UK, and Northampton, MA, USA: Edward Elgar.

Young, M. D., and J. C. McColl. 2005. Defining tradable water entitlements and allocations: A robust system. *Canadian Water Resources Journal* 30:65–72.

Zambatis, G., and N. Zambatis. 1997. *Checklist of the vertebrate fauna of the Kruger National Park (excluding avian fauna)*. Skukuza: Kruger National Park, Scientific Services.

Zedler, J. B. 2004. Compensating for wetland losses in the United States. *Ibis* 146:92–100.

Zedler, J. B., and P. Adam. 2002. Saltmarshes. In *Handbook of ecological restoration*, ed. M. R. Perrow and A. J. Davy, 238–66. Cambridge: Cambridge University Press.

Zhao, B., U. Kreuter, B. Li, Z. J. Ma, J. K. Chen, and N. Nakagoshi. 2004. An ecosystem service value assessment of land-use change on Chongming Island, China. *Land Use Policy* 21:139–48.

Zimmerman, H. G., V. C. Moran, and J. H. Hoffmann. 2004. Biological control in the management of invasive alien plants in South Africa, and the role of the Working for Water Programme. *South African Journal of Science* 100:34–39.

Zolessi, C., and E. Figueroa. 2002. *Los planes de cierre de faenas mineras: El caso de la división Radomiro Tomic, Codelco-Chile. Tesis de Grado*. Facultad de Ciencias Económicas y Administrativas, Universidad de Chile.

James Aronson is head of the Restoration Ecology group at the Center of Functional and Evolutionary Ecology, of the government research network (CNRS), in Montpellier, France. He is also curator of restoration ecology at the Missouri Botanical Garden, USA. He has worked in projects and programs related to the restoration and rehabilitation of degraded ecosystems for over twenty years, in many parts of the Mediterranean region, as well as in southern South America, southern Africa, Madagascar, and elsewhere. He is editor in chief of the book series *Science and Practice of Ecological Restoration*, published jointly by the Society for Ecological Restoration International and Island Press. He has published over 100 journal papers and chapters, and has coedited or coauthored five books.

Suzanne J. Milton is professor in the Department of Conservation Ecology and Entomology, University of Stellenbosch; an associate professor of the Percy FitzPatrick Institute for African Ornithology, South Africa; and associate editor for both the *Journal of Applied Ecology* and the *Journal of Arid Environments*. She has coedited two books, published ten book chapters and 123 peer-reviewed papers in a wide range of journals, and published fifty-one popular articles; she is also a much-solicited botanical and range assessment consultant throughout the Karoo, South Africa. Sue was recently honored with lifetime membership in the Ecological Society of America. Together with her husband, ecologist Richard Dean, she has managed a long-term ecological research site for twenty years, and is currently launching a natural capital restoration project in the Karoo.

James N. Blignaut is an ecological economist attached to the University of Pretoria, South Africa, and director of two companies aimed at economic development and the restoration of natural capital. He is the lead editor of two books, and author or coauthor of six book chapters and forty peer-reviewed academic papers in a range of journals, editor of the *South African Journal of Economic and Management Sciences*, and a member of various academic societies. He focuses primarily on environmental fiscal reform, natural resource accounting, payments for ecosystem goods and services, combating invasive alien plant species, restoring natural capital, and local economic development.

Martin R. Aguiar is an agronomist interested in ecology and management of arid lands. Most of his work deals with plant population ecology in Patagonia and the effects of grazing by domestic herbivores. He is professor at the University of Buenos Aires and independent scientist at the National Research Council in Argentina.

Sean Archer is a labor economist with an additional research interest in the ecology and restoration of arid zones. This applies specifically to the Karoo region of South Africa where he was born and continues to visit regularly. He is currently a research associate in economics at the University of Cape Town, South Africa.

Phoebe Barnard is a conservation biologist working on the vulnerability and adaptation of African biodiversity to environmental change, and the links between conservation science and development policy. She was a Millennium Ecosystem Assessment board member and is based at the South African National Biodiversity Institute, Cape Town.

Nathan Berry is a conservationist with a special interest in natural resource economics and conservation finance. He works with Kim Elliman at the Open Space Institute, a land conservation and research organization based in New York City, USA.

Reinette (Oonsie) Biggs studies socioecological system dynamics in support of improved ecosystem management. She previously worked on the Millennium Ecosystem Assessment while based at the Council for Scientific and Industrial Research in South Africa. She is currently a graduate student at the Center for Limnology, University of Wisconsin, Madison, USA.

Keith Bowers is a restoration ecologist and landscape architect with a special interest in the application of landscape ecology, conservation biology, and restoration ecology to the built environment. He is founder and president of Biohabitats Inc., an ecological restoration, conservation planning, and regenerative design consulting firm active throughout the United States.

Mark Buckley is a doctoral candidate in the Department of Environmental Studies at the University of California, Santa Cruz, USA. He combines economics with environmental studies approaches while studying how game theory can be applied to river restoration and management to describe interactions between restorationists and farmers in the Sacramento Valley, northern California.

Andrew Carey consults with forest managers on restoration and maintenance of biological diversity within the context of general sustainability. He has published extensively on forest ecology and management and is an emeritus scientist with the Pacific Northwest Research Station in Olympia, Washington, USA.

Andre F. Clewell served sixteen years on the faculty at Florida State University in Tallahassee, USA, and owned a consulting company for twenty-two years that specialized in ecological restoration design and implementation. He served two years as president of the Society for Ecological Restoration International.

Richard M. Cowling is a conservation scientist and plant ecologist, with a special interest in turning conservation (including restoration) plans into action. He is professor in the Department of Botany at the Nelson Mandela Metropolitan University, South Africa.

John Craig trained as an ecologist and now specializes in sustainability, including social and economic components. He is active in restoration and has received national and international honors for his contributions to biodiversity and the environment. John is currently professor of Environmental Management at the University of Auckland, New Zealand.

Jim Crosthwaite is an economist in Victoria's Department of Sustainability and Environment, Australia, who has recently managed large research projects on farm businesses and biodiversity, and drivers of land-use change. He also managed a conservation management network program, assisted in developing a biodiversity policy framework for a production agency. He will now input to several high-level initiatives, including a review of Victoria's biodiversity strategy.

Georgina Cundill is a social ecologist specializing in governance and natural resource management in communal areas. She is currently a PhD candidate in the Department of Environmental Science at Rhodes University, South Africa.

Gretchen C. Daily is an ecologist by training, taking integrated, interdisciplinary approaches to making conservation mainstream—economically attractive and commonplace around the world. She is a professor of biological sciences and a senior fellow of the Woods Institute for the Environment at Stanford University, USA, and director of several environmental education and conservation efforts.

Rudolf de Groot is a landscape ecologist with special interest in integrated assessment and valuation of ecosystem services. He is associate professor with the Environmental Systems

Analysis group at Wageningen University, The Netherlands, and founder of an international network on nature valuation and financing.

Martin de Wit is an economist and specializes in managing the risks of, and finding solutions to, environmental problems. He is director of De Wit Sustainable Options (Pty) Ltd, a South Africa–based research and consultancy firm in this field.

W. Richard J. Dean is an ornithologist with a strong interest in biogeography, particularly in the effects of historical and present land-use changes on the distributions of birds. He is currently a research associate with the DST/NRF Centre of Excellence at the Percy FitzPatrick Institute of African Ornithology at the University of Cape Town, South Africa.

Narayan Desai is a restoration ecologist and student of Vedic ecology with special interest in the cultural and spiritual aspects of ecological restoration. He is a founding member of Society for Ecological Restoration–India and of Restoring Natural Capital–South Asia.

Christopher (Kim) Elliman works in land conservation, with particular interest in regulating and valuing natural capital and their ecosystem services in real estate transactions. He has worked in finance, real estate, and conservation. Currently he serves as CEO of the Open Space Institute, USA, and serves on a number of national and regional NGOs and foundations.

Christo Fabricius is a systems ecologist who specializes in community-based natural resource management. He is professor in the Department of Environmental Science at Rhodes University, South Africa.

Joshua Farley is an ecological economist in the department of Community Development and Applied Economics and a fellow at the Gund Institute for Ecological Economics at the University of Vermont, USA. His special interest is ecosystem services, policy, and the theory and practice of restoring natural capital.

Eugenio Figueroa B. is an economist with interest in environmental and natural resource economics, economic development, tourism economics, rural development, international businesses, information and telecommunication technologies. He is a professor in the Department of Economics, and executive director of the National Center for the Environment, at the University of Chile.

Erica J. Brown Gaddis is a watershed scientist and modeler with a special interest in the restoration of freshwater ecosystems. She is currently a doctoral student at the Gund Institute for Ecological Economics at the University of Vermont, USA.

Coert J. Geldenhuys is a forest ecologist with a focus on sustainable multiple resource use by diverse stakeholders from natural forests and woodlands in Africa, and on forest rehabilitation. He runs his own small consulting company, Forestwood. He is associate professor in Forest Science at the University of Stellenbosch, South Africa.

Joshua H. Goldstein works in the field of conservation finance with an interest in developing the financial and institutional capacities to expand restoration projects in working landscapes. He is currently a PhD student at Stanford University, California, USA, in the Interdisciplinary Graduate Program in Environment and Resources.

Jennifer Gouza is project manager in the Working for Water Program, St. Francis Bay, Eastern Cape, South Africa. She is currently a graduate student pursuing a master's degree in Sustainable Development, Planning and Management at the University of Stellenbosch.

Pippa Haarhoff is a paleontologist with a special interest in fossil birds. She is manager of the West Coast Fossil Park on the West Coast of South Africa, which she helped found in 1998.

Stefan Hajkowicz is an environmental economist with research interests in decision support, investment appraisal, policy evaluation, and sustainability metrics. He works with CSIRO and is based in Brisbane, Australia.

Karen D. Holl studies the effect of local- and large-scale processes on the restoration of tropical forests in Costa Rica, and grassland and riparian forests in California. She is a professor of Environmental Studies at the University of California, Santa Cruz, USA.

Louise Holloway is involved in catalyzing ecological restoration of rain forest and rehabilitation of degraded environments in Madagascar, in order to enhance human and ecosystem well-being simultaneously. The ecosystem services market (specifically in carbon) is being piloted as an enabling mechanism.

Patricia M. Holmes is a plant ecologist with a special interest in restoration ecology in fire-prone ecosystems. Until recently she was a freelance consultant and researcher in the Western Cape, South Africa, but now works as biophysical specialist for the City of Cape Town.

Graham I. H. Kerley is a conservation ecologist with interests in animal–plant interactions, desertification, and sustainable land-use options. He is director of the Center for African Conservation Ecology and a professor in the Department of Zoology at Nelson Mandela Metropolitan University, South Africa. He also serves on the board of directors of the Eastern Cape Parks Board.

Carolyn Kousky is involved in two areas of research: public policies for increasing the provision of ecosystem services, and natural disaster policy. She is a doctoral student in public policy at Harvard University's John F. Kennedy School of Government, USA.

Suzanne M. Langridge is an ecologist with a particular interest in the social and ecological interface between natural and agricultural areas, and how restoration can enhance agricultural productivity. She is currently a graduate student in Environmental Studies at the University of California, Santa Cruz, USA.

Patrick Lavelle is a soil ecologist with special interest in sustainable management practices of tropical soils based on soil biological processes. He is a professor at Paris VI University,

France, and heads a project in Amazonia to describe and model the links among socioeconomic conditions, landscapes, and the provision of ecosystem goods and services. He was a co-chair of the Nutrient Cycling working group of the Millennium Ecosystem Assessment.

Richard G. Lechmere-Oertel is a landscape-scale ecologist with a strong interest in conservation planning and management in mountainous areas. He is currently employed as the biodiversity ecologist for the Maloti–Drakensberg Transfrontier Programme in South Africa and Lesotho.

Christina (Tina) E. Loxton is an economist with a special interest in uncertainty and risk analysis, also applied to environmental resource management. She was formerly attached to the Department of Economics, University of Pretoria, South Africa, and is currently a graduate student at the University of Toronto, Canada.

Roy A. Lubke is a botanist and plant ecologist with an interest in restoration ecology and coastal and dune systems. He has been involved in restoration projects in many parts of Africa and Madagascar. He is a founding director of Coastal and Environmental Services and is associate professor emeritus in the Department of Botany, Rhodes University, Grahamstown, South Africa.

John Ludwig is a landscape ecologist with a special interest in the restoration of rangelands and mine sites. He is a researcher with the Tropical Savannas Cooperative Research Centre and is based at Australia's CSIRO laboratory in Athertjameon, Queensland.

Stephanie Mansourian is an environmental consultant based in Switzerland with a special interest in the sustainable management and restoration of forested landscapes. For five years she managed WWF International's forest landscape restoration program.

Christo Marais is an ecologist and natural resource manager specializing in watershed, fire, and natural resource restoration management and its economic consequences. He is responsible for strategic partnerships aimed at mainstreaming invasive alien plant management and natural resource restoration in the Working for Water program in South Africa.

William (Willie) McGhee is a forest ecologist whose pioneering work in social and environmental forestry has influenced the direction of community and native woodland initiatives in the UK. He is the executive director of the Borders Forest Trust, and supply manager, Greenergy Bioenergy Ltd. in Edinburgh, Scotland.

James S. Miller is a research botanist at the Missouri Botanical Garden, USA, with interests in tropical floristics, systematics of Boraginaceae, natural products discovery, community-based conservation, and ecological restoration. He is the head of the William L. Brown Center for Plant Genetic Resources and manages the economic botany programs.

Anthony J. Mills is an ecologist and soil scientist with a special interest in soil–plant relationships. He is a researcher in the Department of Soil Science at the University of Stellenbosch, South Africa.

Richard B. Norgaard is an ecological economist currently doing research on how scientists collectively understand complex systems. He is a professor in the Energy and Resources Group at University of California, Berkeley, USA.

Liba Pejchar is a conservation biologist with an interest in rare birds and interdisciplinary problem solving. Her current research explores the potential of frugivorous birds to speed the reforestation of ranchlands in Hawaii and Costa Rica. She is a postdoctoral fellow with the Center for Conservation Biology at Stanford University, USA.

Shirley M. Pierce has long held an interest in spreading awareness of the real value of nature. Her experience as a researcher has been wide ranging, from marine systems to coastal and arid terrestrial systems. She operates as a freelance plant ecologist from her home at Cape St. Francis, South Africa.

Mike Powell is a botanist and restoration practitioner for subtropical thicket and woodland vegetation. He is technical advisor to the Subtropical Thicket Rehabilitation Project and the Working for Woodlands Program, Eastern Cape, South Africa.

P. S. Ramakrishnan is an emeritus professor working in the School of Environmental Sciences at Jawaharlal Nehru University, New Delhi, India. Taking an integrative socioecological system approach, he has effectively used "knowledge systems" as the basis for community participatory natural resource management–linked sustainable regional development.

Gérard Rambeloarisoa is a forester and ecologist with a special interest in physiology and silviculture. He is active in promoting forest restoration in Madagascar. He is the forest program officer of the WWF Madagascar and a part-time lecturer at the water and forest department at the school of agronomy, University of Antananarivo, Madagascar.

William E. Rees is a human ecologist and ecological economist best known as the originator of ecological footprint analysis. He is a former director and currently professor in the University of British Columbia's School of Community and Regional Planning, Canada, where his current research focuses on the psycho-behavioral roots of human (un)sustainability.

David M. Richardson is an invasion ecologist with a special interest in tree invasions. He is professor of ecology at the Centre for Invasion Biology at Stellenbosch University, South Africa, and is editor in chief of the journal *Diversity and Distributions*.

Chris J. Roche is a historical ecologist with an interest in historical ecological phenomena and the impact an understanding of this can have on modern conservation. He works for a leading southern African ecotourism company as an environmental consultant.

Marcela E. Román is an agronomist specialized in economy interested in development project management with sustainable goals. Her research focuses on family farms and small farmers' strategies in developing countries. She is professor in the Department of Economy at the Faculty of Agronomy, University of Buenos Aires, Argentina.

Gabriella Roscher has a background in economic history and a keen interest in forest conservation and land-use planning. She currently works as the international coordinator of WWF's Forest Conversion Initiative based at WWF Switzerland.

Robert J. Scholes is a systems ecologist employed by the Council for Scientific and Industrial Research in South Africa. His interests include the ecology of African savannas, global biogeochemistry and earth observation systems. He was a co-chair of the Condition and Trend working group of the Millennium Ecosystem Assessment.

Kirsten Schuyt is an environmental economist with a special interest in economic values of tropical forests and freshwater ecosystems and instruments, such as payments for environmental services. She currently works as head of the Forests Program of WWF–Netherlands and has a part-time position as associate researcher at the Erasmus University of Rotterdam in the Netherlands.

Ayanda M. Sigwela is a restoration ecologist for South African National Parks. His area of operation is in arid and savannah parks. Although his interest has always been animal–plant interactions, he is currently developing a research program on human and tourism impacts on biodiversity.

David Tongway devised a method for rapid assessment of soil productive potential that incorporates micromorphology, chemistry, physics, pedology, land-system mapping, and biology. The method is robust enough to work over a range of climates and land-forms in Australia and elsewhere. David is currently a visiting fellow at Australia National University, Canberra, Australia, after thirty-eight years with CSIRO.

Jane K. Turpie is a resource economist with a background in ecology and ornithology. She is a senior lecturer in Conservation Biology and Economics at the Percy FitzPatrick Institute, University of Cape Town, South Africa, and director of Anchor Environmental Consultants.

Katrina Van Dis has been trained in community development and applied economics at the University of Vermont, USA. She has worked for a variety of organizations focused on environmental and youth education, restoration, and sustainable development.

J. Deon van Eeden is a horticulturist and restoration practitioner with a keen interest in arid zones and coastal dunes. He heads Vula Environmental Services, South Africa, and specializes in natural landscaping and rehabilitation work, including seed and plant production. He is currently completing his master's studies at Rhodes University, South Africa, under Professor Roy Lubke.

Éva-Terézia Vesely is an economist with an interest in interdisciplinary research to which she contributes from resource, environmental, and ecological economics perspectives. She is a researcher at Manaaki Whenua, a crown research institute in New Zealand.

Cathy Waters is a research officer with the Department of Primary Industries, Australia, where she has been involved with the restoration of Australian native grasslands in semiarid zones. More recently she has had a special interest in landscape genetics and is currently a postgraduate researcher with Charles Sturt University, Australia.

Steve G. Whisenant is a restoration ecologist with broad international interests and experience. He is professor and head of the Department of Rangeland Ecology and Management at Texas A&M University in College Station, Texas, USA.

Paddy Woodworth is a freelance author, investigative journalist, and former foreign desk editor at the *Irish Times*. He has published widely and is currently working on a book on ecological restoration projects worldwide.

Andrew G. Young is a plant population geneticist. His research focuses on understanding how genetic and ecological processes interact to determine population viability in fragmented habitats in southeastern Australia. He is leader of the biodiversity and sustainable production research program at CSIRO Plant Industry, Australia.

Mike D. Young holds a research chair in water economics and management at the University of Adelaide, Australia, and established CSIRO's Policy and Economic Research Unit. He was awarded the Land and Water Australia Eureka Prize for his research on the design of robust water entitlement and trading systems. In 2003, he was awarded an Australian Centenary medal for his contribution to environmental economics.